SLEEP AND ALERTNESS
Chronobiological, Behavioral, and Medical Aspects of Napping

Sleep and Alertness

Chronobiological, Behavioral, and Medical Aspects of Napping

Editors

David F. Dinges, Ph.D.

Unit for Experimental Psychiatry
The Institute of Pennsylvania Hospital
and
University of Pennsylvania
Philadelphia, Pennsylvania

Roger J. Broughton, M.D., Ph.D.

Division of Neurology
University of Ottawa
and
Ottawa General Hospital
Ottawa, Ontario, Canada

Raven Press  New York

Raven Press, 1185 Avenue of the Americas, New York, New York 10036

Made in the United States of America

Library of Congress Cataloging-in-Publication Data

Sleep and alertness : chronobiological, behavioral, and medical
   aspects of napping / editors, David F. Dinges, Roger J. Broughton.
      p.   cm.
   Includes bibliographies and index.
   ISBN 0-88167-524-5
   1. Napping (Sleep) I. Dinges, David F. II. Broughton, Roger J.
   [DNLM: 1. Circadian Rhythm. 2. Sleep. 3. Wakefulness. WL
108 S605]
QP427.S57   1989
612'.022—dc19   7b29 4
DNLM/DLC
for Library of Congress   86-45981
                                                                CIP

9 8 7 6 5 4 3 2 1

Dedication

Because of his unique pioneering contributions as the father of modern sleep research, his particular interest in napping, which was the theme on the cover of his landmark *Sleep and Wakefulness* (1963), and his exemplary scholarship with its combination of careful data gathering, theoretical innovation, and citation of contributions from many lands and in many languages, we are very pleased and honored to dedicate this volume with his permission to Professor Nathaniel Kleitman.

Preface

This volume springs from our individual and collective experiences studying sleep. Roger J. Broughton has been involved in sleep research since undertaking early studies with Henri Gastaut from 1962 to 1964 and has been particularly interested in the parasomnias, the two-per-day sleep rhythm, and daytime sleep episodes in patients with sleep disorders. David F. Dinges has been studying napping and daytime sleepiness since 1977, together with Martin T. Orne and colleagues, who began pioneering research on napping in 1970.

The first organized symposium on napping was held by sleep researchers outside the field of sleep research, in 1976, at the annual meeting of the American Psychological Association in Washington, D.C. (Chairperson, Hal L. Williams: "Who Needs a 24-hr Day? Efficiency of Napping and Fragmented Sleep"). Five years later napping was, for the first time, the theme of a special session at the 1981 annual meeting of the Sleep Research Society on Cape Cod, Massachusetts (Organizers, David F. Dinges and Martin T. Orne: "Napping: Compensatory, Recreational or Biphasic Sleep?"). The topic gained acceptance in 1983 at a symposium of the 4th International Congress of Sleep Research in Bologna (Chairperson, Roger J. Broughton: "The Siesta: Social or Biological Phenomenon?").

Out of these efforts arose a core group of investigators interested in publishing a volume that reviewed what was known about the significance and utility of naps. This book represents the culmination of those efforts by providing broad state-of-the-art reviews from an international cast of investigators who have carried out seminal work in the area. It is our hope that the volume will fill a critical void in the understanding of sleep and wakefulness, stimulate new insights and experiments, and offer potential solutions to problems of human alertness in a 24-hour world. This book will be especially useful not only to scientists studying sleep, sleep disorders, and chronobiology, but also to psychologists, nurses, engineers, managers, and anyone concerned with maintaining human alertness and optimum performance under circumstances of sleep loss or of shifted or irregular work schedules.

DAVID F. DINGES, PH.D.
ROGER J. BROUGHTON, M.D., PH.D.

Foreword

The general population does not fully understand that the level of alertness in the daytime is inevitably and solely linked to sleep at night. Although the knowledge has been accumulated by researchers, it has not been effectively transmitted to the public. People do not understand what makes them sleepy in the daytime. It may seem a bit self-serving for sleep researchers, but in my opinion, nothing is more important to us as human beings than our level of daytime alertness.

Approximately 10 years ago, I looked through the indices of a number of psychology textbooks and other books, including the Kleitman monograph, *Sleep and Wakefulness*, for the word "sleepiness," and the word failed to appear even once. Even practicing sleep researchers shared the guilt of ignoring this crucially important variable. It remained for clinicians to put it on the map, through studies of narcolepsy (and the unintended sleep episodes associated with this illness) and sleep apnea syndromes.

Falling asleep in the daytime is dependent on some degree of underlying sleep tendency. Carskadon and I, in our studies of children and adolescents, found that children getting an optimal amount of sleep at night were not able to fall asleep during the daytime, under any circumstances (i.e., they could not take naps). We also found, using the Multiple Sleep Latency Test, that a marked tendency to fall asleep develops in the middle of the day from adolescence onward. This tendency reversed itself toward evening whether or not sleep occurred. Thus, it seems nature definitely intended that adults should nap in the middle of the day, perhaps to get out of the midday sun.

It is my strong belief that there should be a clear professional/societal/cultural position on napping. I recently gave a lecture and asked people to raise their hands if they thought that having to take a nap was a little bit on the negative side of the image they wished to project. Nearly half the hands were raised. The current public consensus is: naps at home in bed, good; naps at the office, bad.

We need further research and public awareness on the issue of daytime sleepiness and alertness, and a logical extension of that need is investigation of the topic of napping. At long last, here is a book on this important subject and its many ramifications. The editors are to be congratulated for assembling such a rich fare and presenting it to us.

WILLIAM C. DEMENT, M.D., PH.D.
Lowell W. and Josephine Q. Berry Professor
of Psychiatry and Behavioral Sciences
Director, Sleep Disorders Clinic and
Research Center, Stanford University

Acknowledgments

Much help and support was received while the volume was being developed. We are deeply appreciative of the many hours of skilled work by Emily Carota Orne, Joanne M. Rosellini, Kelly A. Gillen, Stacia A. Bates, and Stephen R. Fairbrother in bringing the volume chapters to a common format, of Wilse B. Webb for his unflagging support of the volume's appearance, and of the typing skills of Barbara Reynolds for the Ottawa manuscripts. We thank the contributors for their standards of excellence and their patience, and the editorial staff of Raven Press for their helpfulness. During the time we worked on the volume, support was received from grant MT3219 and from a Career Investigator Award both of the Medical Research Council of Canada (R.J.B.), and from grants MH19156 and MH44193 of the National Institute of Mental Health, U.S. Public Health Service (D.F.D.), and a grant from the Institute for Experimental Psychiatry Research Foundation. Finally, the effort would not have been possible without the support, encouragement, and understanding of the colleagues working with us in our laboratories, of our staff and students, and, most especially, of Christine and Wendy.

D.F.D. and R.J.B.

Contents

Contributors

Torbjörn Åkerstedt
IPM and Stress Research
Karolinska Institute
National Institute for Psychosocial
Factors and Health
S-10401 Stockholm, Sweden

Robert G. Angus
Defence and Civil Institute of
Environmental Medicine
Downsview, Ontario, Canada

Roger J. Broughton
Division of Neurology
University of Ottawa
Ottawa General Hospital
Ottawa, Ontario, Canada K1H 8L6

Scott S. Campbell
Institute for Circadian Physiology
Boston, Massachusetts 02215, USA

Mary A. Carskadon
Department of Psychiatry and
Human Behavior
Brown University
E. P. Bradley Hospital
East Providence, Rhode Island 02915,
USA

David F. Dinges
Unit for Experimental Psychiatry
The Institute of Pennsylvania
Hospital and
University of Pennsylvania
Philadelphia, Pennsylvania
19139-2798, USA

Mats Gillberg
National Defense Research Institute
Stockholm, Sweden

Peretz Lavie
Sleep Laboratory
Faculty of Medicine
Technion-Israel Institute of
Technology
Haifa 32 000, Israel

Paul Naitoh
Naval Health Research Center
San Diego, California 92318, USA

Claudio Stampi
Human Neurosciences Research Unit
University of Ottawa
Ottawa, Ontario, Canada K1H 8M5

Irene Tobler
Institute of Pharmacology
University of Zurich
CH-8006 Zurich, Switzerland

Lars Torsvall
IPM and Stress Research
Karolinska Institute
Stockholm, Sweden

Wilse Webb
Department of Psychology
University of Florida
Gainesville, Florida 32611, USA

Juergen Zulley
Max-Planck Institute for Psychiatry
Munich, Federal Republic of
Germany

Sleep and Alertness: Chronobiological, Behavioral, and Medical Aspects of Napping, edited by
D. F. Dinges and R. J. Broughton.
Raven Press, Ltd., New York © 1989.

1

Napping: A Ubiquitous Enigma

*Roger J. Broughton and †David F. Dinges

*Division of Neurology, University of Ottawa and Ottawa General Hospital,
Ottawa, Ontario, Canada K1H 8L6, and †Unit for Experimental Psychiatry, The
Institute of Pennsylvania Hospital and University of Pennsylvania,
Philadelphia, Pennsylvania 19139-2798, USA*

Sleep is a central mystery of evolution, occupying a major portion of the life span of most vertebrate species on our planet. As such, it is one of the most pervasive behavioral controls in nature. Sleep is a process that not only alternates with wakefulness, but one that can dramatically affect the quality of wake functioning. Its disorders afflict, often seriously, at least 10% of humans at all ages and present an enormous medical challenge.

Unfortunately, misconceptions about sleep and its disorders are still common, and the importance of sleep for effective wake functioning is often ignored or trivialized. This is especially so for a variant of sleep called napping. Although sleep may be judged a necessary part of life, napping has often been considered to be a deviant and unwanted form of sleep, indicative of laziness, irresponsibility, immaturity, or senility. This negative connotation is captured in the expression "caught napping," which refers to someone being off guard or not vigilant at a time when vigilance is required. In the fields of sleep research and sleep disorders medicine, even voluntary napping has not always been considered healthy or appropriate.

Yet the universal presence of napping across species and ages indicates an extreme biological importance for it and furthermore raises a myriad of issues, some fundamental and theoretical, others very practical, relevant to the fields of general biology, chronobiology, phylogenetic evolution, ontogeny, anthropology, human ergonomics, psychology, and clinical medicine. Animals in their natural habitats are regularly killed by predators when caught napping, and the inability of people to sustain alertness and attention because of sleepiness results in untold accidents in industry and deaths on our highways. In short, the topic is anything but trivial.

This volume attempts to review our current knowledge concerning the nature and importance of napping, isolate and focus on the issues raised, and indicate what mysteries of basic and human sleep/wake biology might be clarified in the future. It is evident that it is meaningless to treat naps outside the context of the overall pattern of sleep/wake oscillations and the continuum of alertness-drowsiness within the nycthemeron—hence the volume's chosen title.

Napping has been sleep's orphan in that it has received little coherent or integrated analysis in the more than 30 years since the benchmark discoveries of Nathaniel Kleitman and his students began modern sleep research. This may be owing in part to the diverse ways in which napping has been defined, especially in the human literature. Characteristics used to describe naps include sleep during the daytime, sleep taken when the person is not in bed or prepared for bed, involuntary sleep, light sleep, and the sleeper's perception that he or she napped regardless of its duration or depth. Most often, however, the term is used simply to denote a short period of sleep, the actual length of which may range from a few minutes to 4 hr. This has prompted some authors to argue that a nap is any sleep period with a duration of less than 50% of the average major sleep period of an individual (7). Accordingly, in humans a nap may consist of anything from a very brief "microsleep" to a very substantial sleep period lasting several hours.

The need to define a nap is evident in the many questions posed about napping. It perhaps becomes most critical in characterizing the human ontogenetic transition between multiple naps to a monophasic sleep pattern. At what age do humans develop beyond a polyphasic sleep/wake (nap) pattern and into a monophasic (nonnap) pattern, if ever? Is it reasonable to describe the polyphasic sleep of human infants and many animal species as nap sleep? Is napping merely the product of a heavy noon meal or the torrid heat of midday, as is often assumed? What distinguishes siesta cultures from nonsiesta cultures? Is there a biological basis for the siesta or midday nap? At what point should sleep episodes of increasing length no longer be considered a nap? What determines the sleeper's perception that he or she napped, even if the sleep was many hours and of considerable depth? What distinguishes nappers from nonnappers? What benefits and consequences derive from napping? What are the major determinants of whether someone will nap? When is the most likely time for a nap and what influences this timing? When is napping pathological?

Such questions are attacked from many perspectives in the volume, based on data from a broad range of scientific technologies. They include laboratory and field-based behavioral studies supplemented or not by filming, videotaping, or video telemetry; macroelectrode or microelectrode recordings in lower animals immobilized or free moving; and surface electroencephalographic and polygraphic recordings of individual naps or sleep throughout the 24 hours including ambulatory monitoring, altered sleep schedules with efforts to increase time in wakefulness (sleep deprivation), enforced bed rest with or without instructions to sleep (as in sleep satiation), altering the timing of the major sleep period (as in studying or replicating the effects of shift work and jet lag), dispersing sleep into shorter sleep episodes throughout the 24 hours (ultrashort sleep schedules), and assessing sleep under conditions in which subjects lack temporal knowledge (time-free studies) or in which all possible further temporal signals are also totally removed (disentrained state).

The effects of these manipulations on sleep structure, waking alertness levels, performance measures, indices of health, and other variables relevant to napping are analyzed. Such outcomes are critically important not only for understanding the biology of sleep and wakefulness, but also for assessing the effects on humans resulting from the increasing need for unusual sleep/ wake patterns, especially those that deviate from the human pattern of mon-ophasic nocturnal sleep. Do napping protocols have the potential to minimize the adverse effects of unusual sleep/wake regimes on human alertness and functioning? Research related to napping and alertness is obviously imper-ative for reasons of industrial safety, optimizing performance during sus-tained operations (such as in coping with disasters), and exploration of the depths of the sea and space—situations in which sleep in crowded quarters with altered zeitgebers and unusual sleep/wake demands are the rule.

What then is the full scope of the relevance of napping? From the stand-point of general biology, the presence of varying nap patterns across species at different points of the evolutionary process and with different lifestyles (nocturnal, diurnal, split activity rhythms; carnivores, herbivores), habitats (secure, insecure; terrestrial, avian, aquatic), and even body mechanics (squirrels, giraffes) raises many interesting issues. What survival trade-offs exist for different sleep lengths in different situations? What is the relative value for environmental monitoring of multiple naps broken by brief arousals versus prolonged periods of sustained wakefulness? Are short naps equally efficient to equivalent duration longer sleep episodes in meeting an animal's sleep need per 24 hr? Are the functions of naps across phylogeny identical to those (yet largely unknown) of sleep itself? In Chapter 2, Tobler focuses on sleep/wake patterns in mammals and gives information relevant to a num-ber of these questions.

Human ontogenetic aspects of napping have played a significant role in the history of sleep research. Pioneering studies by Gesell (10) and above all by Kleitman (11) and by Kleitman and Engelmann (12) analyzed the pre-dictable and lawful evolution from an infantile polyphasic pattern toward increasing consolidation of night sleep with multiple daytime naps, then a single daytime nap, and finally the emergence of the common monophasic adult sleep pattern. Whether this tendency toward a sustained nighttime sleep period reflects maturation of the basic neurobiology of sleep/wake rhythms, a changing sleep need, or a passive response to social pressures for prolonged and sustained daytime wakefulness remains uncertain, especially since in the elderly or when opportunity exists at most ages, daytime napping frequently returns. The orderliness of these developmental changes into adulthood are reviewed from two perspectives in the first part of the volume. In Chapter 3, Webb presents the data on human sleep, and in Chapter 4, Carskadon discusses parallel evidence from studies of human sleepiness using sleep latency tests. Although there is clearly a maturational push toward fewer sleep episodes per 24-hr period, with preferential placement of sleep in the nocturnal phase, the potential for daytime sleepiness in the midafternoon is a remarkably consistent feature of this process leading to consideration of a chronobiologically regulated nap process.

The chronobiological features of endogenous sleep/wake rhythms are obviously genetically coded and structure bound. They are preeminent modulators of the timing of naps and exert a major influence beyond that attributable to sleep need, the absence of arousing stimuli, the availability of a suitable sleep habitat, and the voluntary decision to take a nap. But what are the major biorhythmic aspects of polyphasic sleep or napping? This basic question is attacked from a number of ingenious perspectives and protocols in the second section of the volume. Evidence is reviewed for circa 12-hr, 6-hr, and 3- to 4-hr rhythms, as well as the well-known circadian rhythm. This numerical pattern suggests that there is fundamental biological significance for rhythms with fixed integer ratios (3). Are these periodicities generalizable across all species? If so, what is their survival value? The chronobiological aspects of sleep/wake rhythms relevant to napping are extensively reviewed in Chapter 5 by Broughton, in which a compelling case is made for a circasemidian rhythm underlying the daytime nap tendency in human adults. Robust and remarkably consistent findings on the timing and determinants of naps are then presented in the reviews of two approaches to document the nature of naps taken throughout the circadian cycle. In Chapter 6, Lavie presents findings from his studies of the ability of adults to sleep and to resist sleep while on very short ultradian sleep/wake regimes. In Chapter 7, Campbell and Zulley discuss work on factors related to the timing and perception of naps while subjects live in temporal isolation (a crucial paradigm for understanding the chronobiological dynamics of sleep) including their novel attempts to disentrain the sleep/wake rhythm altogether. The section concludes with Chapter 8 by Stampi, who reviews literature on ultrashort sleep/wake schedules and describes the use of such polyphasic nap regimes by transoceanic solo yacht racers.

Differences in sleep need and patterning among adult humans and even within the same person over time are striking and worthy of detailed study not only because they raise important issues about the limits of plasticity in the human sleep system, but also because they provide information on the practical limits of human sleep/wake manipulations. Variability of sleep patterns among adult humans is high. For example, there are long and short sleepers, larks and owls, and nappers and nonnappers, as well as nap and nonnap days for a given napper. In addition to these constitutional or behavioral-preference differences, there is considerable variability among persons in the ability to adapt to new sleep/wake patterns, such as ultrashort sleep schedules and rotating shift work; some people appear to adjust easily, whereas others appear incapable of adapting. Moreover, there are apocryphal tales of great persons who reputedly survived solely on naps (e.g., Napoleon, Edison, Churchill), although no one has yet been documented to exist solely by napping.

Is napping an adaptive sleep process in humans, and, if so, what are the biological and behavioral differences between those who do nap and those who do not? In terms of their effects, naps have often been alleged to be refreshing disproportionate to their length. What promise does napping hold for sustaining alertness, and what are its consequences for health and functioning, especially when it is used in work/rest scenarios that push the limits of human adaptation? Issues centering on variations in napping behavior and the adaptive use of naps by human adults, including their causes and consequences, are addressed in the third section of the volume.

Part of the variability of napping in human adults appears to relate to the reasons why naps are taken. They may be taken to reduce the effects of an existing sleep debt. The siesta, for instance, is generally present in societies in which people eat very late in the evening, go to bed in the early morning hours, and yet must still get up early to go to work (2). Alternatively, they may be taken to "bank" sleep in anticipation of a foreseeable period of sleep loss, such as in shift work or sustained work. Naps also may be taken simply for the pleasure of sleeping or as part of a habitual biphasic sleep cycle. These three behavioral tendencies have been referred to as replacement, prophylactic, and appetitive napping, respectively (5,6,8,9).

In Chapter 9, Dinges discusses the extent to which these napping patterns are present in healthy adults who are not challenged by rigid work/rest schedules and considers the positive and negative effects of naps, the differences between nappers and nonnappers, and the relative contributions of sleep need and circadian timing to the control of napping. Two subsequent chapters review data vital to determining the potential of naps to maintain effective human functioning in altered work/rest scenarios—a major issue in sleep logistics and occupational health. In Chapter 10, Åkerstedt and colleagues discuss napping patterns among shift workers, as well as the causes and consequences of naps in this population. In Chapter 11, Naitoh and Angus present data on the nature and efficacy of napping during prolonged work, particularly sustained military operations. Chapter 12 by Webb and Dinges concludes the section with a novel review of data on cultural variability in daytime sleep and napping, as well as the factors associated with these differences.

Beyond individual, occupational, and cultural differences in napping, there are differences associated with the health of the napper, particularly the presence of sleep disorders. Patients with various forms of hypersomnia with excessive daytime sleepiness exhibit a continuum of phenomena from brief microsleeps, short sleep episodes, prolonged sleep episodes or "twilight" states (including amnesic automatisms), to remarkable increases of sleep per 24 hr. Sleepiness may become so marked that naps are always semivoluntary or totally involuntary, the patient having little choice but to try to relieve the pathological sleepiness or be overwhelmed by it. To what extent taking short daytime sleep episodes in the patient population improves alertness appears to depend on their duration, structure, and timing, as well as on the specific diagnosis of the sleep disorder.

The pathophysiological cause of daytime naps or overt "sleep attacks" in patients, in fact, appears to differ in different diagnoses. It may include fragmentation of night sleep with no total sleep reduction; nocturnal sleep reduction; selective increase in pressure for nonrapid eye movement, rapid eye movement, or both sleep states; impaired mechanisms of sustained daytime wakefulness ("subwakefulness syndrome"); alterations in the chronobiological aspects of sleep and sleepiness, particularly circadian alterations; and even nocturnal cerebral hypoxia. It is therefore reasonable that different types and timing of naps will have different effects in different medical conditions.

Also of clinical significance is the phenomenon of sleep inertia, which occurs immediately after daytime sleep episodes. Awakening from diurnal naps may, in fact, at times lead to marked behavioral and cognitive effects. The severity of the postnap inertia appears to vary with the depth of sleep in healthy individuals (4). It can result in extreme "sleep drunkenness" in patients with idiopathic hypersomnia (13). In predisposed individuals, somnambulism and sleep terrors may also arise during arousals from daytime deep slow wave sleep (1).

In summary, the information and research on napping pose a challenge to our understanding of sleep and its impact on our waking lives. Whether as a component of a polycyclic sleep schedule, an expression of an inherent response to sleep loss, a diagnostic sign of excessive daytime sleepiness, a way to maintain human performance in sustained operations, or a key to chronobiological dynamics of the sleep/wake cycle, napping touches virtually every aspect of sleep science and medicine. The absence of any cohesive coverage in the literature led us to believe that a *mise au point* was timely. There is a caveat, however; most, but not all, of the aspects of polyphasic sleep that we wished to cover are represented in this volume. We would have liked to include chapters on the fundamental brain mechanisms relating to the nature and timing of polyphasic sleep and on the relevance of napping to theories about sleep function theory. For a variety of reasons, we were unable to cover these two important topics. Our efforts, however, will be more than fully rewarded if this volume helps to catalyze increased interest in the important and ubiquitous enigma of napping.

REFERENCES

1. Broughton, R. J. (1968): Sleep disorders: Disorders of arousal? *Science*, 159:1070–1078.
2. Broughton, R. J. (1983): The siesta: Social or biological phenomenon? *Sleep Res.*, 12:28.
3. Broughton, R. J. (1985): Three central issues concerning ultradian rhythms. In: *Ultradian Rhythms in Physiology and Behavior*, edited by H. Schulz and P. Lavie, pp. 217–233. Springer-Verlag, Berlin/Heidelberg/New York/Tokyo.
4. Dinges, D. F., Orne, M. T., and Orne, E. C. (1985): Sleep depth and other factors associated with performance upon abrupt awakening. *Sleep Res.*, 14:92.
5. Dinges, D. F., Orne, M. T., Orne, E. C., and Evans, F. J. (1980): Voluntary self-control of sleep to facilitate quasi-continuous performance. U.S. Army Medical Research and Performance Command, Maryland (NTIS N. AD-A102264).
6. Dinges, D. F., Orne, M. T., Orne, E. C., and Evans, F. J. (1981): Behavioral patterns in habitual nappers. *Sleep Res.*, 10:136.
7. Dinges, D. F., Orne, M. T., Whitehouse, W. G., and Orne, E. C. (1987): Temporal placement of a nap for alertness: Contributions of circadian phase and prior wakefulness. *Sleep*, 10:313–329.
8. Evans, F. J., Cook, M. P., Cohen, H. D., Orne, E. C., and Orne, M. T. (1977): Appetitive and replacement naps: EEG and behavior. *Science*, 197:687–689.
9. Evans, F. J., and Orne, M. T. (1975): Recovery from fatigue. U.S. Army Medical Research and Performance Command, Maryland (NTIS N. A100347).
10. Gesell, A., and Amatruda, C. S. (1945): *The Embryology of Behavior. The Beginnings of the Human Mind*. Harper Brothers, New York.
11. Kleitman, N. (1963): *Sleep and Wakefulness*. University of Chicago Press, Chicago.
12. Kleitman, N., and Engelmann, T. G. (1953): Sleep characteristics of infants. *J. Appl. Physiol.*, 6:269–282.
13. Roth, B., Nevsimalova, S., Sagova, V., Paroubkova, D., and Horakova, A. (1981): Neurological, psychological and polygraphic findings in sleep drunkenness. *Arch. Suisses Neurol. Neurochir.*, 129:209–222.

Sleep and Alertness: Chronobiological, Behavioral, and Medical Aspects of Napping, edited by D. F. Dinges and R. J. Broughton. Raven Press, Ltd., New York © 1989.

2

Napping and Polyphasic Sleep in Mammals

Irene Tobler

Institute of Pharmacology, University of Zurich, CH-8006 Zurich, Switzerland

MAMMALIAN SLEEP/WAKE PATTERNS

In the early 1900s Szymanski (89) examined the rest/activity behavior in a variety of animal species. He found that several small mammals, including cats, rats, rabbits, and mice, exhibited 10 to 12 cycles of activity per day. Szymanski applied the term *polyphasic* to a behavioral pattern in which bouts of activity and rest alternated with a higher frequency than once per day (see Fig. 1, e.g., rabbit, hamster, pig). This type of cycle is now known to be a fundamental characteristic of the sleep/wake pattern of most mammals, excluding primates.

In contrast, the *monophasic* activity pattern consists of one consolidated period of sleep (short waking episodes are disregarded) followed by a continuous period of activity in which no sleep takes place. This pattern is well-known for the adult human. Between these two extremes lies the *biphasic* sleep/wake pattern typically present in many mammals. It consists of two activity periods that often coincide with dawn and dusk, separated by two troughs: (a) a longer sleep period during night or day, depending on the nocturnal or diurnal placement of activity and (b) a second quiescent period of shorter duration that takes place at approximately midday or midnight (6). A biphasic rest/activity pattern also has been documented for many nonmammalian species (e.g., several species of birds, reptiles, fish, and insects) (6).

Numerous studies have concerned comparisons of sleep structure in mammals without regard for sleep/wake patterning; they have rarely included 24-hr recordings (18). The sleep pattern refers to the temporal relation between sleep and waking, whereas sleep structure refers to characteristics within sleep. Furthermore, several polygraphic sleep recordings have been carried

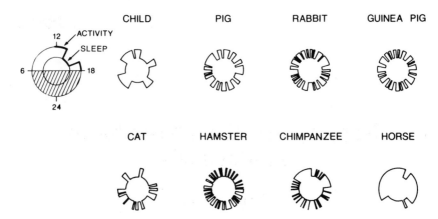

FIG. 1. Actograms of biphasic and polyphasic species. Data for child and rabbit were modified from Szymanski (89). Others were derived as follows: pig (80); cat (62,86); guinea pig (62); hamster (92); chimpanzee (54); and horse (85). Sleep is represented on the inner circle, activity periods on the outer circle, as illustrated on the left. Daytime refers to 12 hr from 0600 to 1800 hr; nighttime covers the 12 hr from 1800 to 0600 hr. Sleep in the child, chimpanzee, cat, pig, rabbit, guinea pig, and hamster are typically polyphasic. The horse is a nocturnal sleeper, which in this example, exhibits a biphasic sleep/wake pattern owing to the occurrence of an afternoon nap. The polyphasic species exhibit a large variability in the 24-hr distribution of the sleep periods and the length of the sleep/wake episodes.

out under constant dim illumination, and therefore diurnal variations in the sleep/wake pattern were not noted. In the following, those studies in mammals that have provided sleep pattern data have been considered. The vast literature on rest/activity rhythms has been mostly disregarded because in these studies sleep was not specifically evaluated.

Nonhuman Primates

Apes (family Pongidae) and Old World monkeys (family Cercopithecidae) are phylogenetically close to humans (family Hominidae). They have been investigated to trace those aspects of sleep that may have developed in the course of primate evolution. Studies of 24-hr sleep/wake cycles in nonhuman primates are particularly rare. Recordings have usually concentrated on the main sleep period (i.e., on the 12-hr dark period). There is general agreement in that the sleep structure of nonhuman primates markedly resembles that of humans.

Sleep in the chimpanzee, the only species of the ape family whose electroencephalogram (EEG) has been recorded, has been found to be more similar to human sleep than to that of other mammals (2,13). The sleep episodes could clearly be divided into nonrapid eye movement (NREM) sleep stages 1 through 4 and rapid eye movement (REM) sleep. However,

based on telemetry recordings of chimpanzees in the wild, several differences were found between sleep in chimpanzees and that in humans. The
apes slept from sunset until sunrise, and total sleep time (TST) comprised
approximately 9.66 hr. The sleep pattern showed short episodes of each
stage and frequent stage shifts as well as many short awakenings (17% waking was found within a night's sleep) (13). The duration of waking varied
from extremely short episodes to others lasting as long as an hour. In this
respect, the chimpanzee sleep pattern resembled that of nonprimates. Similar frequent changes between stages and many short waking bouts were
found in juvenile chimpanzees recorded by telemetry (30).

As noted previously, data on sleep patterns in nonhuman primates are
scarce. The existence of inactivity during the hot hours of the day between
1100 and 1700 hr often has been observed in primates living in the wild in
tropical regions. Thus, mountain gorilla (*G. Gorilla beringei*) built crude
platforms mostly on the ground but also in trees to spend the night. Nests
were built also for resting during the day (82). Orangutan (*Pongo pygmaeus*)
built nests in trees only to sleep during the dark, and napping was never
observed (82). Chimpanzees living in the Gombe Reserve exhibited a consolidated sleep period in tree nests during the night and tended to rest between 0930 and 1500 hr, often in "day nests" on the ground. Behavioral
sleep comprised approximately 30 min of the 5.5-hr rest time (33,34).

A propensity to nap at noon also has been seen in the laboratory. Thus,
chimpanzees recorded by telemetry exhibited a relatively consolidated sleep
period at night. Several naps, in which REM sleep was present, occurred
during the day, usually after a midday meal (29,30). Similarly, an individual
juvenile chimpanzee recorded by telemetry for 30 days exhibited a consolidated sleep period at night that extended into the morning hours and two
activity periods separated by a nap during the day (see Fig. 1). REM sleep
occurred infrequently during the naps (54). In contrast, Bert et al. (13), wrote
that chimpanzees recorded by telemetry "slept hardly at all" during the day.

In monkeys, whose sleep has been studied more extensively than that of
apes, four sleep stages also have been clearly identified. The early studies
were not concerned with sleep patterns. Thus *M. mulatta* were recorded for
30 or 71 nights (46,110), but the animals were kept awake during the day.
This procedure obscured the natural propensity for sleep. Observations of
M. mulatta in a "natural" social environment revealed that they nap (im-
mobility with eyes closed) approximately 6% of the day (88). Moreover,
studies that comprised 24-hr recordings of undisturbed, restrained *M. mulatta* and a closely related species, the pigtail macaque monkey (*M. ne-
mestrina*), revealed frequent daytime napping. Naps occurred usually at
noon and at approximately 1500 to 1700 hr (20) and comprised deep sleep
and REM sleep (8). Naps were absent when the latter species was recorded
in a noisy environment (67) and never occurred in *M. arctioides* (48).

Baboons (*Papio anubis*) in Amboseli Park rested and napped several times

per day (5). When recorded in relatively monotonous laboratory surroundings, baboons exhibited a sleep period in the 12-hr dark period that included
102 min of waking. One animal also slept (stages 1 and 2) for approximately
40 min during the day (103). In baboons recorded by telemetry in Africa,
napping comprised 8.1% of the day and consisted mainly of stage 1 sleep,
but stages 2 and 3 and REM sleep also were present (14).

Several species of New World monkeys (family Ceboidea) have been investigated. Of the 57 species only the owl monkey (*Aotus trivirgatus*) is
nocturnal. In this species, as well as in the squirrel monkey (*Saimiri sciureus*)
NREM sleep was not clearly subdivided into substages although two to three
EEG patterns could be discerned in this sleep stage (63). Squirrel monkeys
recorded under restrained conditions did not nap and spent 17.6% of the 12-
hr dark period in wakefulness (1). Recently, this species has been more
extensively studied. A large amount of waking was present in the dark period
(14%), and several naps occurred during the 12-hr light period (113).

There is little information on sleep in the primate order Prosimia. This
category includes both nocturnal (e.g., *Phaner furcifer, Indri, Propithecus,
Galago senegalensis*), and diurnal (e.g., howlers, slow loris—*Nycticebus cou-
cang*) species. Carpenter (19) observed the howlers on Barro Colorado Island
and reported that whereas infants played during most of the daytime hours
the adults often rested and slept. Bert et al. (12) recorded 15 adult galagos
during the day in darkened boxes. NREM sleep in these animals was shown
to be well developed in that it could be subdivided into four stages. However,
in regard to sleep pattern, the cyclic organization was weaker than in simians
since in galago the four NREM sleep stages and REM sleep followed each
other randomly. A lemur (*Lemur macacus fulvus*) recorded by telemetry
exhibited a polyphasic sleep pattern and well-developed NREM sleep stages
1 through 4 and REM sleep. Two major sleep episodes (both of which included REM sleep) could be distinguished: one main episode of 6.17 hr in
the dark phase and a shorter episode of 3.33 hr during the day (102).

In conclusion, all primates, including relatively primitive lemurs, exhibited
several sleep stages. A well-developed NREM sleep stage 4 was seen in
most species. The main differences among species lie in the duration and
distribution of the sleep stages. Apes, in particular the chimpanzee, exhibit
a consolidated sleep period at night as well as a tendency for short naps
during the day. Interestingly, the occurrence of naps was not restricted to
monkeys recorded in isolation in the laboratory but also was present in most
apes and monkeys in the wild. In Prosimia the monophasic and biphasic
sleep patterns do not occur; their sleep/wake pattern is typically polyphasic.
Thus, in respect to sleep pattern, the monkey is closer to chimpanzees and
humans than it is to lower mammals. This also is true for sleep cycle length
in monkeys (47 min), which is closer to the value in humans (65–110 min)
than that in the cat or rat (30 and 12 min, respectively). However, apes and
monkeys markedly differ from humans in the frequent interruptions of sleep

in the nighttime sleep period. It is possible that human sleep has diverged phylogenetically from that of apes and monkeys. However, it is difficult to discriminate phylogenetic factors from environmental influences.

Rodents

The presence of cyclic changes of sleep stages and their interruption by short waking bouts in the rat, as well as the similarity to various other mammals, was first shown by Roldan et al. (70,109). Thereafter, sleep in rodents has usually been divided into two stages, NREM sleep and REM sleep, although several substages of NREM sleep have been defined in the rat (36). Although most rodents are nocturnal and have a diurnal placement of sleep, they exhibit intervals of sleep and waking throughout the 24-hr period. The typical nocturnal pattern of the rat in which waking predominates during the dark period and sleep during the light period has been well established (15,31,53,55). The light/dark ratio of sleep varied from 1.5 to 2.5, and TST comprised 50% to 55% of the 24-hr period (Table 1).

Similar polyphasic sleep/wake patterns have been established for the Mongolian gerbil (*Meriones unguiculatus*) (87), the Syrian hamster (*Mesocricetus auratus*) (92) (Fig. 2), the mouse (*Mus musculus*) (58,97,112), ground squirrels (*Citellus beldingi, C. lateralis* and *C. tridecemlineatus*) (99,104), and the chinchilla (*Chinchilla laniger*) (99). The main difference in sleep patterns between these species was the light/dark ratio of TST (Table 1). With the exception of ground squirrels, all rodents slept predominantly during the light period. A comparison of wild mice and an inbred mouse strain, A/J, resulted in similar values for TST in the two groups (46% and 51%), but in significantly different light/dark ratios of sleep (1.5 and 1.1, $p < 0.05$) (32).

A comparative study in 14 species of muroid rodents based on observations that were conducted every other hour within 24 hr resulted in large species variations of TST (range, 29.2%–64.0%) and light/dark ratios (0.5–11.85) (9). The guinea pig (*Cavia porcellus*), recorded under natural lighting conditions, slept as little as 28% of 24 hr. Short and numerous sleep and waking episodes were evenly distributed throughout 24 hr so that no diurnal sleep/wake rhythm was apparent (62) (Fig. 1). Similar low values for TST were obtained in guinea pigs (photoperiod unspecified) when they were not well adapted to the recording conditions (25%–30%). In contrast, in well-adapted animals TST comprised 52.5% of 24 hr (42). The lack of a circadian sleep/wake rhythm in guinea pigs raised in a light/dark 12:12 hr photoperiod was confirmed by Ibuka (38).

In summary, the sleep/wake cycle in rodents can be considered a typical example of a polyphasic pattern (Fig. 2). Most sleep occurs during the 12-hr light period, and waking is predominant in the nighttime activity period. There are marked differences, however, among species in TST and the light/dark ratio of sleep.

TABLE 1. *Total sleep time (TST) percentage and the light/dark (L:D) sleep ratio observed in a selection of mammalian species*

Species	Average TST (% of 24 hr)	L:D (12:12 hr) sleep ratio[a]	Ref.
Chimpanzee	57.0	0.11	54
Lemur	39.6	0.54	102
Rat	55.2	1.47	99
	49.5	2.53	95
Mouse	54.8	1.43	99
	45.7	2.66	112
	50.4–56.4[b]	1.1–2.11	97
Hamster	60.1	1.72	99
	64.7	1.53	92
Gerbil	54.4	1.10	87
Chinchilla	52.2	1.23	99
Squirrel	57.5	0.91	99
	66.4	0.58	104
Guinea pig	28.4	~1.00	62
	34.7	~1.00	38
Rabbit	47.5	~1.03	64
Cat	50.6	~1.00	16
Dog	52.6	0.81	90
	45.6	0.48	35
	34.8	0.35	93
Horse	12.0	0.00[c]	73
Donkey	13.0	0.00	72
Cow	16.4	0.003	74
Pig	32.6	0.53	74
Sheep	16.0	0.17	74
Rock hyrax	48.0	1.05	84
	56.0	1.22	84
Tree hyrax	29.9	0.90	84
Hedgehog	72.6[d]	1.36[e]	28
Tree shrew	65.8	0.56	11
Rat kangaroo	44.6	3.87	7

[a] Derived by dividing the TST in the 12-hr light period by the TST in the 12-hr dark period. A value greater than 1.0 indicates that more sleep took place in the light period; a value less than 1.0 indicates that more sleep took place in the dark period.
[b] Several mouse strains were studied.
[c] A ratio of 0.00 indicates that virtually all sleep took place in the dark period.
[d] Includes drowsiness (approximately 25%).
[e] Light/dark ratio was 13:11 hr.

Rabbit

A recent study has provided normative values for the rabbit (*Oryctolagus cuniculus*) (64). In animals well adapted to the recording conditions, TST comprised 47% (including 12% drowsiness) of the light/dark 12:12-hr cycle. NREM sleep was evenly distributed throughout the light/dark period, and waking slightly predominated during the 12-hr dark period.

SYRIAN HAMSTER

VIGILANCE STATES AND EEG SLOW WAVE ACTIVITY

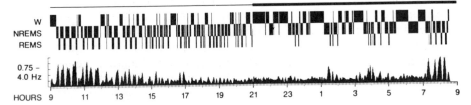

FIG. 2. Vigilance states and EEG slow wave activity in the delta band (0.75–4.0 Hz) of a Syrian hamster for 12 hr light and 11.5 hr dark (delimited by the black bar on top). Values are plotted for 1-min epochs.

Carnivores

In most mammals other than primates, only two sleep stages, NREM versus REM sleep, have been scored. A notable exception among the carnivores is the domestic cat (*Felis catus*), for which two distinct stages of NREM sleep have been described, "light and deep SWS [slow wave sleep]" (96). Drowsiness in cats has usually been considered to belong to the waking state (86). In reference to sleep/wake distribution, early behavioral studies reported that cats were diurnal and monophasic, since they exhibited an almost unbroken rest period during the night (89). More recent studies of cats in the laboratory have shown a large variation in the sleep/wake pattern depending on laboratory conditions and a large interindividual variability. Sleep was divided into two main periods, one occurring in the morning after early morning activity and the other in the afternoon or late evening, indicating a bimodal pattern (16,52). The sleep pattern of the cat could be considered polyphasic, however, if one considers the relatively long bouts of waking (20 min) between sleep episodes (80 min).

Considerable individual variability was present in the actograms of cats recorded for several days in total isolation (47,66). Of 15 cats, 12 were nocturnal, 2 random, and 1 diurnal. A clear daily sleep/wake distribution was present in 16 cats maintained in a light/dark, 12:12-hr ratio with *ad libitum* feeding and drinking, adapted to recording conditions and isolated from laboratory noise (47). More sleep occurred during the light period (light, 58.6%; dark, 46.7%) and the polyphasic sleep/wake pattern included both more NREM and REM sleep and fewer awake episodes in the light period than in the dark period. Several sleep bouts occurred during the night, but awake episodes predominated and became most prominent toward the end of the dark period. Interestingly, five cats were arrhythmic (47).

The dog (*Canis familiaris*) appears to differ from the cat in TST and in the temporal and quantitative distribution of sleep and waking. A comparative study reported that waking episodes between sleep bouts were longer in the dog. Sleep comprised 36.7% of 24 hr in a light/dark ratio of 9:15 hr in dogs and 51.7% in cats (50). A polyphasic sleep/wake pattern with a nocturnal placement of the main sleep bouts was recorded in several studies in mongrel and pointer dogs in 12:12 hr light/dark ratio (35,51,90). TST comprised 35% to 53% of 24 hr, and most sleep occurred in the second third of the night. This result was confirmed in long-term ambulatory recordings of motor activity in 15 mongrel dogs in which rest predominated during the 12-hr dark period (74% of total rest time) and several rest episodes (26%) also occurred during the 12-hr light period (93).

Two dog breeds, Labrador retriever and Doberman pincher, slept for 57.4% of 24 hr in 12:12 hr light/dim (43,57). The daily placement of sleep was not specifically studied, but sleep and waking seemed to be randomly distributed across the 24-hr period. EEG recordings of sleep in three captive foxes (*Vulpes vulpes*) showed a predominance of sleep during the day, with a major sleep period between 1600 and 2000 hr. However, several sleep bouts occurred also during the night (21).

In view of the scarce electrographic data on carnivores, as well as the lack of data from wild animals, several behavioral studies providing normative data of rest/activity are also included. Radio-tracking of feral dogs revealed that they were most active at night, in spite of large daytime activity (83). Three species of arctic carnivores (wolf, wolverine, and fox) retained a nocturnal activity pattern during the 82 days of arctic summer in Alaska (27). Rest usually occurred between 1430 and 2000 hr in the three species. In the arctic wolves, activity gradually declined during the day, with troughs at 2100 and 0100 hr, followed by a sharp peak of activity at 0300 hr. The three arctic wolverines also demonstrated a predominantly nocturnal activity pattern with an activity trough either between 1600 and 2100 hr or 1000 and 1700 hr. Arctic foxes exhibited a major period of rest between 1400 and 1800 hr. However, some foxes appeared to have a second period of rest that usually occurred between 2200 and 0100 hr. Wolf records also showed a drop in activity around midnight (27).

Wheel-running activity was measured in single specimens of several medium-sized American carnivores for 17 to 59 days under a regime that consisted of 12-hr light, 1-hr twilight, 10-hr dim light, and 1-hr twilight (44). The activity pattern of the grison (*Galictis vittatus*), which is related to the weasel, was biphasic; in addition to sleeping during the night, the animal typically took a midday sleep at noon. The tayra (*Eira barbara*) generally "retired several hours before dusk" and became active several hours before dawn, and on some days interrupted the otherwise fairly continuous activity bout by a short midday sleep. No activity was present in either species during the dark. However, the presence of quiet waking during this time cannot be excluded. The ringtail (*Bassariscus astutus*) was never active during the day. Wheel-running episodes of a single bobcat (*Lynx rufus*) occurred at irregular intervals during a cycle of 12-hr light, 10-hr dim, and 2-hr dusk. The rest periods during the day were longer and more frequent (161 min) than those at night (67 min).

Ungulates

The sleep patterns of several domestic and wild ungulates (ruminants and nonruminants) have been investigated. Four vigilance states have been distinguished: waking, NREM sleep, REM sleep, and drowsiness or relaxed wakefulness. Drowsiness was considered to be a vigilance state closer to waking than to sleep. It is the most common stage in domestic animals, particularly in the cow. Furthermore, it is usually associated with rumination, which occurs around the clock (in stall) or predominantly at night (in the field) (74).

As early as 1937, Steinhart (85) observed approximately 600 horses, some of them continuously for 14 hr. Individual animals exhibited a relatively consistent individual timing and placement of sleep. TST was found to vary between 3 and 10.25 hr. The sleep/wake pattern was polyphasic. However, the number of sleep episodes varied considerably among animals (3–16 episodes per 24 hr). The variation was attributed to the diverse daily schedules that included stalling as well as intense activity in a military regiment for 8 to 12 hr per day. When the horses remained in the stall for the entire 24-hr period, eight sleep periods were usually present, five of which occurred during the night. The longest and most consistent sleeping periods occurred in the early morning hours. Many years later, these observations were confirmed by electrographic recordings of stalled horses (77,78). Sleep comprised only 3 to 5 hr a day, and its distribution was nocturnal and monophasic, with the main occurrence between 0200 and 0600 hr. Three to four sleep bouts were present at night. Each NREM sleep episode was interrupted by short EEG arousals (30–50 sec) without change in behavior.

In the pony, sleep was distributed in two main periods, one between 2100 and 2300 hr and the other between 0200 and 0400 hr (darkness occurred from 2000 to 0800 hr) (22). Electrographic recordings and photographs of a single donkey recorded for several 24-hr periods also showed a nocturnal placement of sleep (72). In contrast to the horse, usually none or only two short sleep bouts occurred during the day, whereas five to six bouts of 4- to 5-min duration occurred during the night.

The cow is the domestic animal whose sleep has been most extensively studied. In his early study, Szymanski (89) considered rest to be identical to sleep. This simplification may be adequate for animals such as the rat and mouse in which little quiet waking occurs. In the cow, however, drowsiness has been shown to occupy a large part of each 24-hr period, making EEG recordings necessary to establish sleep/wake values. Normative values for the cow have been provided by Ruckebusch (73,74). Sleep in cows in the stall tended to be monophasic and nocturnal and was thus comparable to sleep in horses. The nocturnal sleep episodes tended to begin after a period of rumination. Cows recorded in a shielded chamber with food *ad libitum* exhibited sleep episodes throughout the entire 24-hr period (78).

Goats recorded and observed for many months distributed their sleep episodes irregularly during the nycthemeron (71). Polygraphic records of daytime hours revealed a polyphasic sleep/wake pattern with 0 to 6 sleep episodes per 12 hr (45). Sleep in Saanen goats and Merino sheep was found to be polyphasic, and the sleep/wake pattern independent from rumination (10). A comparison of horse, cow, and pig showed that the latter spent by far the longest time lying down and that during the day as well as the night pigs slept more than the other animals. The sleep pattern was polyphasic and nocturnal (80) (Fig. 1). Similarly, the incidence of sleep in juvenile pigs was highest at night (69).

The majority of studies of sleep placement have been carried out in captive and domestic ungulates. Nevertheless, several species also have been observed in the wild. Under these conditions it is difficult to determine if an animal lying down is awake or sleeping. In view of this difficulty, only observations that include descriptions of "lying without ruminating" are discussed.

Several 24-hr observations (on full-moon nights) of a wild ruminant ungulate, the impala (*Aepycerus melampus*), revealed little recumbency of individual animals in a herd during the daytime and major bouts of rest (lying and lying/ruminating) during the night. Sleep bouts (lying without ruminating) were nocturnal, scarce, and of less than 10-min duration (41). Thus, during the 12 daylight hours, lying down without rumination only comprised between 0.11 and 0.47 hr. In contrast, during 12-hr dark it comprised between 5.45 and 7.01 hr, depending on the season. A similar rest/activity pattern was observed for the waterbuck and several herd-forming antelopes.

Thomson's gazelle (*Gazella thomsoni*) was observed in one study lying down and sleeping between 0900 and 1100 hr, after a sustained period of activity that began after sunrise and, in another study, resting several times (10–11) during the day in alternation with bouts of activity (105). During the daily rest period, several short sleep episodes (5–10 min) were infrequently observed at irregular intervals (based on observations every 15 min in 10 24-hr periods). TST was estimated at 0.5 to 2 hr. Sleep was clearly polyphasic, but rest episodes were longer and included sleep more often during the night. Also, during the night more individuals (as much as 100%) of a herd engaged in lying down, whereas during the day the number of animals resting comprised 50% to 90% and seldom exceeded 90%. In contrast, Gerenuks (*Litocranius walleri*), observed only during daytime, were seen lying down (sleep not mentioned) mainly during the middle of the day (49).

Male giraffes (*Giraffa camelopardalis*) showed the most marked temporal pattern for lying down, relative to other activities (e.g., locomotion, feeding). It occurred in two short periods, one in the late morning and the other in the early afternoon. Females were not seen lying down (49).

Seven female African elephants (*Laxodonta africana*) have been observed for 24 hr in the wild. They showed a main sleep period (defined as lying down) during the small hours of the morning (between 0400 and 0700 hr) and a shorter rest period in the early afternoon (1200–1400 hr), when the animals appeared to be dozing (114). Lying down occurred only during the night. From these data it can be concluded that elephants in the wild are monophasic with a nocturnal placement of sleep. EEG recordings would be needed to determine if the midday rest period involves sleep.

Three species belonging to the order Hydracoidea were recorded by telemetry. The rock hyrax (*Procavia*) was polyphasic and lacked a circadian sleep/wake distribution, whereas among the representatives of two closely related genera, the rock hyrax (*Heterohyrax*) slept more at night and the tree hyrax (*Dendrohyrax*) slept predominantly during the day (84) (Table 1).

Edentates

Polygraphic features of waking, NREM sleep, and REM sleep in edentates were typically mammalian. The armadillo (*Dasypus novemcinctus*) slept for most of a 24-hr recording period (72.5%–77%, consisting of 11–21 sleep epochs), with the exception of an active period in the early evening lasting 3 to 5 hr (65,100). The sleep periods often consisted of many brief awakenings. EEG recordings have been made in the three-toed sloth (*Bradypus tridactylus*) for 90 sec of each 15-min period during 24 hr, with natural light during the day and dim light during the night. Light sleep, deep sleep, and waking were present throughout the entire 24-hr period (23), and TST comprised 66% of recording time. Although the sleep/wake pattern was polyphasic, the longest sleep period occurred between 0600 and 1000 hr.

Insectivores

Sleep in the hedgehog (*Erinaceus europaeus*) recorded in light/dark ratio of 13:11 was similar to that of other nocturnal mammals. It comprised 72.6% of 24 hr and predominated in the light period (28). A substantial amount of drowsiness (25%) was included in this value. In contrast, in the mole (*Scalopus aquaticus*) five EEG states were distinguished: waking, drowsy, spindle sleep, NREM sleep, and REM sleep. TST was 35.2% of a 24-hr period. A mean of 24 sleep periods occurred at irregular intervals during the 24-hr recording period. Periods of sleep lasting 1 to 3 hr alternated with periods of active waking; however, the mean duration of uninterrupted sleep epochs was 22 min (3).

Tree Shrew

The vigilance states of the tree shrew (*Tupaia glis*) have been scored into waking, drowsiness, light and deep NREM sleep, and REM sleep. TST comprised 65.8% of the 24-hr period (26% drowsiness was not included). Sleep predominated during the night (11).

Marsupials and the Echidna

From the running-wheel activity data obtained in two Tasmanian devils recorded in light/dark 12:12 hr, it can be deduced that they sleep mainly during the light period, when there is little running activity (61). In the echidna (*Tachyglossus aculeatus*) two patterns of sleep/wake behavior were seen. Some animals became active toward the end of the day, whereas others tended to be polyphasic. Sleep comprised 35.8% of 24 hr, and drowsiness comprised a similar amount; no REM sleep was observed (4). The American opossum (*Didelphis marsupialis*) exhibited polyphasic sleep with relatively long (1–4 hr) uninterrupted sleep bouts (100). TST was extremely high, comprising 80% of 24 hr. The rat kangaroo (*Potorous apicalis*) showed a diurnal placement of sleep (70.5% of sleep in 12-hr light) that consisted of long sleep bouts interrupted by short (90 sec) wakefulness. Only 18.2% of sleep, which was primarily short episodes of NREM sleep, took place during the dark (7).

In summary, in nonplacental mammals, sleep stages were found to be similar to those found in placental mammals; with the exception of the echidna, two stages (NREM and REM sleep) were discriminated. Generally the sleep pattern tended to be polyphasic.

STABILITY OF SLEEP PATTERN AND SLEEP PLACEMENT

There are marked differences in the sleep/wake patterns across species. However, the interindividual variation within each species is relatively small. The sleep/wake distribution, particularly in rodents, is robust and resistant to manipulation, indicating a stable and organized sleep/wake pattern (101,106). This appears to be the case in many other species as well, including human (107,108), cat (52), horse, cow, and pig (73,74). For example, cows recorded in light/dark 12:12 hr were prevented from lying down during the last 2 hr of the light period and during the entire dark period for 4 weeks, but nevertheless adhered to a nocturnal placement of sleep. Two thirds of NREM sleep occurred during the night in spite of enforced standing (TST in control dark was 17.4% and in light was 0%; TST in "restricted" dark was 8.1% and in light was 5.9%) (75).

On the other hand, several factors do influence the sleep/wake pattern.

Effects of the laboratory versus natural surroundings as well as of restrained versus unrestrained recording conditions on sleep patterns in several apes and monkeys have been extensively documented (14,17). Furthermore, it has been established for several species that the sleep/wake pattern is susceptible to the amount and availability of food. Limitation of food to the light period of the light/dark 12:12-hr cycle induced rats to redistribute their sleep so that equal amounts occurred in light and dark (60). Large effects of feeding and stall conditions were reported in cows and horses (73,74,78). Thus drowsiness in the cow was reduced from 42% to 9% by changing food from hay to pelleted grass (73). The distribution of sleep episodes in free and foraging cattle was found to be biphasic, occurring at dawn and dusk after the two major grazing periods (94). When the same animals were recorded in the stable, TST increased by 30% and the sleep pattern became monophasic and nocturnal (76). The nocturnal sleep episodes tended to begin after a period of rumination. When the animals were isolated in a chamber with little external influence and given food *ad libitum*, the sleep episodes were equally distributed throughout the 24-hr period (78). Thus, in herbivores, in which sleep is particularly labile, the influence of the environment is marked.

The cat also exhibits a clear susceptibility to the laboratory environment and feeding. The largest circadian difference in the sleep/wake pattern was observed when food was reduced to the basic amount necessary to avoid weight gain and offered as three meals (during the light period of the light/dark 12:12 photoperiod). Under these conditions the wake pattern included little drowsiness, and most sleep occurred during the 12-hr dark period (79). The authors concluded that a cat, like stalled cattle, may exhibit a "luxury form of sleep" owing to the ease of food acquisition and excessive consumption. Also, the distribution of sleep and waking within a sleep cycle was found to differ from control when cats were trained to expect a food reward. The average sleep/wake cycle duration was shortened (from 104 to 93 min) and waking between sleep periods was prolonged (from 26 to 35 min) (52). The reduction of external stimuli, which were mainly present during the light period, induced an increase of sleep during the 12-hr light period and the 24-hr recording period (79).

A large effect on sleep/wake consolidation was observed in mice when they were provided with a running wheel (111). The ratio of sleep in light/dark 12:12 hr was changed in that sleep was enhanced in the light period when a running wheel was available. Chipmunks (*Tamias striatus*) exhibited a more nocturnal placement of sleep when they were housed in a large complex cage (25). Similar effects of enclosure size on sleep placement were found in a diurnal rodent (*Octodon degus*) (26).

On the basis of these results it is questionable that the sleep/wake distribution obtained in the laboratory reflects natural behavior. In the field, the conditions are more similar to those in the wild and the sleep/wake patterns may more closely reflect the natural behavior of animals. On the other hand, data based on long-term records of sleep/wake polygraphy under constant condition and in isolation have served, at least to some extent, to exclude the effects of the environment.

ENDOGENOUS NATURE OF THE SLEEP/WAKE PATTERN

The endogenous nature of the sleep/wake pattern has been demonstrated in several species. Sleep placement, as well as the sleep pattern, has been shown to be influenced by the light/dark rhythm. Squirrel monkeys were shown to maintain a stable consolidated sleep/wake rhythm in constant light. More sleep was present during the subjective day (i.e., the inactive half of the circadian period, 8.3%) than in the 12-hr light period of the light/dark 12:12-hr regime. This change was followed by a larger amount of waking during the subjective night (21.8% versus 14.0% in the dark period of light/dark) (113). The sleep/wake rhythm was dampened when mice were recorded in constant dark or constant light (58,112), and sleep became more fragmented (112).

On the basis of observations in one cat it has been reported that cats maintained in constant conditions lack a circadian rest/activity cycle (37). In contrast, a weak but significant circadian sleep/wake rhythm was found in preliminary studies of cats kept for 8 days in constant dim illumination (59). Cats recorded in isolation in dark/dark for several weeks or months maintained a free-running rest/activity rhythm, but in light/light the pattern became random (66). Also, sleep in cats recorded in isolation in dark/dark for 12, 15, and 60 days predominated during the subjective night. Thus, TST decreased during the subjective day and conversely increased during the subjective night. The circadian sleep/wake pattern persisted and free-ran with a period of 24.2 hr for approximately 7 to 12 days and then gradually disappeared (47).

Rodents recorded under constant conditions exhibited free-running sleep/wake rhythms. For example, mice kept in dark/dark for 60 to 280 days still exhibited clear, free-running circadian rhythms for each of the three stages: waking, NREM sleep, and REM sleep (68). Mice strains were shown to differ in the daily rhythm of sleep and waking as well as in TST values (e.g., CBA strain, 791 min; C57BR strain, 705 min) (97,98). The CBA strain exhibited a smaller difference between the 12-hr light and 12-hr dark period than the C57BR strain. Both strains exhibited more sleep in the dark period (see also Table 1). Such data provide evidence that genetic factors may be involved in the organization of the sleep/wake pattern.

Complete bilateral lesions of the nuclei suprachiasmatichi resulted in a disruption of sleep/wake pattern in the rat, mouse, and Siberian chipmunk (*Eutamias sibiricus*) (24,39,40,56,81,91). The NREM/REM sleep cycle was not affected. On the basis of these results it can be concluded that there is a strong circadian component that regulates sleep distribution.

EEG SPECTRAL ANALYSIS AND SLEEP/WAKE PATTERN

To determine the species-specific type of sleep/wake pattern, it seems relevant to assess whether long napping episodes are occasional or if they recur in a predictable manner. Carnivores such as the dog and cat exhibit a nocturnal placement of sleep but also have a strong propensity for it during daylight hours. However, the amount of daytime napping in these animals is largely dependent on the environment, a finding that also has been reported for human subjects (see Chapter 7). To collect data under natural conditions would be more relevant for understanding the significance of the sleep/wake pattern. However, natural conditions are unstable and virtually impossible to simulate in the laboratory. The proportion of sleep between light and dark seems to be an adequate measure to characterize sleep distribution. Thus, in rodents, in which a large amount of sleep also takes place in the active period of the light/dark cycle, the sleep pattern is considered to be typically polyphasic. It is, therefore, in these animals that the significance of the polyphasic sleep/wake pattern should be examined.

EEG frequency analysis of continuous 24-hr recordings of two rodent species (rat and hamster) have revealed processes that are superimposed on the polyphasic sleep/wake pattern. In both species, EEG slow wave activity showed a marked consistent decreasing trend in the course of the 12-hr light period, the circadian rest phase of the animals (92,95) (Figs. 2 and 3). Comparison of the slow wave activity decrease in NREM sleep in the course of the sleep period (i.e., the difference between maximum and minimum percentages for 12 hr in the rat and hamster and 8 hr in the human) revealed similar values for the three species (rat, 61%; hamster, 63%; human, 58%) (92). This finding is of interest because the decrease of EEG slow wave activity does not seem to depend on (a) sleep distribution (monophasic or polyphasic), (b) sleep continuity, (c) sleep-cycle length, which differs greatly between rodents and humans (rat, 12 min; hamster, 11.5 min; human, 90 min), and (d) the amount of sleep that takes place in the active part of the circadian light/dark cycle (rat, 28.0%; hamster, 50.4%; human, none). The independence of the progressive decline of slow wave activity during the sleep period with waking interruptions reveals the presence of a stable, continuous process underlying the sleep/wake pattern.

Since the main difference between monophasic, biphasic, and polyphasic rhythms appears to be the proportion of sleep in the active period, a comparison of sleep in the two periods may help to elucidate their function. In several species (e.g., ground squirrel, rat, rabbit) the mean episode durations of waking, NREM sleep, and REM sleep have been found to differ markedly

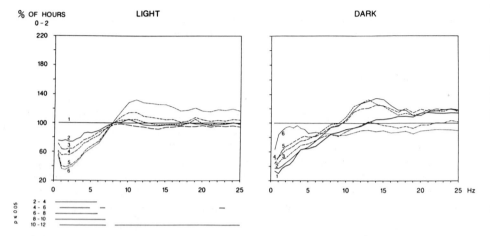

FIG. 3. Spectral distribution of relative EEG power density in NREM sleep computed for consecutive 2-hr periods. The curves connect mean values of six Syrian hamsters for the 12-hr light period and four hamsters for the 12-hr dark period of the light/dark 12:12 hr. For each animal the mean value of the first 2 hr of the light period was defined as 100% (*horizontal line*). The values for the subsequent 2-hr periods are plotted relative to the value for hours 0 to 2. Numbers designate consecutive 2-hr periods. Lines below the abscissa indicate those frequency bands in which the spectral values differ significantly from those of hours 0 to 2 ($p < 0.05$; two-tailed, Wilcoxon paired signed ranks test). (From ref. 92.)

between the 12-hr light and 12-hr dark period (15,64,104). In other species, such as the guinea pig and the hamster, no differences were found (38,92). In addition, spectral analysis of the sleep EEG has revealed qualitative differences between sleep in the 12-hr light period and 12-hr dark period. NREM sleep EEG frequencies in the delta band decreased progressively in the course of the light period (i.e., the main sleep period). The level remained low during most of the dark period and rose sharply at the dark-light transition. In contrast, the EEG frequencies above 7 Hz remained on a relatively stable level during most of the light period, increased during the last hours, and remained on a high level during most of the dark period (92,95). Thus, circadian factors seem to play an important role in the distribution of sleep/wake activity and the length of sleep/wake episodes, as well as the EEG spectrum in NREM sleep.

CONCLUSION

Sleep distribution has been shown to vary between two basic patterns: monophasic sleep placement within a 24-hr period and polyphasic sleep placement. Biphasic sleep placement (i.e., two sleep periods within 24 hr) is considered here to be a variant of monophasic placement. In humans, apes, and monkeys, the sleep pattern is monophasic, although all three occasionally exhibit napping (biphasic pattern). Of the three, however, only humans have a consolidated sleep period, whereas apes and monkeys exhibit a large degree of sleep fragmentation.

Frequent interruptions of sleep by wakefulness are present in most mammalian species, independent of the 24-hr sleep/wake distribution. These short waking episodes may subserve a function of vigilance (e.g., environmental scanning). It is possible that animals have a need for vigilance during the sleep period that may have become unnecessary for the survival of humans.

The environment influences not only TST and sleep pattern, but also sleep placement. Ruckebusch (76) concluded that the species differences in these parameters may be attributed to "adaptive situational responsivity." Thus, within species there is a clear response to different situations, such as sleep deprivation, noise, light, anxiety, hunger. These responses include shifts of the sleep/wake cycle in time, lengthening and shortening of both sleep and waking, sleep fragmentation, and sleep intensification as responses to the exigencies of various situations. In conclusion, napping and polyphasic sleep patterns allow a large degree of flexibility of adaptation to rapid changes in the environment.

ACKNOWLEDGMENTS

I thank Dr. Alexander Borbély for critical reading of the manuscript. The study was supported by the Swiss National Science Foundation, grant 3.234-0.85.

REFERENCES

1. Adams, P. M., and Barratt, E. S. (1974): Nocturnal sleep in squirrel monkeys. *Electroencephalogr. Clin. Neurophysiol.*, 36:201–204.
2. Adey, W. R., Kado, R. T., and Rhodes, J. M. (1963): Sleep: Cortical and subcortical recordings in the chimpanzee. *Science*, 141:932–933.
3. Allison, T., and Van Twyver, H. (1970): Sleep in moles, Scalopus aquaticus and Condylura cristata. *Exp. Neurol.*, 27:564–578.
4. Allison, T., Van Twyver, H., and Goff, W. R. (1972): Electrophysiological studies of the echidna, Tachyglossus aculeatus. *Arch. Ital. Biol.*, 110:119–144.
5. Altmann, S. A., and Altmann, J. (1970): Baboon ecology: African field research. In: *Bibliotheca Primatologica, Vol. 12*, edited by H. Hofer, A. H. Schultz, and D. Starck. S. Karger, Basel.
6. Aschoff, J. (1966): Circadian activity pattern with two peaks. *Ecology*, 47:657–662.
7. Astic, L., and Royet, J. P. (1974): Sommeil chez le rat-kangourou, Potorous apicalis. Etude chez l'adulte et chez le jeune un mois avant la sortie definitive du marsupium. Effets du sevrage. *Electroencephalogr. Clin. Neurophysiol.*, 37:483–489.
8. Batini, C., Radulovacki, M., Kado, R. T., and Adey, W. (1967): Effect of interhemispheric transection on the EEG patterns in sleep and wakefulness in monkeys. *Electroencephalogr. Clin. Neurophysiol.*, 22:101–112.
9. Baumgardner, D. J., Ward, S. E., and Dewsbury, D. A. (1980): Diurnal patterning of eight activities in 14 species of muroid rodents. *Animal Learn. Behav.*, 8:322–330.

10. Bell, F. R., and Itabisashi, T. (1973): The electroencephalogram of sheep and goats with special reference to rumination. *Physiol. Behav.*, 11:503–514.
11. Berger, R. J., and Walker, J. M. (1972): A polygraphic study of sleep in the tree shrew (Tupaia glis). *Brain Behav. Evol.*, 5:54–69.
12. Bert, J., Collomb, H., and Martino, A. (1967): L'electroencephalogramme de sommeil d'un pro-simien. Sa place dans l'organisation du sommeil chez les primates. *Electroencephalogr. Clin. Neurophysiol.*, 23:342–350.
13. Bert, J., Kripke, D. F., and Rhodes, J. (1970): Electroencephalogram of the mature chimpanzee: Twenty-four hour recordings. *Electroencephalogr. Clin. Neurophysiol.*, 28:368–373.
14. Bert, J., Balzamo, E., Chase, M., and Pegram, G. V. (1975): The sleep of the baboon, papio papio, under natural conditions and in the laboratory. *Electroencephalogr. Clin. Neurophysiol.*, 39:657–662.
15. Borbély, A. A., and Neuhaus, H-U. (1979): Sleep-deprivation: Effect on sleep and EEG in the rat. *J. Comp. Physiol.*, 133:71–87.
16. Bowersox, S. S., Baker, T. L., and Dement, W. C. (1984): Sleep-wakefulness patterns in the aged cat. *Electroencephalogr. Clin. Neurophysiol.*, 58:240–252.
17. Breton, P., Gourmelon, P., and Court, L. (1986): New findings on sleep stage organization in squirrel monkeys. *Electroencephalogr. Clin. Neurophysiol.*, 64:563–567.
18. Campbell, S. S., and Tobler, I. (1984): Animal sleep: A review of sleep duration across phylogeny. *Neurosci. Biobehav. Rev.*, 8:269–300.
19. Carpenter, C. R. (1965): The howlers of Barro Colorado Island. In: *Primate Behavior. Field Studies in Monkeys and Apes*, edited by I. DeVore, pp. 250–277. Holt, Rinehart and Winston, New York.
20. Crowley, T. J., Kripke, D. F., Halberg, F., Pegram, G. V., and Schildkraut, J. J. (1972): Circadian rhythms of Macaca mulatta: Sleep, EEG, body and eye movement and temperature. *Primates*, 13:149–168.
21. Dallaire, A., and Ruckebusch, Y. (1974): Rest-activity cycle and sleep patterns in captive foxes (Vulpes vulpes). *Experientia*, 30:59–60.
22. Dallaire, A., and Ruckebusch, Y. (1974): Sleep patterns in the pony with observations on partial perceptual deprivation. *Physiol. Behav.*, 12:789–796.
23. De Moura, G. A., Espe Huggins, S., and Gomes Lines, S. (1983): Sleep and waking in the three-toed sloth, Bradypus tridactylus. *Comp. Biochem. Physiol. [A]*, 76:345–355.
24. Eastman, C. I., Mistlberger, R. E., and Rechtschaffen, A. (1984): Suprachiasmatic nuclei lesions eliminate circadian temperature and sleep rhythms in the rat. *Physiol. Behav.*, 32:357–368.
25. Estep, D. Q., Fischer, R. B., and Gore, W. T. (1978): Effects of enclosure size and complexity on the activity and sleep of the eastern chipmunk Tamias striatus. *Behav. Biol.*, 23:249–253.
26. Fischer, R. B., Meunier, G. F., O'Donoghue, P. J., Rhodes, D. L., and Schafenaker, A. M. (1980): Effects of enclosure size on activity and sleep of a hystricomorph rodent (Octodon degus). *Bull. Psychonom. Soc.*, 16:273–275.
27. Folk, G. E., Jr. (1964): Daily physiological rhythms of carnivores exposed to extreme changes in arctic daylight. *Fed. Proc.*, 23:1221–1228.
28. Fourré, A. M., Rodriguez, F. L., and Vincent, J. D. (1974): Etude polygraphique de la veille et du sommeil chez le Hérisson (Erinaceus europaeus L.). *C R Soc. Biol.*, 168:959–964.
29. Freemon, F. R., McNew, J. J., and Adey, W. R. (1969): Sleep of unrestrained chimpanzee: Cortical and subcortical recordings. *Exp. Neurol.*, 25:129–137.
30. Freemon, F. R., McNew, J. J., and Adey, W. R. (1971): Chimpanzee sleep stages. *Electroencephalogr. Clin. Neurophysiol.*, 31:485–489.
31. Friedman, L., Bergmann, B., and Rechtschaffen, A. (1979): Effects of sleep deprivation on sleepiness, sleep intensity, and subsequent sleep in the rat.. *Sleep*, 1:369–391.
32. Friedmann, J., Estep, D., Huntley, A., Dewsbury, D., and Webb, W. (1975): On the validity of using laboratory mice versus wild mice for sleep research. *Sleep Res.*, 4:142.
33. Goodall, J. M. (1962): Nest building behavior in the free ranging chimpanzee. *Ann. NY Acad. Sci.*, 102:455–467.
34. Goodall, J. M. (1965): Chimpanzees of the Gombe Stream Reserve. In: *Primate Behavior.*

Field Studies of Monkeys and Apes, edited by I. DeVore, pp. 425–449. Holt, Rinehart and Winston, New York.

35. Gordon, C. R., and Lavie, P. (1984): Effect of adrenergic blockers on the dog's sleep-wake pattern. *Physiol. Behav.,* 32:345–350.
36. Gottesmann, C. (1967): Recherche sur la psychophysiologie du sommeil chez le rat. Thesis, CNRS, Paris.
37. Hawking, F., Lobban, M., Gammage, K., and Worms, M. (1971): Circadian rhythms (activity, temperature, urine and microfilarie) in dog, cat, hen, duck, Thamnomys and Gerbillus. *J. Interdisc. Res.,* 2:455–473.
38. Ibuka, N. (1984): Ontogenesis of circadian sleep-wakefulness rhythms and developmental changes of sleep in the altricial rat and in the precocial guinea pig. *Behav. Brain Res.,* 11:185–196.
39. Ibuka, N., and Kawamura, H. (1975): Loss of circadian rhythm in sleep-wakefulness cycle in the rat by suprachiasmatic nucleus lesions. *Brain Res.,* 96:76–81.
40. Ibuka, N., Nihonmatsu, I., and Sekiguchi, S. (1980): Sleep-wakefulness rhythms in mice after suprachiasmatic nucleus lesions. *Waking Sleeping,* 4:167–173.
41. Jarman, M. V., and Jarman, P. J. (1973): Daily activity of impala. *East Afr. Wildlife J.,* 11:75–92.
42. Jouvet-Mounier, D., and Astic, L. (1966): Etude du sommeil chez le cobaye adulte et nouveau-né. *C R Soc. Biol.,* 160:1453–1457.
43. Kaitin, K. I., Kilduff, T. S., and Dement, W. C. (1986): Sleep fragmentation in canine narcolepsy. *Sleep,* 9:116–119.
44. Kavanau, J. L. (1971): Locomotion and activity phasing of some medium-sized mammals. *J. Mammals,* 52:386–403.
45. Klemm, W. R. (1966): Sleep and paradoxical sleep in ruminants. *Proc. Soc. Exp. Biol. Med.,* 121:635–638.
46. Kripke, D. F., Reite, M. L., Pegram, G. V., Stephens, L. M., and Lewis, O. F. (1968): Nocturnal sleep in rhesus monkeys. *Electroencephalogr. Clin. Neurophysiol.,* 24:582–586.
47. Kuwabara, N., Seki, K., and Aoki, K. (1986): Circadian, sleep and brain temperature rhythms in cats under sustained daily light-dark cycles and constant darkness. *Physiol. Behav.,* 38:283–289.
48. Leinonen, L., and Stenberg, D. (1986): Sleep in Macaca arctoides and the effects of Prazosin. *Physiol. Behav.,* 37:199–202.
49. Leuthold, B. M., and Leuthold, W. (1978): Daytime activity patterns of gerenuk and giraffe in Tsavo National Park, Kenya. *East Afr. Wildlife J.,* 16:231–243.
50. Lucas, E. A., Powell, E. W., and Murphee, O. D. (1976): A comparison of dog and cat baseline sleep-wake (S-W) patterns. *Sleep Res.,* 5:98.
51. Lucas, E. A., Powell, E. W., and Murphee, O. D. (1977): Baseline sleep-wake patterns in the pointer dog. *Physiol. Behav.,* 19:285–291.
52. Lucas, E. A., and Sterman, M. B. (1974): The polycyclic sleep-wake cycle in the cat: Effects produced by sensorimotor rhythm conditioning. *Exp. Neurol.,* 42:347–368.
53. Matsumoto, J., Nishisho, T., Suto, T., Sadahiro, T., and Miyoshi, M. (1967): Normal sleep cycle of male albino rats. *Proc. Jpn. Acad.,* 43:62–64.
54. McNew, J. J., Burson, R. C., Hoshizaki, T., and Adey, W. R. (1972): Sleep-wake cycle of an unrestrained isolated chimpanzee under entrained and free running conditions. *Aerospace Med.,* 43:155–161.
55. Michel, F., Klein, M., Jouvet, D., and Valatx, J. L. (1963): Etude polygraphique du sommeil chez le rat. *C R Soc. Biol.,* 155:775–785.
56. Mistlberger, R. E., Bergmann, B. M., and Rechtschaffen, A. (1983): Recovery sleep following sleep deprivation in intact and suprachiasmatic nuclei lesioned rats. *Sleep,* 6:217–233.
57. Mitler, M. M., and Dement, W. C. (1977): Sleep studies in canine narcolepsy: Pattern and cycle comparisons between affected and normal dogs. *Electroencephalogr. Clin. Neurophysiol.,* 43:691–699.
58. Mitler, M. M., Lund, R., Sokolove, P. G., Pittendrigh, C. S., and Dement, W. C. (1977): Sleep and activity rhythms in mice: A description of circadian patterns and unexpected disruptions in sleep. *Brain Res.,* 131:129–146.
59. Moore-Ede, M. C., Gander, P. H., Eagan, S. M., and Martin, P. (1981): Evidence for weak circadian organization in the cat sleep-wake cycle. *Sleep Res.,* 81:298.

60. Mouret, J. R., and Bobillier, P. (1971): Diurnal rhythms of sleep in the rat: Augmentation of paradoxical sleep following alternations of the feeding schedule. *Int. J. Neurosci.,* 2:265–270.
61. Packer, W. C. (1966): Wheel-running behavior in the marsupial sarcophilus. *J. Mammals,* 47:698–701.
62. Pellet, J., and Béraud, G. (1967): Organisation nycthemerale de la veille et du sommeil chez le cobaye (Cavia porcellus). *Physiol. Behav.,* 2:131–137.
63. Perachio, A. A. (1971): Sleep in the nocturnal primate, Actus trivirgatus. *Proceedings of the 3rd International Congress on Primates, Vol. 2,* Zurich, 1970. Karger, Basel, pp. 54–60.
64. Pivic, R. T., Bylsma, F. W., and Cooper, P. (1986): Sleep-wakefulness rhythms in the rabbit. *Behav. Neur. Biol.,* 45:275–286.
65. Prudom, A., and Klemm, W. (1973): Electrographic correlates of sleep behavior in a primitive mammal, the Armadillo Dasypus novemcinctus. *Biol. Behav.,* 10:275–282.
66. Randall, W., Johnson, R. F., Randall, S., and Cunningham, J. T. (1985): Circadian rhythms in food intake and activity in domestic cats. *Behav. Neurosci.,* 99:1162–1175.
67. Reite, M., Rhodes, J., Kavan, E., and Adey, W. (1965): Normal sleep patterns in macaque monkey. *Arch. Neurol.,* 12:133–144.
68. Richardson, G. S., Moore-Ede, M. C., Czeisler, C. A., and Dement, W. C. (1985): Circadian rhythms of sleep and wakefulness in mice: Analysis using longterm automated recording of sleep. *Am. J. Physiol.,* 248:R320–R330.
69. Robert, S., and Dallaire, A. (1986): Polygraphic analysis of the sleep-wake states and the REM sleep periodicity in domesticated pigs (Sus scrofa). *Physiol. Behav.,* 37:289–293.
70. Roldan, E., Weiss, T., and Fifkova, E. (1963): Excitability changes during the sleep cycle of the rat. *Electroencephalogr. Clin. Neurophysiol.,* 15:775–785.
71. Ruckebusch, Y. (1962): Activité corticale au cours du sommeil chez la chèvre. *C R Soc. Biol.,* 156:867–870.
72. Ruckebusch, Y. (1963): Etude EEG et comportementale des alternances veille-sommeil chez l'ane. *C R Soc. Biol.,* 157:840–844.
73. Ruckebusch, Y. (1972): Comparative aspects of sleep and wakefulness in farm animals. In: *The Sleeping Brain. Perspectives in the Brain Sciences,* edited by M. Chase, pp. 23–28. Brain Information Service/Brain Research Institute, UCLA, Los Angeles.
74. Ruckebusch, Y. (1972): The relevance of drowsiness in the circadian cycle of farm animals. *Animal Behav.,* 20:637–643.
75. Ruckebusch, Y. (1974): Sleep deprivation in cattle. *Brain Res.,* 78:495–499.
76. Ruckebusch, Y. (1977): Environment and sleep patterns in domestic animals. In: *Sleep 1976,* edited by W. P. Koella, pp. 159–163. Karger, Basel.
77. Ruckebusch, Y., Barbey, P., and Guillemot, P. (1970): Les états de sommeil chez le cheval (Equus caballus). *C R Soc. Biol.,* 164:658–665.
78. Ruckebusch, Y., Dallaire, A., and Toutain, P. L. (1975): Sleep patterns and environmental stimuli. In: *Sleep 1974,* edited by P. Levin and W. P. Koella, pp. 273–276. Karger, Basel.
79. Ruckebusch, Y., and Gaujoux, M. (1976): Sleep patterns of the laboratory cat. *Electroencephalogr. Clin. Neurophysiol.,* 41:483–490.
80. Ruckebusch, Y., and Morel, M. T. (1968): Etude polygraphique du sommeil chez le porc. *C R Soc. Biol.,* 162:1346–1354.
81. Sato, T., and Kawamura, H. (1984): Effects of bilateral suprachiasmatic nucleus lesions on the circadian rhythms in a diurnal rodent, the Siberian chipmunk (Eutamias sibiricus). *J. Comp. Physiol. [A],* 155:745–752.
82. Schaller, G. B. (1965): The behavior of the mountain gorilla. In: *Primate Behavior. Field Studies of Monkeys and Apes,* edited by I. DeVore, pp. 324–367, 474–481. Holt, Rinehart and Winston, New York.
83. Scott, M. D., and Causey, K. (1973): Ecology of feral dogs in Alabama. *J. Wildlife Management,* 37:253–265.
84. Snyder, F. (1974): Sleep-waking patterns of hydracoidea. *Sleep Res.,* 3:87.
85. Steinhart, P. (1937): Der Schlaf des Pferdes. Seine Dauer, Tiefe, Bedingungen. *Z. Veterinarkunde,* 6:193–232.
86. Sterman, M. B., Knauss, T., Lehmann, D., and Clemente, C. D. (1965): Circadian sleep

and waking patterns in the laboratory cat. *Electroencephalogr. Clin. Neurophysiol.,* 19:509–517.

87. Susic, V. T., and Masirevic, G. (1986): Sleep patterns in the Mongolian gerbil, Meriones unguiculatus. *Physiol. Behav.,* 37:257–261.
88. Swett, C. (1969): Daytime sleep patterns in freeranging Rhesus monkeys. *Psychophysiology,* 6:227–228.
89. Szymanski, J. S. (1920): Aktivitat und Ruhe bei Tieren und Menschen. *Z. Allg. Physiol.,* 18:105–162.
90. Takahashi, Y., Ebihara, S., Nakamura, Y., Nishi, C., and Takahashi, K. (1978): Circadian sleep and waking patterns in the laboratory dog. *Sleep Res.,* 7:144.
91. Tobler, I., Borbély, A. A., and Groos, G. (1983): The effect of sleep deprivation on sleep in rats with suprachiasmatic lesions. *Neurosci. Lett.,* 42:49–54.
92. Tobler, I., and Jaggi, K. (1987): Sleep and EEG spectra in the Syrian hamster (Mesocricetus auratus) under baseline conditions and following sleep deprivation. *J. Comp. Physiol.,* 161:449–459.
93. Tobler, I., and Sigg, H. (1986): Long-term motor activity recording of dogs and the effect of sleep deprivation. *Experientia,* 42:987–991.
94. Toutain, P. L., and Ruckebusch, Y. (1973): Sommeil paradoxal et environment. *C R Soc. Biol.,* 167:550–555.
95. Trachsel, L., Tobler, I., and Borbély, A. A. EEG analysis of nonREM sleep in the rat. *Am. J. Physiol. (in press).*
96. Ursin, R. (1971): Differential effect of sleep deprivation on the two slow wave sleep stages in the cat. *Acta Physiol. Scand.,* 83:352–361.
97. Valatx, J. L., and Bugat, R. (1974): Facteurs genetiques dans le determinisme du cycle veille-sommeil chez la souris. *Brain Res.,* 69:315–330.
98. Valatx, J. L., Bugat, R., and Jouvet, M. (1972): Genetic studies of sleep in mice. *Nature,* 238:226–227.
99. Van Twyver, H. (1969): Sleep patterns of five rodents. *Physiol. Behav.,* 4:901–905.
100. Van Twyver, H., and Allison, T. (1970): Sleep in the opossum Didelphis marsupialis. *Electroencephalogr. Clin. Neurophysiol.,* 29:181–189.
101. Vivaldi, E., Pastel, R. H., Fernstrom, J. D., and Hobson, A. (1984): Long term stability of rat sleep quantified by microcomputer analysis. *Electroencephalogr. Clin. Neurophysiol.,* 58:253–265.
102. Vuillon-Cacciuttolo, G., Balzamo, E., Petter, J. J., and Bert, J. (1976): Cycle veille-sommeil étudié par télémetrie chez un lemurien (Lemur macaco fulvus). *Rev. Electroencéphalogr. Neurophysiol. Clin.,* 6:34–36.
103. Vuillon-Cacciuttolo, G., and Seri, B. (1978): Effets de la section des nerfs optiques chez le baboun sur l'activité à type de pointes genouillées et corticales au cours des divers états de vigilance. *Electroencephalogr. Clin. Neurophysiol.,* 44:754–768.
104. Walker, J. M., Glotzbach, S. F., Berger, R. J., and Heller, H. C. (1977): Sleep and hibernation in ground squirrels (Citellus spp) electrophysiological observations. *Am. J. Physiol.,* 233:R213–R221.
105. Walther, F. R. (1973): Round-the-clock activity of Thomson's gazelle (Gazella thomsoni Gunther 1884) in the Serengeti National Park. *Z. Tierpsychol.,* 32:75–105.
106. Webb, W. B. (1971): Sleep behavior as a biorhythm. In: *Biological Rhythms and Human Performance,* edited by W. P. Colquhoun, pp. 149–177. Academic Press, London.
107. Webb, W. B. (1981): Some theories about sleep and their clinical implications. *Psychiatr. Ann.,* 11:415–422.
108. Webb, W. B., and Agnew, H. W., Jr. (1965): Sleep: Effects of a restricted regime. *Science,* 150:1745–1746.
109. Weiss, T., and Roldan, E. (1964): Comparative study of sleep cycles in rodents. *Experientia,* 20:280–281.
110. Weitzman, E. D., Kripke, D. F., Pollack, C., and Dominguez, J. (1965): Cyclic activity in sleep of macaca mulatta. *Arch. Neurol.,* 12:463–467.
111. Welsh, D. K., Richardson, G. S., and Dement, W. C. (1985): Effect of running wheel availability on the circadian pattern of sleep and wakefulness in the mouse. *Sleep Res.,* 14:316.

112. Welsh, D. K., Richardson, G. S., and Dement, W. C. (1986): Effect of age on the circadian pattern of sleep and wakefulness in the mouse. *J. Gerontol.,* 41:579–586.
113. Wexler, D. B., and Moore-Ede, M. C. (1985): Circadian sleep-wake cycle organization in squirrel monkeys. *Am. J. Physiol.,* 248:R353–R362.
114. Wyatt, J. R., and Eltringham, S. K. (1974): The daily activity of the elephant in the Rwenzori National Park, Uganda. *East Afr. Wildlife J.,* 12:273–289.

Sleep and Alertness: Chronobiological, Behavioral, and Medical Aspects of Napping, edited by
D. F. Dinges and R. J. Broughton.
Raven Press, Ltd., New York © 1989.

3

Development of Human Napping

Wilse B. Webb

*Department of Psychology, University of Florida,
Gainesville, Florida 32611, USA*

Developmental status, generally indexed by age, is a major determinant of the amount, structure, and pattern of sleep. This is most apparent in human napping, which is the focus of this chapter. Napping in humans displays a systematic development from the polyphasic pattern of the neonate. There is a consolidation into two daytime naps, then a single nap, a decline of napping in early childhood, and a return of napping with advancing age. This chapter reviews this process.

The chapter is divided into the age groupings of the first year, the second through fifth year, childhood and adolescence, and advancing age. The characteristics of adult napping are detailed in Chapter 9. This chapter focuses on the pattern or presence of naps within the 24-hr period.

THE FIRST YEAR

Noting the small number of 24-hr observational studies of early infant sleep, Parmelee et al. (29) reported on the sleep patterns of 75 infants in the first 3 days after birth. In 1964, Parmelee and co-workers (30) extended their observations on 46 infants from birth to 16 weeks of age. The focus of these studies was on sleep amounts and their primary unit of analysis was the longest and shortest sleep and waking episodes. In an earlier study, Kleitman and Engelmann (21) reported on the sleep of 19 infants from 3 weeks to 26 weeks of age. The data of these studies consisted primarily of sleep charts kept by mothers, although the Kleitman and Engelmann study supplemented the observations with motility measures.

These studies differentiated between day sleep and night sleep [7:00 A.M.– 7:00 P.M. and 8:00 A.M.–8:00 P.M. defined the day period in the studies of Parmelee et al. (30), and Kleitman and Engelmann (21), respectively]. Their

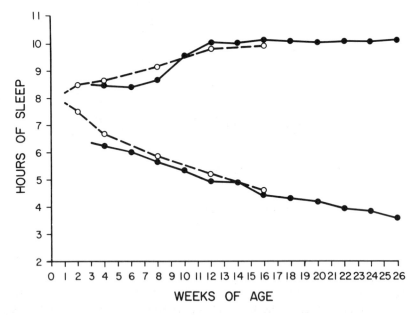

FIG. 1. Hours of nighttime sleep (*top lines*) and daytime sleep (*bottom lines*) during infancy. Data from refs. 21 (●) and 30 (○).

findings for total sleep amounts are in remarkable accord. Figure 1 displays the data of these studies. Although sleep was approximately equally distributed at birth, an early circadian tendency is present. In the first 3 days, Parmelee et al. (29) reported that the longest sleep period occurred 24% of the time between 7:00 A.M. and 3:00 P.M., 18% of the time between 3:00 P.M. and 11:00 P.M., and 57% of the time between 11:00 P.M. and 7:00 A.M. The authors pointed out that this may be partially attributable to the mothers' sleeping during this period. However, Parmelee et al. (30) also noted that daytime sleep averaged 7.8 hr during the first week compared with 8.3 hr during night sleep (standard deviation equaled "about 1 hr"). This early night/day difference presages the systematic shift seen in Fig. 1 of a slow-rising consolidation of a long sleep period in the night hours and a declining amount of sleep in the daytime hours.

The declining daytime sleep shows a systematic tendency to consolidate into two nap periods. Table 1 presents data detailing this evolution from the Kleitman and Engelmann (21) study. This table is derived from a chart of the percentage of time awake by hour of the day for the group of infants and shows the data of daytime sleep from 7 to 10 weeks of age and 23 to 26 weeks of age in 2-hr intervals. There are two "notches" of sleep tendency that have developed in the later age range, between 10:00 A.M. and noon and 2:00 P.M. and 4:00 P.M.

This development of a binary daytime pattern is also seen in the detailed

TABLE 1. *Percentage of time asleep from 8:00 A.M. to 7:00 P.M. during the 7th to 10th and 23rd to 26th week of age*

	8:00 A.M. 9:00 A.M.	10:00 A.M. 11:00 A.M.	12:00 P.M. 1:00 P.M.	2:00 P.M. 3:00 P.M.	4:00 P.M. 5:00 P.M.	6:00 P.M. 7:00 P.M.
7–10 weeks	40	47	49	50	53	45
23–26 weeks	20	43	28	41	28	20

Adapted from ref. 21.

reports of single infants. Three such reports are those of Gesell and Ilg (16), Gesell and Amatruda (15), and an infant reported by Parmelee (28). Table 2 extracts the number of sleep periods occurring between 6:00 A.M. and 6:00 P.M. for these infants during the first year.

The most detailed analysis of the development of this binary tendency is reported by Meier-Koll et al. (25). An infant was observed round the clock from birth through the first 4 months. By use of a computerized multiple averaging procedure, the "waking density" of these observations was developed and is presented in Fig. 2. Here are clearly displayed the decreasing waking tendencies in the night period and the emergence of two sleep troughs during the daytime (e.g., week 15) from multiple and irregular tendencies in the earlier periods.

The issue of an endogenous circadian tendency in the development of sleep patterns and its entrainment by social factors has been addressed by two early uses of the time-free or "free-running" biological rhythm design. Both Gesell and Amatruda (15) and Kleitman and Engelmann (21) reported on the development of sleep patterns in single infants who were on demand feeding schedules. Gesell and Amatruda (15) specifically stated the question: "it is interesting to inquire whether there is a natural diurnal rhythm of sleep and wakefulness. . . . Is the infant born with a diurnal pattern which favors activity by day and quiescence by night?"

TABLE 2. *Number of daytime sleep periods by weeks of age*

Study[a]	Weeks of age[b]									
	1	2	4	12	14	16	20	40	48	52
Gesell and Ilg (16)			4	4	3	3	3	2	1	2
Gesell and Amatruda (15)	4	4	2			2		2		1
Parmelee (28)	3	3	4	3	2	2				

[a] Information adapted from studies.
[b] Note scale is irregular.

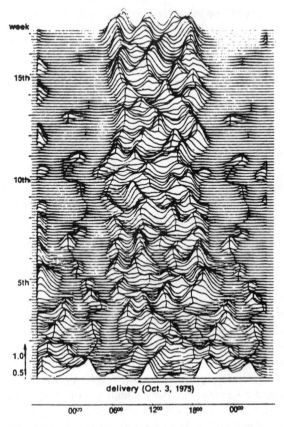

FIG. 2. Distribution of "wake density" across 24 hr in a single infant from birth to 4 months of age. (From ref. 25.)

Gesell and Amatruda (15) noted in their infant that as early as day 3 there was an emerging block of sleep from 7:00 A.M. to 4:00 P.M. By day 12 there was a continuous sleep period from 9:00 P.M. to 3:00 A.M. and by 16 weeks of age one from 6:00 P.M. to 5:00 A.M. This period was shifted to 9:00 P.M. to 8:00 A.M. by 40 weeks of age. They concluded

> Consolidation is essentially an embryological phenomenon. This does not mean that the infant is immune from acculturation. . . . The span of consolidation is determined through maturation. The culture simply takes the incidence of the consolidated blocks to conform to its own customs.

This process is yet more dramatically demonstrated by Kleitman and Engelmann (21). Figure 3 displays the sleep/wake pattern of an infant, to quote the authors, "[whose] parents were sufficiently indulging to permit her to set her own sleep-wakefulness patterns and feeding schedule." [This figure

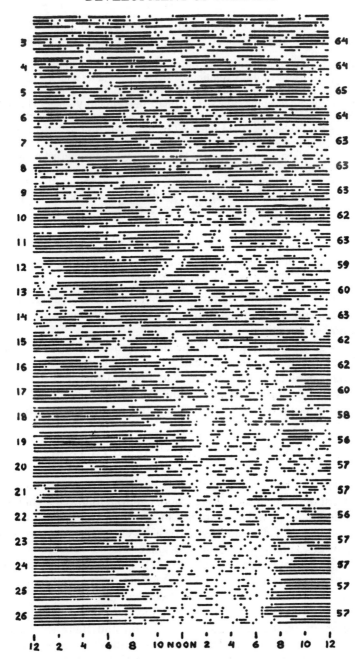

FIG. 3. Sleep (*solid lines*), wakefulness and feedings (*dots*) for one infant from birth to 26 weeks of age. Weeks are indicated on the left; percentage of time asleep is indicated on the right. (From ref. 21.)

also appears on the cover of Kleitman's (20) *Sleep and Wakefulness*.] The description of the emergent sleep patterns is described by the authors:

> The top and the bottom . . . do not differ from any other infant's chart. They show an initial haphazard distribution of sleep . . . and a terminal adjustment to an uninterrupted 10-hour sleep period during the night, . . . [and] short naps . . . during the intervening hours. But there the similarity ends. This infant began its adjustment by developing a 25-hour periodicity. This is evident from the successive displacements to the right of the clusters of daily feedings, the white spaces of wakefulness and the bands of "long" sleep. . . . The second turn is . . . slightly steeper, because of the shortening of the diurnal period from 25 hours to a mean of 24.86 hours. . . . The third spiral . . . has a much greater slope . . . the mean "diurnal" period being down to 24.5 hours. . . . From the middle of the 18th week to the end of 21st week, the consolidated long sleep hours, as well as the much widened light space, appear stationary, indicating that the diurnal periodicity is now down to the astronomically correct 24 hours. However, community-wise the adjustment is far from perfect, as the long sleep period starts at midnight or later. A secondary adjustment, . . . of the diurnal period to 23.86 hours, causing a leftward displacement of the night sleep to the 8 P.M. to 6 A.M. interval, occurred during weeks 22–24. . . .

Figure 3 and the authors' comments provide early dramatic evidence of a human endogenous circadian free-running sleep/wake rhythm with its gradual entrainment over time.

To briefly review the pattern of the development of daytime sleep and waking in infants during the first year, we see clear evidence of a decrease in the amount of sleep during the day and a consolidation of this sleep into a binary pattern during the first 6 months. The reduction in total sleep time (TST) per 24 hr is almost exclusively owing to a reduction in daytime sleep.

FROM THE SECOND TO THE FIFTH YEAR

Despert (10), in her review of sleep in preschool children, cited 54 references. Most of these studies of preschool children followed an initial major study by Flemming (12). These studies included an extensive survey of reports by parents of 1,816 children (13), a 3-year study of nursery school observations of 73 children by Dales (8), and a summer school observational study of a residential group across 24 hr (32). This section reviews the findings of these major studies with additional data from Chant and Baltz (6), Shinn (34), Moore and Ucko (26), and Beckman (2).

Few data are reported on 1-year-old children. Foster et al. (13) reported a daytime sleep average of nearly 2 hr. Chant and Baltz (6) reported an average of 72 min for 11 children. Only one study refers to the two-nap pattern in the 1-year group. Moore and Ucko (26) noted that at the age of 2 years, 25% of their sample were taking two naps, 68% were taking one nap regularly, and 8% were napping irregularly.

The overall circadian pattern of sleep can be seen as a systematic extension

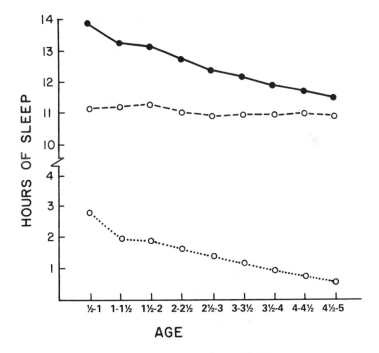

FIG. 4. Sleep in early childhood. Total sleep (●—●), night sleep (○—○), day sleep (○·○). (From ref. 13.)

of the pattern seen in Fig. 1. Figure 4 presents the data from the questionnaire study of Foster et al. (13). There is a continued decline of sleep through the first 5 years, but it can be seen that this decline is almost exclusively attributable to the decline in daytime sleep.

A study by Dales (8) sheds light on the nature of this daytime sleep change. She reported the duration and latencies of daytime sleep episodes in young children and compared them with those of Reynolds and Mallay (32). These data are shown in Table 3 along with Beckman's (2) results. There is considerable agreement between the results of Dales and the findings reported by Reynolds and Mallay. Both indicated an increasing resistance to napping as evidenced by a 30% increase in nap sleep latency and an 11% to 36% decrease in mean duration of nap sleep. Other investigators noted similar drops in nap sleep durations between the second and fourth years of age: Flemming (12), 27%; Foster et al. (13), 21%; and Beckman (2), 13%.

Dales' (8) findings emphasize the substantial individual differences among subjects in daytime sleep episodes. She divided her subjects into mean age groups of 2.9 years ($n = 31$), 3.7 years ($n = 51$), and 4.6 years ($n = 46$). Daytime sleep length standard deviations and ranges of these age groups were, respectively, 11 min (56–107), 11 min (52–100), and 9 min (48–90).

TABLE 3. *Daytime sleep latencies and sleep lengths (min) derived from three studies of young children*

	Years[b]				
	2.0–3.0	2.4–3.0	3.0–3.11	4.0–4.11	4.3–5.6
Latency[a]					
Dales (8)	24		31	35	
Reynolds and Mallay (32)	27		35	38	
Beckman (2)		27			32
Length[a]					
Dales (8)	77		75	69	
Reynolds and Mallay (32)	89		77	57	
Beckman (2)		78			68

[a] Information adapted from studies.
[b] Note scale is irregular.

A major contribution to the decreased amount of daytime sleep, however, is the deletion of the nap. Table 4 presents the data from five studies reporting on the percentage of days by age groups during which the children did not take a nap. It should be emphasized in considering these data that all of the children were encouraged to take naps and required to lie down during a specified nap period. Although there are exceptional figures, which undoubtedly reflect differences in conditions, there is a consistent trend for the nap to occur in most 2-year-olds, whereas one-third to one-half of the children had abandoned the nap by 5 years of age.

A striking figure on the incidence of napping is reported by Reynolds and Mallay (32). In their 2-year-old group, 50% took naps every day and all took naps at least 3 or 4 days each week. Among the 4-year-old children, none napped every day and 50% napped only once or twice a week.

Dales (8) reported on a subgroup of 10 children that she observed from 2

TABLE 4. *Percentage of young children not napping daily as derived from five studies*

	Years		
Study[a]	2.5–3.0	3.0–4.0	4.0–5.0
Dales (8)	20	17	40
Flemming (12)	2	45	50
Reynolds and Mallay (32)	4	10	30
Shinn (34)	4	42	42
Foster et al. (13)	4	14	33
Median	4	17	40

[a] Information adapted from studies.

through 4 years of age. The data from this subgroup typify and detail the general trends during these years. There was a slight decline of mean nap duration (75 to 71 min). There was an increase in presleep latencies (26 to 34 min). The percentage of days without sleep for the group rose from 4% to 31%. Seven 2-year-olds took a nap every day or only skipped 1 day during some 50 days of observations. However, one child did not nap approximately 20% of the days. At the age of 3, six children continued to nap daily. However, three children skipped 13%, 14%, and 33% of their naps. At 4 years of age, only two children continued the daily nap pattern and three children skipped 69%, 73%, and 84% of their naps.

The individual data reported by Dales (8) on these subjects permit a further analysis with striking results. The individual sleep tendencies, although showing wide interindividual differences, show remarkable within-subject consistency. The rank order correlations of the average nap durations were, respectively, 2 to 3 years = 0.74, 3 to 4 years = 0.80. Furthermore, the individual subject that skipped the greatest percentage of naps at age 2 also skipped the greatest percentage of naps at 3 and 4 years of age. Conversely, the two children who consistently maintained their naps at age 4 did so as 2- and 3-year-olds.

A more recent study conducted in France by Koch et al. (22) affirms and extends the more general findings of the earlier studies. They analyzed the duration of nap and night sleep during a 9-month period (September through June) of 107 kindergarten children between 2.5 and 4.5 years of age. Teachers in the eight classes reported the presence and duration of daytime naps, and parents reported when the children went to sleep at night and awakened in the morning.

Koch et al. (22) found that the duration of naps showed no difference across the school year. They were remarkably similar to the data reported in Table 3 (e.g., mean duration of 79 min). However, the frequency of napping showed a significant decline during the 9 months (90% in the first third of the period, 81% in the last third). There was also a slight but significant decline in the duration of night sleep (673 min versus 660 min). The authors concluded that "the tendency of the sleep-waking rhythm is mainly toward fewer naps and not toward shorter naps." There were, of course, substantial individual differences: "some children at 40 months had naps lasting .4 to 1.5 hours and night sleep lasting from 9.7 to 11.9 hours."

This study also reports significant negative correlations both between the duration of daytime naps and the following night's sleep duration and between the duration of night sleep and the subsequent daytime nap. These correlations were highly significant across all subjects. The nap-to-night correlations were significant within six of eight classes, and the night-to-subsequent-nap duration correlations were significant for three of the classes. The authors (22) cite these results as affirming the earlier studies of Klackenberg (19) and Basler et al. (1). It should be noted that the correlations of

Koch et al. (22) can be interpreted as reflecting "compensatory" sleep processes, which perhaps override more generalized sleep tendencies that would be reflected in positive correlations.

In summary, from 2 through 4 years of age, the amount of daytime sleep shows a continuous decline, and this serves as the major contributor to the decline in TST per 24 hr. By the age of 2 years, daytime sleep tends to occur in a single period during the afternoon. Although there appears to be a limited decline in the length of this period, generally the period begins to take an all-or-none character approximately 1 hr in length. From 2 through 5 years of age, there is evidence of increasing presleep latencies. Consistently, there is an increase in the tendency not to nap with fewer individuals napping every day. There is, nevertheless, a wide range of individual differences within this overall pattern.

CHILDHOOD AND ADOLESCENCE

From age 6 to 16, there is a continued slow decline in the amount of TST. From an earlier report on TST (41), two figures of sleep amounts for the ages 8 to 12 and 13 to 17 years from two large samples of school children were given. Terman and Hocking (38) gave means of 625 and 560 min for the two groups. Fifty years later O'Connor (27) reported means of 550 and 475 min. The differences between these studies may reflect cultural changes in sleep habits that have occurred during this century.

Only one study could be located that described the prevalence of naps across this age range. Simonds and Parraga (35) obtained questionnaire responses from 309 mothers of children in community schools whose ages ranged from 5 to 20 years. They found that 17% of a 5- to 8-year-old group ($n = 75$), 9% of a 9- to 11-year-old group ($n = 65$), 9% of a 12- to 14-year-old group ($n = 58$), and 23% of a 15- to 18-year-old group ($n = 110$) were reported to take naps. The decline in napping tendencies in the middle age ranges is in accord with findings obtained by Carskadon and associates using the Multiple Sleep Latency Test (MSLT) (see Chapter 4). The return of a nap tendency in the midteen years is also consistent with MSLT results and studies of napping in late adolescence and young adulthood (see Chapter 9).

NAPS WITH ADVANCING AGE

Napping appears to increase systematically with advancing age. In his classic study of the sleep patterns of 509 normal adults, Tune (39) used sleep charts during an 8-week period. Reported naps rose linearly from an average of one nap across the 8-week reporting period in the 20- to 30-year-old group to an average of 14 naps in the 70- to 80-year-old group. Webb and Swinburne (47) reported an observational study across 4 days of 19 subjects whose ages ranged from 66 to 96 years (mean = 77 years). Only one subject was without a nap during the 4 days and one took one nap. Four persons averaged three or more naps.

In spite of this increasing number and later ubiquity of naps, few systematic studies have been reported. Those available show napping to be a "volatile" event with a range of consistent individual differences and a limited relationship to night sleep. Tune (39) found the mean number of naps for his oldest group (70–80 years) to be 14 across the 8 weeks (or approximately 1 of every 4 days). There was a small but significant relationship between frequency of napping and frequency of nighttime awakenings ($r = 0.188$). Spiegel (36) reported on the naps of a group of healthy Swiss retirees with an age range of 53 to 70 years (mean = 63). With no difference in frequency between men and women, he found that of his subjects, 34% reported never napping, 18% occasional napping, and 48% always napping. There was no relationship between these categories and night sleep duration, sleep latency, and nighttime awakenings.

An increase in dozing or sleeping during the day in elderly subjects was also found by Gerard et al. (14). They compared questionnaire responses of young adults (16–39 years) with those of 103 elderly persons (65–98 years). Of the younger subjects, 41% reported daytime dozing compared with 65% of the elderly. The younger subjects reported an average of 3.1 days per month (SD = 5.6) compared with 12.1 days (SD = 12.8) for the elderly ($p < 0.001$).

The most extensive analysis of napping in older subjects was reported by Webb (43), who analyzed the napping patterns of 40 men and 40 women between 50 and 60 years of age. The data were drawn from 2-week sleep diaries. He noted that, although naps contributed an average of only 15 min to the average TST, they revealed idiosyncratic individual patterns and significant gender differences.

Table 5 presents the basic data from this study. It can be seen that approximately an equal number of men and women reported no naps (25% and 27%, respectively). However, the men reported more frequent napping and shorter naps. The men reported 161 naps compared with 100 naps reported by the women; 71% of the women reported four or fewer naps, whereas 45% of the men reported seven or more naps. In terms of the length of naps, 26% of the men's naps were less than 30 min compared with 13% of the women's naps, but 29% of the women's naps were longer than 90 min compared with 17% of the men's naps. Fifty-one percent of the naps exceeding 90 min were on weekends, and no subject reported more than two such long naps during the 2 weeks. No relationship was found between napping and total bedtime ($r = 0.03$).

Zepelin (50) reported that, in 60- to 70-year-old subjects, naps accounted for an average of 4 min in men and 11 min in women of total daytime sleep, when nap days and nonnap days were averaged together.

TABLE 5. *Percentage of 50–60-year-old men and women reporting naps during a 2-week period*

	Men (n = 40)	Women (n = 40)
Number of naps		
0	25	27
1–4	30	50
5–8	30	22
9+	15	0
Mean length (min)		
0–29	26	13
30–60	36	34
60+	38	54

These limited data thus indicate that, although there is an increase in napping in older subjects, it is not a universal development. One-fourth to one-third of older adults do not nap at all and a large number of the remaining persons nap only occasionally.

DISCUSSION

Throughout this presentation, time or, in particular, age has generally served as an index for developmental changes. As I heard a wise, early teacher, John McGeoch, say, "Time, in and of itself, means nothing. It is events within time that must be understood." To try to better comprehend the factors underlying these changes in napping patterns, three sleep-associated and age-associated factors must be considered: sleep demand levels, environmental demands and associated behavioral control, and endogenous circadian tendencies. Let us begin by considering these factors in young children.

Through the age of 5 years, sleep demand is approximately indexed by the total sleep obtained in the 24-hr period. During this age span, we have seen that sleep amount declines from an average of 16 to 10 hr and that this decline is almost exclusively attributable to the decline in daytime sleep. Further, as a process, the general pattern was the deletion of two naps and then the abandonment of the single nap. To this is added the finding reported by Parmelee and colleagues (30): During the rapidly changing levels of TST in the first 6 weeks, the length of the longest waking period systematically increased from approximately 2.5 to approximately 4 hr at 16 weeks.

It can be argued that a primary determinant of the amount and pattern of napping in this age range is attributable to a decreasing sleep demand and an increasing ability to sustain wakefulness. From this perspective, the consolidated night sleep period is not sufficient to meet the 24-hr sleep demands and the daytime sleep, or nap, serves to meet these needs. Further, as this demand declines to a level that can be met within the nighttime period, the tendency for the daytime nap declines and is finally eliminated. The two-nap/one-nap/no-nap pattern may be considered an expression of the initial inability of nighttime sleep to meet the sleep demand level during the daytime (i.e., to sustain wakefulness in the face of high levels of sleep demand).

The apparent continued decrease in napping tendencies in the prepubertal period can be viewed as a simple extension of the decreasing sleep need, which is being met fully during the nighttime sleep period.

It is clear, however, that, although the 24-hr sleep demands may continue to decline or stabilize, at pubescence and beyond there is a return of napping tendencies. At this point, one must consider further the issue of sleep demand and sleep amounts and take note of additional age-related changes, in particular those changes associated with the behavioral control of sleep. As the child matures and enters primary school, behavioral demands and control begin to change significantly. Because sleep, within limits, can be voluntarily and involuntarily delayed and truncated, the amount of sleep obtained does not necessarily reflect the level of sleep need or demand. This point is illustrated by self-report data obtained by Carskadon (4). Table 6 presents her questionnaire results for nighttime sleep in 10-, 11-, 12-, and 13-year-old children. A most striking aspect of these results is the relatively limited decline in TST on nonschool nights but the increasing discrepancy between nocturnal sleep lengths on school nights across the prepubescent age range. To this may be added questionnaire data cited by Carskadon (4): 50% of the 10-year-old children had their night bedtime hour controlled by the parents,

TABLE 6. *Sleep habits reported by 218 early adolescents*

Parameter	10-year-olds (n = 35)		11-year-olds (n = 58)		12-year-olds (n = 78)		13-year-olds (n = 47)	
	Mean	SD[a]	Mean	SD	Mean	SD	Mean	SD
Bedtime (hr)								
School night	2122	(34)	2137	(39)	2146	(34)	2214	(40)
Nonschool night	2222[b]	(46)	2239[b]	(60)	2300[b]	(49)	2332[b]	(74)
Rising time (hr)								
School night	0704	(71)	0705	(26)	0703	(28)	0656	(26)
Nonschool night	0807[b]	(70)	0808[b]	(73)	0840[b]	(65)	0855[b]	(64)
Sleep time (min)								
School night	587	(30)	569	(38)	557	(47)	522	(49)
Nonschool night	587	(84)	566	(82)	580[b]	(68)	562[b]	(88)

[a] Standard deviation (SD) in min.
[b] $p < 0.02$ by t-test for related means.
From ref. 4.

and 19% of the 13-year-old children gave that response and "the 12 and 13 year olds tended to be awakened . . . on school mornings; the younger children tended to awake spontaneously on school mornings more frequently than the older children." Carskadon concluded: "These data lead to the speculation that the need for sleep may not be changing across this span of years, but that school and social pressures on older children may decrease the amount of time they have available for sleep."

Within the perspective of sleep demand and behavioral control variables, some portion of the return of naps in late adolescence may be considered, therefore, attributable to unmet sleep demand needs.

As one turns to young adulthood, the most widely studied population is the college group. In this population, napping is at a high level (see Chapter 9). There is ample evidence of a discrepancy between sleep demand levels and sleep amounts seen in the discrepancy between weekday and weekend sleep (45) and the environment is more schedule free than the typical job-bound population. Certainly some portion of the burst of daytime napping in this population also appears to be attributable to sleep need and behavioral control factors.

Let us turn to the increase in napping associated with aging within the perspective of associated sleep demand and behavioral factors. Two aspects of changes in sleep need considered as a function of age deserve notice.

The most substantial data are derived from questionnaires from more than 1 million respondents obtained in a 1959 to 1960 survey by the American Cancer Society. The sleep data from this survey were analyzed and reported by Kripke et al. (23). Sleep amount was assessed from the response to the question, "How many hours of sleep do you usually get per night?" For men, the average rose continuously from the age of 40 (7.8 hr) through the ninth decade (8.7 hr). For women, the sleep amounts showed no change from 40 to 70 years and then rose sharply to 8.6 hr in the ninth decade. A subanalysis of the data reported by Kripke et al. (23) reveals an accelerated increase in the short and long sleep tendencies. These are shown in Fig. 5, which plots by age the percentage of persons reporting more than 9 hr and less than 6 hr of sleep. Some portion of those with short sleep patterns may reflect an inability to sleep during the primary sleep period with a resultant increase in sleep need that in turn finds expression in naps. The long sleep group can be considered to reflect an increase in sleep demand levels that makes them particularly vulnerable to behaviorally based restrictions on sleep.

A second age-associated sleep change that may contribute to the sleep debt (and in turn to nap tendency) is the increased inability to sustain sleep throughout the night. Measurements of the amount of time awake after sleep onset in a large group of healthy men and women between 50 and 70 years of age found that 26% had average awake times of more than 30 min (46). This deconsolidation of the nighttime sleep period may contribute to an

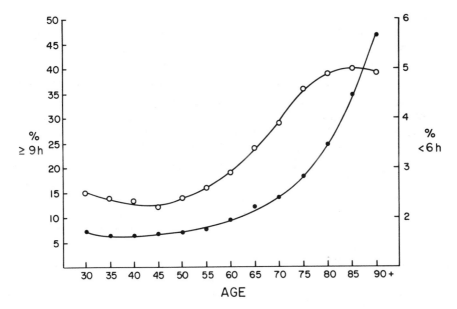

FIG. 5. Long and short sleep tendencies as a function of age. Percentages for ≥9-hr (●) and <6-hr sleepers (○). (From ref. 23.)

uncompleted sleep demand and in turn result in an increased nap tendency in the daytime.

In considering these factors in the older group, it should be noted that, although these age-associated changes involve only a portion of the older subjects, the level of napping to be accounted for is not high and affects only a portion of these adults. The question that remains unanswered is whether these are the same subset of elderly adults; that is, are those with the greater nocturnal awake time also more likely to engage in daytime napping?

Aside from the changes in nighttime sleep per se, some portion of the increased napping in older adults may be attributable to behavioral facilitation or increased opportunities for napping. For example, as already noted, levels of napping of Swiss retirees were higher than that of an actively employed sample of older subjects (36).

In summary, with regard to the sleep demand factor, it can be argued that sleep demand may play a major role in determining the amount and presence of napping across the age dimension. In infancy and early childhood, the high and then decreasing sleep demand may contribute to the high initial level and the decreasing level of napping. With increasing age, the increased nap tendency may be, at least partially, a function of change in the demand level as well as a function of increased demand levels resulting from inabilities to sustain sleep.

In regard to behavioral factors, it can be argued that some portion of the return of the nap in adolescence may be attributable to voluntary and involuntary limitations of night sleep resulting in a sleep demand pressure. Certainly, this is arguable in the case of the college student who not only may have the increased sleep demand but also a more unscheduled environment, which may be more permissive of the nap.

Let us turn to the third component to which napping tendencies may be ascribed: circadian biological rhythm tendencies. We should first briefly define these biological rhythm tendencies. A chronobiological or biological rhythm is seen in the repetition of a biological event at a regularly timed interval. A circadian biological rhythm is the repetition of the event within a 24-hr (circa) matrix. There is ample evidence, in infancy and early childhood, of tendencies to sleep briefly (nap) at approximately the same time each day. This occurs, at least in some individuals, through the age range and there is an increase in this with aging.

However, the establishment of a repetitive event is not sufficient to attribute the tendency to a biological rhythmic tendency. As discussed elsewhere (44), five classical sources of behavioral responding may be cited: reflex responding, instinctive responses, learned responding, homeostatic responding, and maturational changes. All of these may display temporal regularity. The first three of these (reflex, instinctive, and learned) merely require a regularity of stimuli in the timing of environmental sequences. The latter two (homeostatic and maturational events) are inherently organized, often in a timed sequence. How then do these paradigms differ from the biological rhythm model?

> Although the other response determinants may modulate or even override biological rhythms, they are conceived of as endogenous, inherent systems whose sole phase determination is the passage of time *per se*. Simply, a mature organism in a physiologically "balanced" (homeostatic) state in a constant stimulus environment would, in a biological rhythm model, continue to display a rhythmic response at repeated time points (44).

What evidence is there for a biological rhythm determination of naps, as proposed earlier (3,42)? This question is the focus of Chapters 5 to 8, but it needs to be addressed in the context of development as well. In particular, one might ask: Is there evidence to indicate an ontological course for a biological rhythm underlying the tendency to nap?

There is, first, much evidence that nighttime sleep displays a major biological rhythm component. This is clearly seen in the systematic development of the continuous period of nighttime sleep in the infant (see above). Using the chronobiological techniques of time-free environments, sleep displacement, and sleep deprivation, biological rhythm aspects of the major sleep period have been amply demonstrated in both humans and animals [cf. (18,44)]. These biological rhythm tendencies have been elegantly incorporated into contemporary sleep theory [cf. (7)]. It would be unlikely that this variable is confined to nighttime sleep.

At this point, to consider more specifically the circadian aspects of naps throughout development, we must introduce two other important characteristics of biological rhythms: period and amplitude. Period refers to the frequency of a rhythm and amplitude refers to its magnitude. In terms of frequency, biological rhythms may be ultradian, circadian, or infradian (i.e., they may be less than, approximately, or more than 24 hr). In the instance of naps, the rhythm may be considered ultradian or a two-phase sleep rhythm of a circadian event. In this chapter we have arbitrarily treated it as a circadian rhythm. In terms of amplitude, the rhythm may be strong (of high amplitude) or weak. Amplitude, of course, may vary within and between subjects.

In this light, our discussion of sleep demand and behavioral factors associated with napping throughout the life span involved primarily consideration of them as modulators of amplitude of nap tendency. We now turn specifically to the circadian phase aspects of naps.

As we have noted, a timed presence is a necessary but not sufficient condition for a biological rhythm. If brief sleep periods occurred randomly in the daytime as extensions of (or anticipation of) nighttime sleep or as random intermittent bursts of sleep of varying duration, there would be no case for a circadian rhythm. This is not the pattern seen, however. The data reviewed previously reveal a systematic appearance of two sleep periods in the morning and the afternoon in infancy and, thereafter, a remarkably systematic occurrence of a brief sleep period in the midafternoon.

There is considerable evidence to indicate that these are endogenous self-generated tendencies. From intensive studies of infants, the interrelationship between feeding and sleep periods has been shown to be largely independent [cf. (21)]. Similarly, dips in afternoon performance in adults, which are interpreted as associated with nap tendencies, have also been established as being independent of postprandial effects (17). Of particular relevance are the demand-feeding studies (see the section on the first year), which for the infant approximate the classical chronobiological time-free design. The developmental patterns are consistent with the conclusion that daytime sleep (naps) are endogenously timed events within a free-running biological rhythm.

We have seen that, throughout adolescence and adulthood, although there are potential amplitude changes, the phase relationship of a nap is firmly stable in the midafternoon. Particularly impressive in this regard are the results from studies of daytime sleepiness using the MSLT in which prior wakefulness was manipulated. Although there are systematic magnitude effects on the MSLT function, the phase relation (i.e., the midafternoon dip) is stable.

The data on circadian rhythms, aging, sleep, and naps are of particular interest. There is experimental evidence to indicate that the circadian system is affected by aging. The amplitudes of rhythms are diminished with aging in rodents (33,49). Of particular pertinence is a report by von Gool and Mirmiran (40) on the sleep of rats. In this study, the light/dark ratio changed from 0.59 in younger rats to 0.71 in older rats (where 1.0 is equal distribution). This flattened amplitude of biological rhythms has also been well demonstrated in humans [cf. (11,24)]. There is also evidence that there is a reduced capacity for entrainment. Rats resynchronized more slowly after a phase shift with increased age (31). Wever (48) has noted a sharply increased tendency for internal desynchronization of biological rhythms in older human subjects.

Reduced biological rhythm control in advanced age is consistent with the overall pattern of the sleep of older persons. They are less capable of sustaining the long night phase of sleep [cf. (44)]. The higher tendency for daytime napping is readily interpreted, therefore, as being, at least partially, a retrogression toward the more polyphasic pattern of early childhood.

A final speculative but provocative argument for the endogenous biological rhythm during the day can be made. For most biological rhythms, a sensible, adaptive functional purpose can be found. Simply put, biological rhythms typically are nature's coding, through evolutionary processes, of appropriate timings of organisms' functionings. What could be the functional significance of naps? Elsewhere (42) I wrote

> I propose that, in man in particular, and perhaps in many other beasts, it was simply a behavioral control system to avoid the torrid heat of midday. Over simplified, those of the species which could suppress activity during these times [napped] survived and reproduced and those who maladaptively continued functioning had less probability of surviving. This can be labelled the MD and E notion—"only mad dogs and Englishmen go out in the noon day sun."

SUMMARY AND CONCLUSIONS

The data present a systematic change in the patterns of napping associated with development and age. Within the first 6 months of life, daytime sleep consolidates into two periods, midmorning and midafternoon, and by the first year there is usually only a single afternoon nap. Nap tendency declines throughout childhood into puberty but returns postpubescence. With advancing age there is a linear increase in napping. Within this overall pattern there is at any given age a wide range of individual differences in the time of development and the level of napping.

In examining the factors that modulate this pattern, there is evidence of an endogenous sleep tendency that underlies the phase of nap tendency. Within this phase-related tendency the likelihood of a nap appears to depend on two age-associated factors: changes in the 24-hr sleep demand levels and behavioral control. In infancy, the amount of daytime sleep is primarily determined by a high level of sleep demand, and the decline in this demand with age is related to the decline in daytime sleep. As the amount of sleep that occurs at night approximates the total sleep demand, as in the prepubescent child, daytime napping becomes minimal. The return of naps in puberty and in certain young adult populations arguably reflects the presence of sleep demands resulting from voluntary and involuntary behavioral limitations on nighttime sleep. In older persons, napping tendency may be increased by associated age changes in sleep demand levels, inabilities to sustain sleep during the night because of age-associated biological rhythm changes, and more permissive behavioral schedules.

REFERENCES

1. Basler, K., Largo, R., and Molinari, L. (1980): Die entwicklung des schlafverhaltens in der ersten funf lebensjahren. *Acta Paediatr. Scand.,* 35:211–213.
2. Beckman, W. B. (1932): Daytime sleep of nursery children. Thesis, University of Chicago.
3. Broughton, R. (1975): Biorhythmic variations in consciousness and psychological functions. *Can. Psychol. Rev.,* 16:217–239.
4. Carskadon, M. (1982): The second decade. In: *Sleeping and Waking Disorders: Indications and Techniques,* edited by C. Guilleminault, pp. 99–125. Addison-Wesley, Menlo Park, CA.
5. Deleted at proofs.
6. Chant, N., and Baltz, W. (1928): A study of the sleeping habits of children. *Genet. Psychol. Monogr.,* 4:13–43.
7. Daan, S., Beersma, D. G. M., and Borbély, A. (1984): Timing of human sleep: Recovery process gated by a circadian pacemaker. *Am. J. Physiol.,* 246:R161–R178.
8. Dales, R. J. (1941): Afternoon sleep in a group of nursery-school children. *J. Genet. Psychol.,* 58:161–180.
9. Deleted at proofs.
10. Despert, J. L. (1949): Sleep in pre-school children: A preliminary study. *Nerv. Child,* 8:8–27.
11. Finkelstein, J. W., Roffwarg, R. P., Boyar, R. M., Krean, J., and Hellman, L. (1972): Age related changes in 24 spontaneous secretion of growth hormone. *J. Clin. Endocrinol.,* 35:665–670.
12. Flemming, B. M. (1925): A study of the sleep of young children. *J. Am. A. Univer. Wom.,* 19:25–27.
13. Foster, J. C., Goodenough, F. L., and Anderson, J. E. (1928): The sleep of young children. *J. Genet. Psychol.,* 35:201–232.
14. Gerard, P., Collins, K. J., Dore, C., and Exton-Smith, A. N. (1978): Subjective characteristics of sleep in the elderly. *Age Ageing,* 7(suppl.):55–63.
15. Gesell, A., and Amatruda, C. S. (1945): *The Embryology of Behavior. The Beginnings of the Human Mind.* Harper Brothers, New York.
16. Gesell, A., and Ilg, F. C. (1937): *Feeding Behavior of Infants. A Pediatric Approach to Mental Hygiene.* Lippincott, Philadelphia.
17. Hockey, G. R. J., and Colquhoun, W. P. (1972): Diurnal variations in human performance: A review. In: *Aspects of Human Efficiency: Diurnal Rhythms and Loss of Sleep,* edited by W. P. Colquhoun, pp. 1–23. English University Press, London.
18. Hume, K. I. (1983): The rhythmical nature of sleep. In: *Sleep Mechanisms and Functions,* edited by A. Mayes, pp. 18–56. Van Nostrand Reinhold, Wokingham.
19. Klackenberg, G. (1971): A prospective longitudinal study of children. Further studies of sleep behavior in a longitudinal follow-up sample. *Acta Paediatr. Scand.* [*Suppl.*], 224.
20. Kleitman, N. (1963): *Sleep and Wakefulness.* University of Chicago Press, Chicago.

21. Kleitman, N., and Engelmann, T. G. (1953): Sleep characteristics of infants. *J. Appl. Physiol.*, 6:269–282.
22. Koch, P., Soussignan, R., and Montagner, H. (1984): New data on the wake-sleep rhythm of children aged from 2½ to 4½ years. *Acta Paediatr. Scand.*, 73:667–673.
23. Kripke, D. F., Simons, R. N., Garfinkel, M. A., and Hammond, C. (1979): Short and long sleep and sleeping pills. *Am. J. Psychiatry*, 36:103–116.
24. Lobban, M. C., and Tredre, C. (1967): Diurnal rhythms of renal excretion and body temperature in aged subjects. *J. Physiol. (Lond.)*, 188:48–49.
25. Meier-Koll, A., Hall, U., Hellwig, U., Kott, G., and Meier-Koll, V. (1978): A biological oscillator system and the development of sleep-waking behavior during early infancy. *Chronobiologia*, 5:425–440.
26. Moore, T., and Ucko, L. E. (1957): Night waking in early infancy. *Arch. Dis. Child.*, 32:333–345.
27. O'Connor, A. L. (1964): Questionnaire responses about sleep. MA thesis, University of Florida.
28. Parmelee, A. H. (1974): Ontogeny of sleep patterns and associated periodicities in infants. In: *Pre- and Postnatal Development of the Human Brain, Mod. Probl. Paediatr.*, 13:298–311. Karger, Basel.
29. Parmelee, A. H., Schultz, H. R., and Disbrow, M. A. (1961): Sleep patterns of the newborn. *J. Pediatr.*, 58:241–250.
30. Parmelee, A. H., Werner, W. H., and Schultz, H. R. (1964): Infant sleep patterns from birth to 16 weeks of age. *J. Pediatr.*, 65:578–582.
31. Quay, W. B. (1972): Pineal homeostatic regulation of shifts in circadian activity rhythms during maturation and aging. *Trans. NY Acad. Sci.*, 34:239–254.
32. Reynolds, M. M., and Mallay, H. (1933): The sleep of children in a 24-hour nursery school. *J. Gen. Psychol.*, 43:322–351.
33. Sacher, G. A., and Duffy, P. H. (1978): Age changes in rhythms of energy metabolism activity and body temperatures in Mus and Peromyscus. In: *Aging and Biological Rhythms*, edited by H. Samis and S. Capobianco. Plenum Press, New York.
34. Shinn, A. F. (1932): A study of the sleep habits of two groups of pre-school children. *Child Dev.*, 3:159–166.
35. Simonds, J. F., and Parraga, H. (1982): Prevalence of sleep disorders and sleep behaviors in children and adolescents. *J. Am. Acad. Child Psychiatry*, 21:383–388.
36. Spiegel, R. (1981): *Sleep and Sleepiness in Advanced Age.* SP Medical and Scientific Books, New York.
37. Deleted at proofs.
38. Terman, L. M., and Hocking, A. (1913): The sleep of school children: Its distribution according to age and its relation to physical and mental efficiency. *J. Educ. Psychol.*, 4:138–147.
39. Tune, G. S. (1969): Sleep and wakefulness in normal human adults. *Br. Med. J.*, 2:269–271.
40. van Gool, W. A., and Mirmiran, M. (1985): Sleep in young and old rats: Influence of increased environmental complexity. *Sleep Res.*, 14:81.
41. Webb, W. B. (1969): Twenty-four hour sleep cycling. In: *Sleep: Physiology and Pathology*, edited by A. Kales, pp. 53–65. Lippincott, Philadelphia.
42. Webb, W. B. (1978): Sleep and naps. *Specul. Sci. Tech.*, 1:313–318.
43. Webb, W. B. (1981): Patterns of sleep in healthy 50–60 year old males and females. *Res. Commun. Psychol., Psychiatr. Behav.*, 6:133–140.
44. Webb, W. B. (1983): *Biological Rhythms, Sleep and Performance.* Wiley, New York.
45. Webb, W. B., and Agnew, H. W. (1975): Are we chronically sleep deprived? *Bull. Psychonomic. Soc.*, 6:47–48.
46. Webb, W. B., and Schneider-Helmert, D. (1984): A categorical approach to changes in latency, awakening, and sleep length in older subjects. *J. Nerv. Ment. Dis.*, 172:291–295.
47. Webb, W. B., and Swinburne, H. (1971): An observational study of sleep of the aged. *Percept. Mot. Skills*, 32:895–898.

48. Wever, R. A. (1979): *The Circadian System of Man: Results of Experiments Under Temporal Isolation*. Springer, New York.
49. Yunis, E., Fernandez, G., Nelson, W., and Halberg, F. (1974): Circadian temperature rhythms and aging in rodents. In: *Chronobiology,* edited by L. Sheving, F. Halberg, and J. Pauly, pp. 358–363. Iguku Shoin, Tokyo.
50. Zepelin, H. (1983): A life span perspective on sleep. In: *Sleep Mechanisms and Functions,* edited by A. Mayes, pp. 126–160. Von Nostrand Reinhold, Wokingham.

Sleep and Alertness: Chronobiological,
Behavioral, and Medical Aspects of
Napping, edited by
D. F. Dinges and R. J. Broughton.
Raven Press, Ltd., New York © 1989.

4

Ontogeny of Human Sleepiness as Measured by Sleep Latency

Mary A. Carskadon

Department of Psychiatry and Human Behavior, Brown University and
E. P. Bradley Hospital, East Providence, Rhode Island 02915, USA

The measurement of daytime sleepiness or sleep tendency using sleep latency tests is of relatively recent vintage, having begun with the Multiple Sleep Latency Test (MSLT) (7). The MSLT measures speed of falling asleep in a standard setting designed to maximize the likelihood of falling asleep (15). The measure was conceived as a way to achieve an objective, repeatable quantification of sleepiness for evaluating the levels and determinants of this behavioral manifestation (3), whether in normal individuals undergoing particular sleep/wake regimens or in patients with sleep disorders.

Much of the normative data using the MSLT was gathered by the group at Stanford University (13). Several other groups, most notably that of the Henry Ford Hospital in Detroit, have added significantly to the normative data pool. A number of variants to the MSLT have been introduced. Most of these were designed to measure ability to stay awake rather than sleep tendency; that is, alertness rather than sleepiness. Although these measures have been used primarily in clinical evaluations, some normative data are available for several of them.

The goal of this chapter is to review the age-related trends in human daytime sleepiness as measured by the MSLT and its contextual variants in order to determine whether the patterning of daytime sleepiness throughout life suggests an inherent tendency to nap. Although maturation and aging have previously been shown to affect the MSLT, several other factors also have significant impact on this measure (13) and must therefore be taken into account. Chief among the other determinants is the amount of sleep on the night(s) before the MSLT is measured. This factor is the largest confound that is likely to occur, and it should remain foremost in the reader's mind when evaluating the data. It should also be kept in mind that the amount of sleep on only the night immediately preceding the MSLT may be misleading, as several experiments have shown a cumulative impact of several nights of sleep restriction (8) or sleep extension (9). Another factor affecting the MSLT manifestation of sleepiness is the continuity or fragmentation of nocturnal sleep, which is particularly relevant when considering sleepiness in the elderly (6).

SLEEP LATENCY TESTS

A standard for performing the MSLT has been described by a task force of the American Sleep Disorders Association (formerly the Association of Sleep Disorders Centers) (15). In brief, electrodes are applied to the subject for recording standard polysomnography, including electroencephalogram (EEG), electro-oculogram (EOG), and electromyogram (EMG). The first sleep latency test of the MSLT begins 1.5 to 3 hr after the end of the nocturnal sleep episode; following this test, the procedure is repeated at 2-hr intervals. For each test the subject lies comfortably in bed in a quiet, dark room and is requested to remain still, keep eyes closed, and try to fall asleep (or not resist sleep). A limit of 20 min is given to fall asleep, at which time the test is terminated. If the subject falls asleep, the test ends when three consecutive 30-sec epochs are identified as sleep (stages 1, 2, 3, 4, or REM sleep).

The score for each test is the elapsed time (in min) from the beginning of the procedure (lights out) to the first 30-sec epoch identified as sleep. MSLT data are generally presented either as test-by-test sleep latencies (particularly when time-of-day effects are studied) or as the mean or median of sleep latencies for an entire day. (In the clinical version of the MSLT, primarily used to diagnose narcolepsy and other Disorders of Excessive Somnolence, each MSLT continues for 15 min beyond the first epoch of sleep.) The underlying assumption of the MSLT is that long sleep latencies indicate relative alertness (or low sleep tendency), whereas short sleep latencies indicate relative sleepiness (or high sleep tendency). Values of less than 5 min indicate a vulnerability for unintentional sleep episodes (13) and have been termed "pathological sleepiness" in patient groups (28,37).

The MSLT relies on a standard environment and conditions that maximize the probability of falling asleep. Thus, subjects are required to stay awake between tests, refrain from vigorous physical or mental activity during the 15 min before each test, refrain from caffeine on the day of testing, refrain from tobacco smoking during the 30 min before each test, remove shoes and loosen constricting clothing during the tests, and be as free as possible of competing needs (e.g., hunger, thirst, urination).

Variants of the MSLT include the Maintenance of Wakefulness Test (MWT) (26) and the Modified Maintenance of Wakefulness Test (36), the Repeated Test of Sustained Wakefulness (RTSW) (24), the Modified As-

TABLE 1. *Methodologies for variants of the Multiple Sleep Latency Test (MSLT)*

Procedure	Variants of the MSLT				
	MWT[a]	MMWT[b]	RTSW[c]	MAST[d]	MRT[e]
Duration	20 min	40 min	30 min	20 min	25 min
Posture	Seated	Seated	Lying	Seated	Lying(?)
Lighting	Dark	Dark	Dark	Low-level	Semidark
Apparatus	None	None	None	Reading	Headset, signal switch
Eye closure instructions	None	None	Closed	Open	N/A
Sleep instructions	Stay awake	Stay awake	Stay awake	Stay awake	Relax
Latency measure	3 epochs stage 1 sleep	3 epochs stage 1 sleep	1 epoch stage 1 sleep	Stage 1 sleep	Stage 1 sleep

[a] Maintenance of Wakefulness Test (26).
[b] Modified Maintenance of Wakefulness Test (36).
[c] Repeated Test of Sustained Wakefulness (24).
[d] Modified Assessment of Sleepiness Test (23), which consists of five periods that alternate between the MSLT format and a seated test.
[e] Multiple Relaxation Test (31), during which the subject wears earphones and listens to the tones of the Wilkinson auditory vigilance test "displayed at a low level . . . the subject might indicate signal detection if by chance recognizing any."

sessment of Sleepiness Test (23), and the Multiple Relaxation Test (31). Table 1 lists these variants and describes procedures used by each.

Relatively few normative data are available from the newer sleep latency tests, thus it is difficult to present age-related data from these tests. Furthermore, whereas a number of factors affecting or determining MSLT values have been studied and described (9,13), no similar normative database is available for each of the various "stay awake" tests. Lacking such comparative data, it is not only unclear whether the same factors that affect the MSLT alter these tests, but also whether differences in test conditions themselves significantly affect test scores. Because a great deal more data on daytime sleep tendency have been gathered using the MSLT, this measure of sleepiness will form the primary focus of the chapter.

EFFECT OF AGE ON SLEEP LATENCY IN HEALTHY VOLUNTEERS

Groups Studied

The MSLT and its variants have not been used to measure sleepiness in healthy children younger than 7 years. The MSLT has been used to characterize alertness in older children, adolescents, and adults under a number of circumstances. The data presented below include volunteers studied by the Stanford group in the "Stanford Summer Sleep Camp" during the course of 10 years. This laboratory provided facilities to study four or five volunteers at a time in a setting that was very comfortable for around-the-clock assessments. Sleep was always recorded the night before MSLTs were obtained. The nocturnal sleep schedule was rigorously observed, and the volunteers remained in the laboratory for at least 2 nights and days.

Maturational status was determined in all children and adolescents by Tanner staging (35), in which a pediatrician examined secondary sexual characteristics. According to this staging system, for example, Tanner stage 1 is prepubescent, stage 3 midpubescent, and stage 5 fully mature. Tanner stage and age in this sample were highly correlated ($r = 0.825$, $df = 126$, $p < 0.001$). Children and adolescents were evaluated for at least 3 consecutive nights and days. Bedtime was 2200 hr and rising time was 0800 hr each night. MSLTs were given at 2-hr intervals each day from 0930 through 1930 hr.

Assessment by MSLT of daytime sleepiness in children and adolescents has involved both cross-sectional and longitudinal evaluations. The youngest children studied included a group between the ages of 7 and 10 years, who were evaluated for 3 consecutive nights and days on a single occasion (17). Older adolescents included a group of 27 volunteers who came to the laboratory twice or more for six summers (3,4,14,16,20). The total number of individual assessments in children and adolescents was 128.

To assess developmental changes, the data have been grouped by Tanner stage, giving the following distribution (see Table 2 for details): 34 subjects recorded at Tanner stage 1, 24 subjects studied at Tanner stage 2, 18 subjects evaluated at Tanner stage 3, 23 subjects recorded at Tanner stage 4, and 29 subjects who were seen at Tanner stage 5.

Volunteers in three adult age groups were studied under conditions identical to those described for the children and adolescents (see Table 2). Sixteen young adults (Young Adults-A), ages 17 to 24 years, were studied with the same 10-hr time in bed (TIB), as part of two separate studies (8,12). Fifty elderly subjects who took part in several studies (5,6,10,21,29) were also recorded for at least 2 nights and days with the 10-hr TIB. These elderly subjects were divided into two groups based on ages. The first group (Elderly-A) included 22 subjects, ages 60 to 69 years; the second group (Elderly-B) included 28 subjects, ages 70 to 79 years.

To provide comparisons among other adult age groups, four additional sets of data—all gathered with a nocturnal TIB of less than 10 hr—were assessed separately (see Table 2). The first group included 22 young adults (Young Adults-B), ages 18 to 22 years. These subjects were recorded for at least 2 consecutive nights and days with TIB from midnight to 0800 hr (9) or 2300 to 0730 hr (19). Two groups of older adults were studied for at least 2 consecutive nights and days with TIB from 2330 to 0800 hr (18) or 2300

TABLE 2. *Subject groups studied with MSLT by Carskadon and colleagues*

Group	F/M ratio[a]	Age (years) Range	Age (years) Mean (SD)	TIB[b]	Mean TST[c] (SD)
Tanner stage 1	10/24	7–11	10.2 (1.7)	10	9.2 (0.5)
Tanner stage 2	12/12	9–14	11.5 (1.5)	10	9.3 (0.5)
Tanner stage 3	10/8	11–16	13.2 (1.3)	10	9.2 (0.3)
Tanner stage 4	11/12	12–16	13.8 (1.2)	10	8.8 (0.9)
Tanner stage 5	10/19	13–18	15.7 (1.4)	10	8.9 (0.8)
Young Adults-A	7/9	17–24	19.4 (1.8)	10	9.1 (0.5)
Elderly-A	11/11	60–69	65.2 (2.2)	10	7.8 (1.1)
Elderly-B	20/8	70–79	73.3 (2.6)	10	7.8 (0.8)
Young Adults-B	8/14	18–22	19.6 (1.4)	8 or 8.5	7.6 (0.6)
Older Adults-A	4/11	26–39	30.9 (4.0)	8.5 or 9	7.4 (0.6)
Older Adults-B	3/6	40–55	45.2 (4.5)	8.5 or 9	8.1 (0.5)
Elderly-C	12/7	61–79	68.0 (5.2)	8 or 8.5	7.0 (0.5)

[a] Ratio of female to male subjects in each group.
[b] Time in bed (hr).
[c] Total sleep time (hr).

to 0800 hr (29). One group (Older Adults-A) included 15 volunteers, ages 26 to 39 years, 13 of whom had an 8.5-hr TIB. The second group (Older Adults-B) consisted of 9 volunteers between 40 and 55 years of age, seven of whom had a 9-hr TIB. The final group of adults included 19 elderly volunteers (Elderly-C), ages 61 to 79 years, who were recorded for at least 2 consecutive nights and days with a TIB either from 2300 to 0730 hr (21) or from midnight to 0800 hr ($n = 13$). (These elderly subjects were not separated into two age groups as was done above, because only five were 70 years of age or older.)

Effects of Age and Nocturnal Sleep Time on MSLT

The data presented here for all groups are from the second night and day of testing (second day followed second night). A median MSLT value was derived for each volunteer from the series of tests given on the second day. Group scores are the median value, averaged across subjects, at each testing time, as well as the median of the group's summary scores. Correlations of median MSLT, total sleep time (TST) on the previous night, and age were performed using the Spearman rank order procedure.

Figure 1 illustrates the median daily MSLT scores for the groups of subjects recorded with a 10-hr TIB. The youngest subjects (Tanner stages 1 and 2) rarely fell asleep and consequently achieved the maximal group daily median MSLT score of 20 min. In the Tanner stage 1 volunteers, the individual median daily MSLT score was less than 20 min in only three subjects

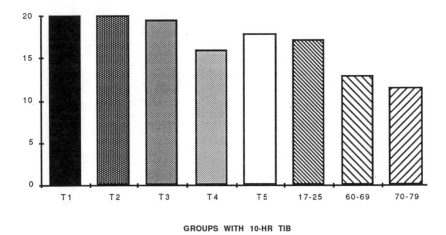

FIG. 1. Group median daily MSLT scores for the second day in subjects given a 10-hr TIB. T1, T2, T3, T4, and T5 designate the five Tanner stage groups. The group labeled 17–25 is Young Adults-A (Table 2), 60–69 is Elderly-A, and 70–79 is Elderly-B. The value at T4 (16 min) was significantly lower than the T3 value. The values for both elderly groups were significantly lower than the other groups. These data suggest a pubertal augmentation of daytime sleepiness, as well as increasing sleepiness in elderly individuals.

(9%); in the Tanner stage 2 group, this was the case in six subjects (25%). At Tanner stage 3, although the overall group daily median MSLT remained near 20 min (19.6), only 50% achieved a maximum score of 20 min. A further reduction of MSLT scores occurred at Tanner stage 4 (group median = 16 min; approximately 35% with a median of 20 min) and remained at approximately this level in the Tanner stage 5 (median = 18 min) and young adult (median = 17.2 min) groups. The reduction of MSLT scores in these adolescents supports previous findings from a subset of these data (16,20), which suggested a midpubertal augmentation of daytime sleepiness.

The MSLT scores of the elderly subjects who spent 10 hr in bed showed a further reduction in sleep latency, with only 14% of both elderly groups having 20-min median daily scores. The group median MSLT score in the 60- to 69-year-old group was 13 min and in the 70- to 79-year-old group was 11.5 min. Analysis of variance showed a significant effect of groups on MSLT scores ($F = 12.55$; $df = 6, 190$; $p < 0.001$) across these eight groups with a 10-hr TIB.

As Fig. 2 suggests, however, the amount of sleep at night also affected MSLT values, particularly in the two oldest groups. In spite of an identical testing schedule, the elderly volunteers slept less by more than an hour on average than the younger groups (see also Table 2). Within the entire group of 197 subjects with a 10-hr TIB, significant correlations were found among MSLT scores, TST values, and age: Age and TST significantly correlated ($r = 0.596$, $p < 0.001$), and both were significantly correlated with median

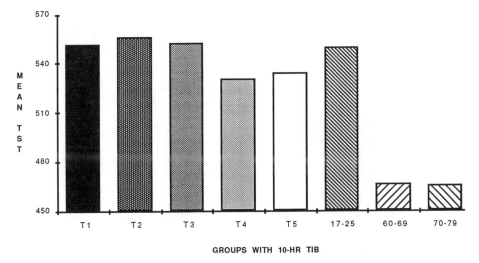

FIG. 2. Group mean TST for the second night in the same subjects as Fig. 1. TST did not differ significantly among the five Tanner stage and young adult groups. The two groups of elderly subjects slept significantly less than the younger subjects.

MSLT scores (MSLT versus TST, $r = 0.157$, $p < 0.05$; MSLT versus age, $r = 0.528$, $p < 0.001$).

The data from groups of subjects with a TIB of less than 10 hr were examined to determine if similar relationships were present and to see whether groups comparable for TST at different ages could be distinguished. Figures 3 and 4 show the MSLT and TST data for the four age groups with shorter sleep allotments. Once again, the oldest group averaged the lowest group median MSLT (5.8 min; 11% with 20 min median) and the shortest mean TST (422 min). From all the data, two groups emerged who were markedly different in age but whose nocturnal TST averages were comparable: the Young Adults-B, in whom mean TST was 456 min (SD = 38); and the Elderly-A group, whose mean TST was 466 min (SD = 64). The sleep times of the younger group overlap 100% with the older group. Furthermore, the distributions of median daily MSLT scores in these groups do not differ significantly. This result highlights the relevance of TST versus age as a determinant of MSLT.

Figure 5 plots the group median MSLT scores as a function of TST. Although this figure exaggerates the relationship by using grouped data, it is nevertheless clear that TST and MSLT are related. The correlation coefficient of median MSLT versus TST across all 262 individual subjects was statistically significant ($r = 0.295$, $p < 0.001$). This relationship has been previously shown in experiments that have tested the assumption directly by manipulating sleep schedules within groups of subjects (15).

The question remains, however, whether the age-related change in day-

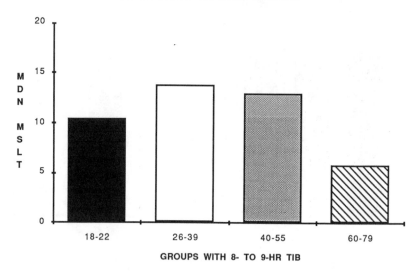

FIG. 3. Group median daily MSLT scores for the second day in subjects given an 8- to 9-hr TIB. The group designated 18–22 corresponds to Young Adults-B (Table 2), 26–39 is Older Adults-A, 40–55 is Older Adults-B, and 60–79 is Elderly-C. The median MSLT for the Elderly-C group (5.8 min) was significantly lower than the other groups. The values for Young Adults-B and Elderly-C were significantly lower than the Tanner stage groups and Young Adults-A, although not Elderly-A and B.

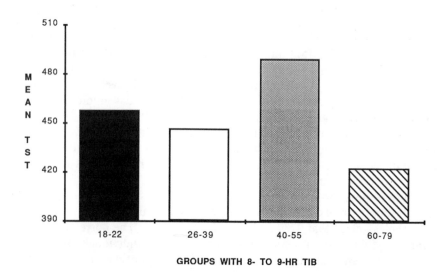

FIG. 4. Group mean TST for the second night in the same subjects as Fig. 3. These values were not significantly different among groups, although they were all significantly reduced as compared with the Tanner stage and Young Adults-A groups.

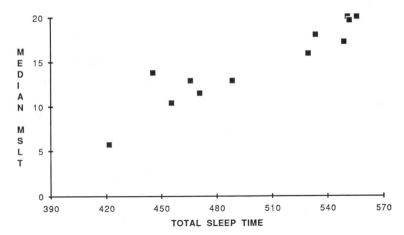

FIG. 5. Group median daily MSLT versus group mean TST in all 12 groups (Table 2). This plot exaggerates the relationship between MSLT and TST, although they were significantly correlated in the raw data of the 262 subjects ($r = 0.295$; $p < 0.001$).

time sleepiness simply results from an age-dependent decline in nocturnal sleep time or if aging also acts directly on waking alertness. In the adolescent groups, it appears that TST was less a factor than was maturational status. Thus, although no significant change in amount of sleep accompanied the Tanner 3 to Tanner 4 transition ($t = 1.55$; $p > 0.10$), MSLT scores showed a significant increase in daytime sleepiness between these stages (chi-square $= 4.19$; $p < 0.05$). A further test of this relationship during maturation awaits studies in children with precocious and delayed puberty.

In the adult volunteers, the critical factor appeared to be amount of nocturnal sleep, which was related to the experimental paradigms and to age. Thus, even when elderly volunteers were given the opportunity to sleep 10 hr at night, their sleep was reduced versus that of the younger groups. When sleep allotments were reduced (Elderly-C), sleep times remained lower relative to younger subjects (Young Adults-B) given the same schedules. Furthermore, we have previously shown that the fragmentation of nocturnal sleep by "microarousals" is generally high in elderly persons and that this fragmentation has a further impact on MSLT scores (6).

A definitive test of the effects of age versus TST on daytime sleepiness may be very difficult to achieve. Nevertheless, correlational data from our subjects recorded with a 10-hr TIB provide strong evidence that both factors are major determinants of MSLT. Thus, partial correlations across these 197 subjects reveal significant relationships: r for age and TST with MSLT partialed out was -0.809 ($p < 0.001$); r for age and MSLT with TST partialed out was 0.783 ($p < 0.001$); and r for TST and MSLT with age partialed out was 0.692 ($p < 0.001$). The changes in daytime sleepiness we have observed in our oldest groups may not be reflecting sleep restriction per se but may instead reflect a weakening of the sleeping and waking systems; to wit, a reduced ability to consolidate either state. The increased nocturnal sleep fragmentation that accompanies aging may also reflect this process.

Several other studies have incorporated comparisons of MSLT values across ages of normal subjects. Roehrs and colleagues (30) reported normative data on 33 healthy adult volunteers ages 30 to 60 years. In another study (25), this group reported MSLT data from a group of 129 younger adults (ages 18–29 years) and four groups of older adults (ages 30–80 years). In general, the MSLT data were collected from four daytime tests (1000, 1200, 1400, and 1600 hr) on the first day following a single night of sleep in the laboratory.

Table 3 summarizes the TST and MSLT data for the eight age groups from the two studies conducted at Henry Ford Hospital (25,30). Although both TST and MSLT data show a decline in the eldest group in the Roehrs et al. (30) study, the authors remarked on the absence of age-related variation in sleepiness in their subjects. In the Levine et al. (25) report, it was found that the young adults (18–29 years) had significantly lower MSLT scores than the older (30–80 years) subjects. Unfortunately, sleep times of the young adults were not reported, although the authors state that those young adults with a high sleep efficiency (≥95%) were significantly sleepier than those with a low sleep efficiency (≤86%). (TIB in this study was reported to be 8 hr.) This suggests that the sleepier volunteers among the young adults were those who had the higher TSTs on an 8-hr schedule.

Seidel and colleagues (32) also reported a negative correlation between mean daily MSLT scores and TST ($r = -0.23$; $p < 0.01$) in a group of 89 normal volunteers (47 men, mean age 26 years, SD = 4.8 years; 40 women, mean age 27 years, SD = 5.5 years). Furthermore, they reported a positive

TABLE 3. *Subject groups studied with MSLT at Henry Ford Hospital*

Study	Age range	N	F/M ratio[a]	TST Mean (SD)[b]	MSLT Mean (SD)[c]
Roehrs et al. (30)	30–39	12	4/8	7.2 (0.6)	11.5 (5.1)
	40–49	10	4/6	7.2 (0.6)	12.1 (4.5)
	50–59	11	4/7	6.7 (1.0)	9.2 (4.5)
Levine et al. (25)	18–29	129		8-hr TIB	10.7
	Decade 4			7.1 (0.6)	13.8 (1.4)
	Decade 5			7.0 (0.7)	12.3 (1.2)
	Decade 6			6.9 (0.7)	12.1 (1.1)
	Decade 7+			6.8 (0.8)	11.9 (1.6)

[a] Ratio of female to male subjects in each group.
[b] TST (hr).
[c] MSLT (min).

correlation between age and mean daily MSLT scores ($r = 0.28$, $p < 0.01$) in this fairly young group of adult subjects. We are unable to explain this age-related finding, although a positive (but nonsignificant) correlation was apparent in several of the subgroups of our data (specifically, Tanner 2 and 3, and Elderly-C groups).

Similarly, a negative correlation between MSLT and TST was found in several of our subgroups, reaching significance in the Tanner stage 2 group ($r = -0.497$; $p < 0.02$) and the Tanner stage 3 group ($r = -0.666$; $p < 0.01$). We concur with the interpretation of Levine et al. (25) that a negative correlation between MSLT and TST may reflect an accumulated sleep debt. Thus, individuals who are chronically sleep deprived and then placed in a standard testing environment will tend to sleep more than non-sleep-deprived individuals, yet the sleep debt is not recovered in a single night of relatively longer sleep and is manifested as greater daytime sleepiness on the MSLT. We predict that this pattern is similar across age groups, although Levine et al. (25) suggest that it is most commonly seen in young adults.

Hartse and colleagues (24) examined whether age was a contributing factor determining daytime sleepiness measured both by the MSLT and the RTSW, in which the instruction is to stay awake rather than to fall asleep. Both tests were given to 20 normal subjects (11 women, 9 men) ages 24 to 70 years. The subjects were divided into two groups based on age: 12 (7 women, 5 men) under 55 years of age and 8 (4 women, 4 men) over 55 years old. The subjects were studied on nonconsecutive nights but with rigidly controlled TIB. As expected, across the four daytime tests (1000, 1200, 1400, and 1600 hr), RTSW scores consistently averaged durations that were approximately 4 min longer than MSLT duration. No significant difference within either MSLT or RTSW scores was found between those subjects under and over 55 years of age.

Effect of Time of Day on MSLT

The time-of-day findings from these MSLT studies emphasize a circadian component in human daytime sleepiness, one that emerges during adolescence and is present thereafter. To illustrate, Fig. 6 shows the median MSLT scores for the prepubescent and postpubescent adolescent groups and the three adult groups with a 10-hr TIB. A midafternoon trough in MSLT scores is present in each group, with the exception of the Tanner stage 1 group (as well as Tanner stages 2 and 3, which are not shown). By Tanner stage 5, MSLT scores tended to drop in the early afternoon (1330 hr) and midafternoon (1530 hr) tests and rise again in the late afternoon (1730 hr) and early evening (1930 hr) tests. This early-to-midafternoon dip in MSLT values, present from postpubescence to older adulthood, even after 10-hr TIB, corresponds remarkably well to the time zone during which naps occur in healthy young adults (see Chapter 9).

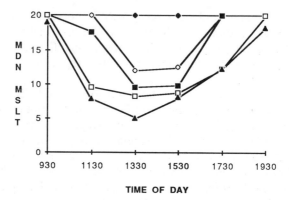

FIG. 6. Group median MSLT values as a function of time of day in subjects given a 10-hr TIB. This figure highlights the diurnal pattern of sleepiness in human subjects. Whereas the prepubescent subjects at T1 (and T2, not shown) rarely fell asleep and consequently received the maximal score of 20 min on each test of the day, all other groups (starting with T4, not shown) had a pattern of MSLT scores that included a midday decline in sleep latency (increase of sleep tendency). (♦) Tanner stage 1; (◊) Tanner stage 5; (■) Young Adults-A; (□) Elderly-A; (▲) Elderly-B.

Interestingly, the daytime dip appears to broaden to include the time from 1130 hr to 1730 hr in the elderly group means depicted in Fig. 6. This suggests a generalized loss of maximum alertness throughout the day in later years. On the other hand, even with increasing age and decreasing nocturnal sleep, the evening rise in MSLT is well maintained. As Fig. 7 shows, the morning MSLT values may not stay at a high level with reduced sleep, but evening scores remain high except in the oldest group, which also has the shortest mean TST (see Table 2).

EFFECT OF AGE ON SLEEP LATENCY IN SLEEP DISORDERS PATIENTS

Insomnia Complaints

A number of studies reporting daytime sleepiness, as assessed by sleep latency tests, in patients with insomnia have also included information on age-related changes in sleep latency in this population. Seidel et al. (32) compared the daytime sleepiness of 89 volunteers without sleep complaints (mean age 26 years) to that of 138 patients with the complaint of chronic insomnia: 43 men (mean age 57 years, SD = 17 years) and 95 women (mean age 52 years, SD = 16 years). Age was not a significant correlate of mean daily MSLT duration in the insomnia complaint group ($r = 0.02$) nor was TST ($r = -0.07$). Although the insomnia complaint group was significantly older than the noncomplaining subjects, mean daily MSLT scores were com-

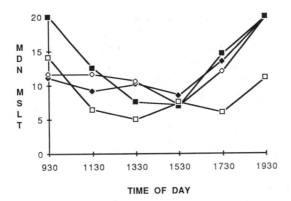

TIME OF DAY

FIG. 7. Group median MSLT values as a function of time of day in subjects given a TIB of 8 to 9 hr. This figure again emphasizes the diurnal sleepiness pattern of postpubescent humans. The reduction of the first morning test score (compare with the 10-hr groups) is thought to be a result of nocturnal sleep restricted below the required sleep need. Reduction of the evening test score, seen only in the Elderly-C group, is thought to result only with a fairly severe reduction of nocturnal sleep. The latter may, however, represent an age-related exaggeration of the sleepiness response. (♦) Young Adults-B; (◊) Older Adults-A; (■) Older Adults-B; (□) Elderly-C.

parable between the two groups (insomnia complaint group averaged 12.4 min, SD = 5.4 min; noncomplaint group averaged 10.9 min, SD = 6.5 min), despite the fact that nocturnal TSTs were generally lower in the insomnia complaint group.

Sleep efficiency in the insomnia group was significantly correlated with mean daily MSLT ($r = -0.25$; $p < 0.01$), which may suggest a cumulative sleep debt in these patients, although Seidel and Dement (33) have previously interpreted such a relationship in insomniacs as reflecting a "chronic activation" process in a subgroup of patients. Stepanski et al. (34) have reported a similar negative correlation between MSLT and TST ($r = -0.56$; $p < 0.01$) in a group of 24 patients (14 women, mean age 50 years, SD = 14.8 years) with complaints of chronic insomnia. Mean daytime MSLT in these insomniacs was 15.1 min (SD = 3.1 min), which was significantly higher than the 12.7 min daily mean (SD = 4.7 min) for a matched control group.

Narcolepsy

Daytime sleepiness in patients with narcolepsy has also been evaluated with regard to age. As part of our longitudinal assessment of sleep and sleepiness in adolescents, we followed nine youngsters with a family history of narcolepsy (4,20). Two of these children developed symptoms of narcolepsy during the study and showed reduced MSLT scores relative to the children with no history of narcolepsy. For example, one child at age 14 (Tanner stage 4) had a mean MSLT score of approximately 7 min; the second, a boy age 16 (Tanner stage 5), had a mean MSLT score of approximately 10 min. Both of these MSLT scores, however, are well above the mean MSLT typically reported for adult patients with narcolepsy, which is approximately 3 min [cf. (29)]. Because of this discrepancy, we postulated that age might be significantly related to excessive sleepiness in patients with narcolepsy (4).

Others have evaluated daytime sleep tendency in narcoleptic children and adults, finding that children with the disorder may be just as sleepy as adult patients. For example, Chisholm et al. (22) reported the case of a 6-year-old girl with narcolepsy and precocious puberty, who fell asleep within 1 min on five daytime nap opportunities. One confound of this case report, however, was that TST on the preceding night was only 6.5 hr, which is well below the sleep requirement of most children this age. Thus, it is unclear in this case whether sleep loss may have contributed to the MSLT score.

Young and colleagues (39) from the Henry Ford Hospital have evaluated daytime sleepiness in a group of eight children and adolescents with narcolepsy (ages 7–15 years, mean age 11.6 years). Table 4 lists the mean MSLT scores for this group along with similar data from a previous study of the effects of age in adult patients with narcolepsy (38). These data show no significant differences in MSLT scores as a function of age, although TST declines significantly across age groups. Thus, it does not appear that aging is a significant factor related to daytime sleepiness in patients with narcolepsy.

It is possible, however, that such relationships may be obscured in patients with excessive daytime sleepiness owing to the "floor effect" of the MSLT. That is, very short duration MSLT scores (those at or near zero latency) in patients with excessive daytime sleepiness may not reflect further variation in sleepiness because of aging. To obviate this problem, variants of the MSLT have been used to examine the relationship between age and sleep latency in very sleepy patients.

TABLE 4. *Narcoleptic subject groups studied with MSLT at Henry Ford Hospital*

Study	Age range	N	Mean TST[a]	Mean MSLT[b]
Young et al. (39)	7–15	8	9.0	1.4
Young et al. (38)	20–29	17	7.7	3.4
	30–39	60	7.7	2.8
	40–49	79	7.6	2.6
	50–59	40	7.1	3.3
	60+	24	7.1	3.1

[a] TST (hr).
[b] MSLT (min).

Baker and colleagues (1) studied a group of 55 narcoleptic patients (28 men, mean age 41.4, SD = 2.6 years; 27 women, mean age 46.2, SD = 2.4 years) for 2 consecutive nights and days, using both the MSLT and the MWT performed in a random sequence on the 2 days. Like Young et al. (39), they found that the MSLT and age were not significantly correlated; however, they found a significant relationship between age and the MWT ($r = 0.62$; $p < 0.005$). Although their results concerning TST were not reported, several other factors were also correlated with age, including several measures of nocturnal sleep disruption. Thus, although this study supports the notion that aging has a significant impact on alertness in narcoleptic patients, it is likely that age-related changes in nocturnal sleep are also involved.

In studies reported by Mitler et al. (26,27) and Browman et al. (2), the MWT was used to measure sleepiness in 12 patients with narcolepsy (6 women, mean age 45.7 years), in 12 patients with obstructive or mixed sleep apnea syndrome (4 women, mean age 43.8 years), and in 10 control subjects (5 women, mean age 35.7 years). Correlation coefficients computed in these small groups indicated no statistically significant relationship between age and MWT in narcoleptics ($r = 0.030$), sleep apneics ($r = -0.079$), or controls ($r = 0.237$).

CONCLUSIONS

Age is a major determinant of sleep and daytime sleepiness, a relationship that may become muddied in the case of daytime sleepiness because sleep and wakefulness are not only complementary, but also interacting phases in the daily cycle of existence. The nearly universal sleep debt of subjects at all ages, which may indeed be greatest in older adolescents and young adults, as Levine et al. (25) suggest, is likely to be a major confound at most ages. The effect of age on sleep latency is quite obvious over the adolescent span. Thus, MSLT scores in our adolescent volunteers declined at Tanner stage 4 even though the amount of nocturnal sleep remained unchanged from Tanner stage 3.

From the perspective of napping, the diurnal pattern of postpubescent daytime sleepiness, which manifests as an afternoon increase in sleep tendency that dissipates whether or not a nap is taken, suggests that the human brain is "programmed" for a siesta. Although this behavioral program may once have been adaptive for humans, a midday vulnerability to sleep (or sleepiness) does not seem to be adaptive in most modern cultures. Among groups in which napping is possible or desirable, however, a schedule that capitalizes on this diurnal susceptibility might be advantageous. Thus, for example, college students in whom chronic sleep restriction potentiates daytime sleepiness might benefit from a planned midday nap (in contrast to nodding off during afternoon classes). Similarly, the elderly, in whom consolidation of sleep or wake states is problematic, may be better served by reducing rather than extending nocturnal sleep and taking an afternoon siesta. Conversely, the diurnal data seem to suggest that napping in the morning or early evening would be inefficient, if possible, except perhaps in circumstances of severe chronic sleep restriction.

ACKNOWLEDGMENT

Much of the research on which this substantive evaluation was based was supported by grant MH31845 from the National Institute of Mental Health, U.S. Public Health Service.

REFERENCES

1. Baker, T., Leder, R., Colbert, M., Kim, J., and Gujavarty, K. (1987): Periodic leg movements during NREM and REM sleep in narcolepsy syndrome. *Sleep Res.*, 16:302.
2. Browman, C. P., Gujavarty, K. S., Sampson, M. G., and Mitler, M. M. (1983): REM sleep episodes during the maintenance of wakefulness test in patients with sleep apnea and patients with narcolepsy. *Sleep*, 6:23–28.
3. Carskadon, M. A. (1979): Determinants of daytime sleepiness: Adolescent development, extended and restricted nocturnal sleep. Doctoral dissertation, Stanford University.
4. Carskadon, M. A. (1982): The second decade. In: *Sleeping and Waking Disorders: Indications and Techniques*, edited by C. Guilleminault, pp. 99–125. Addison-Wesley, Menlo Park, CA.
5. Carskadon, M. A., Bliwise, D. L., Bliwise, N. G., and Dement, W. C. (1984): Effects of two sleep regimens on sleep and sleepiness in elderly volunteers. *Sleep Res.*, 13:76.
6. Carskadon, M. A., Brown, E. D., and Dement, W. C. (1982): Sleep fragmentation in the elderly: Relationship to daytime sleep tendency. *Neurobiol. Aging*, 3:321–327.
7. Carskadon, M. A., and Dement, W. C. (1977): Sleep tendency: An objective measure of sleep loss. *Sleep Res.*, 6:200.
8. Carskadon, M. A., and Dement, W. C. (1981): Cumulative effects of sleep restriction on daytime sleepiness. *Psychophysiology*, 18:107–113.
9. Carskadon, M. A., and Dement, W. C. (1982): Nocturnal determinants of daytime sleepiness. *Sleep*, 5(S):73–81.
10. Carskadon, M. A., and Dement, W. C. (1985): Sleep loss in elderly volunteers. *Sleep*, 8:207–221.
11. Carskadon, M. A., and Dement, W. C. (1985): Midafternoon decline in MSLT scores on a constant routine. *Sleep Res.*, 14:292.
12. Carskadon, M. A., and Dement, W. C. (1986): Effects of a daytime nap on sleepiness during sleep restriction. *Sleep Res.*, 15:69.
13. Carskadon, M. A., and Dement, W. C. (1987): Daytime sleepiness: Quantification of a behavioral state. *Neurosci. Biobehav. Rev.*, 11:307–317.
14. Carskadon, M. A., and Dement, W. C. (1987): Sleepiness in the normal adolescent. In: *Sleep and Its Disorders in Children*, edited by C. Guilleminault, pp. 53–66. Raven Press, New York.
15. Carskadon, M. A., Dement, W. C., Mitler, M. M., Roth, T., Westbrook, P., and Keenan, S. (1986): Guidelines for the multiple sleep latency test (MSLT): A standard measure of sleepiness. *Sleep*, 9:519–524.
16. Carskadon, M. A., Harvey, K., Duke, P., Anders, T. F., Litt, I. F., and Dement, W. C. (1980): Pubertal changes in daytime sleepiness. *Sleep*, 2:453–460.
17. Carskadon, M. A., Keenan, S., and Dement, W. C. (1987): Nighttime sleep and daytime

sleep tendency in preadolescents. In: *Sleeping and Its Disorders in Children*, edited by C. Guilleminault, pp. 43–52. Raven Press, New York.

18. Carskadon, M. A., Kerr, E. L., and Dement, W. C. (1984): Phase reversal: Effects of flurazepam (30 mg), temazepam (30 mg), and placebo on sleep, sleepiness, performance, and mood. *Sleep Res.*, 13:45.

19. Carskadon, M. A., Mancuso, J., Keenan, S., Littell, W., and Dement, W. C. (1986): Sleepiness following oversleeping. *Sleep Res.*, 15:70.

20. Carskadon, M. A., Orav, E. J., and Dement, W. C. (1983): Evolution of sleep and daytime sleepiness in adolescents. In: *Sleep/Wake Disorders: Natural History, Epidemiology, and Long-Term Evolution*, edited by C. Guilleminault and E. Lugaresi, pp. 201–216. Raven Press, New York.

21. Carskadon, M. A., Van den Hoed, J., and Dement, W. C. (1980): Sleep and daytime sleepiness in the elderly. *J. Geriatr. Psychiatry*, 13:135–151.

22. Chisholm, R. C., Brook, C. J., Harrison, G. F., Lyon, L., and Zukaitis, D. (1986): Prepubescent narcolepsy in a six year old girl. *Sleep Res.*, 15:113.

23. Erman, M. K., Beckham, B., Gardner, D. A., and Roffwarg, H. P. (1987): The modified assessment of sleepiness test (MAST). *Sleep Res.*, 16:550.

24. Hartse, K. M., Roth, T., and Zorick, F. J. (1982): Daytime sleepiness and daytime wakefulness: The effects of instruction. *Sleep*, 5(S):107–118.

25. Levine, B., Roehrs, T., Lamphere, J., Zorick, F., Stepanski, E., and Roth, T. (1987): Daytime sleepiness in young adults. *Sleep Res.*, 16:207.

26. Mitler, M. M., Gujavarty, S., and Browman, C. P. (1982): Maintenance of wakefulness test: A polysomnographic technique for evaluating treatment efficacy in patients with excessive somnolence. *Electroencephalogr. Clin. Neurophysiol.*, 53:658–661.

27. Mitler, M. M., Gujavarty, K. S., Sampson, M. G., and Browman, C. P. (1982): Multiple daytime nap approaches to evaluating the sleepy patient. *Sleep*, 5(2):119–127.

28. Richardson, G. S., Carskadon, M. A., Flagg, W., Van den Hoed, J., Dement, W. C., and Mitler, M. M. (1978): Excessive daytime sleepiness in man: Multiple sleep latency measurement in narcoleptic and control subjects. *Electroencephalogr. Clin. Neurophysiol.*, 45:621–627.

29. Richardson, G. S., Carskadon, M. A., Orav, E. J., and Dement, W. C. (1982): Circadian variation of sleep tendency in elderly and young adult subjects. *Sleep*, 5(S):82–94.

30. Roehrs, T., Zorick, F., McLenaghan, A., Sicklesteel, J., Lamphere, J., Wittig, R., and Roth, T. (1984): Sleep and MSLT norms for middle age adults. *Sleep Res.*, 13:87.

31. Schneider-Helmert, D. (1985): Multiple relaxation test (MRT): An investigation into pathophysiology of chronic insomnia. *Sleep Res.*, 14:211.

32. Seidel, W. F., Ball, S., Cohen, S., Patterson, N., Yost, D., and Dement, W. C. (1984): Daytime alertness in relation to mood, performance, and nocturnal sleep in chronic insomniacs and noncomplaining sleepers. *Sleep*, 7:230–238.

33. Seidel, W. F., and Dement, W. C. (1982): Sleepiness in insomnia and treatment. *Sleep*, 5(S):182–190.

34. Stepanski, E., Zorick, F., Sicklesteel, J., Young, D., and Roth, T. (1986): Daytime alertness-sleepiness in patients with chronic insomnia. *Sleep Res.*, 15:174.

35. Tanner, J. (1962): *Growth at Adolescence*, 2nd ed. Blackwell, Oxford.

36. Timms, R. M., Shaforenko, R., Hajdukovic, R. M., and Mitler, M. M. (1985): Sleep apnea syndrome: Quantitative studies of nighttime measures and daytime alertness. *Sleep Res.*, 14:222.

37. Van den Hoed, J., Kraemer, H., Guilleminault, C., et al. (1981): Disorders of excessive daytime somnolence: Polygraphic and clinical data for 100 patients. *Sleep*, 4:23–37.

38. Young, D., Zorick, F., Lamphere, J., Roehrs, T., Wittig, R., and Roth, T. (1986): Fragmented sleep, daytime somnolence and age in narcolepsy. *Sleep Res.*, 15:186.

39. Young, D., Zorick, F., Wittig, R., Roehrs, T., Stepanski, E., and Roth, T. (1987): Narcolepsy in a pediatric population. *Sleep Res.*, 16:461.

Sleep and Alertness: Chronobiological, Behavioral, and Medical Aspects of Napping, edited by
D. F. Dinges and R. J. Broughton.
Raven Press, Ltd., New York © 1989.

5

Chronobiological Aspects and Models of Sleep and Napping

Roger J. Broughton

*Division of Neurology, University of Ottawa and Ottawa General Hospital,
Ottawa, Ontario, Canada K1H 8L6*

In order to assess the reliability of chronobiological models of sleep/wake regulation, it is first necessary to review the data base for which they should be predictive. To date such models have concentrated on adult human data (i.e., that body of literature for which we have the most empirical facts). Yet it is evident that a comprehensive theory of the biorhythmic aspects of sleep/wake states should eventually be generalizable across all ages and all species that show the states involved. Although studies under time-free, and preferably totally disentrained, conditions are preeminently necessary both to prove an endogenous origin of biorhythmic phenomena and to obtain descriptive data from which the features of the underlying oscillator(s) may be inferred, comprehensive theories must also predict the effects of various zeitgebers on the endogenous rhythms and explain the phenomenology of a variety of conditions of entrainment.

The phenomenon of napping is particularly relevant to the current predictive models of sleep/wake states. There is growing evidence that, even in adult humans, sleep is inherently polyphasic. As we shall see, in both the disentrained and entrained states, naps, when allowed, tend to recur in a temporally lawful manner rather than randomly. The monophasic sleep pattern characteristic of most industrial societies therefore appears to be purely of social origin. Important issues are raised. These include the effects of naps in optimizing waking vigilance (signal detection ability) and the scheduling aspects of daytime naps in those situations in which sleep is reduced (sleep deprivation), becomes totally fragmented (loss of sleep continuity), or is temporally displaced (as in shift work and jet lag). It is precisely in those situations in which sleep is most fragmented that prolonged ongoing performance demands may be most critical and performance ability be affected most by the timing of naps.

This chapter will first provide a brief synoptic review of the evidence for circadian, circasemidian, and ultradian sleep/wake rhythms in both the entrained and disentrained conditions. It will then critically discuss the main current chronobiological models with emphasis on their relative abilities to predict this data base. Finally, it will suggest some further assumptions that might be useful toward elaborating more comprehensive models and will schematize their application to our adult society's typical entrained monophasic nocturnal sleep schedule. It is thought that the current data base is insufficient to propose a more comprehensive model at this time. Because of space restrictions, the review sections will necessarily be highly selective in literature citations, and reference to animal data will unfortunately only be possible where it appears essential for the specific issues raised.

BIORHYTHMS IN SLEEP STATES AND RELATED VARIABLES

Circadian Rhythms

The *major sleep period* itself, which is typically the only sleep obtained in adult humans, recurs once a day. A very large literature now exists concerning these circadian features. Only the outstanding characteristics relevant to modeling will be mentioned.

In the entrained state, with usual evening sleep onset and morning awakening both recurring at fairly regular hours, sleep has essentially a precise 24-hr periodicity. Entrainment to this period is related to the more or less consistent timing both of going to bed and, in particular, of morning awakening often caused by available zeitgebers (alarm clock, sunrise, social activity).

Central body temperature shows a parallel pattern. From a usual early-to-midevening maximum, it begins to fall approximately 1.0 to 1.5 hr prior to sleep onset, reaches a minimum approximately two-thirds of the way through sleep, begins to rise approximately 0.5 to 2.0 hr prior to the hour of morning awakening, and then increases throughout the day to the next evening maximum. There is evidence that the morning increase is earlier and/or more marked in so-called "larks," who perform best in the morning, and later and/or more marked in the evening for "owls," who function best late in the day (65). Because the usual relationships between the circadian central body temperature and sleep/wake patterns are so lawful, a number of authors have suggested that body temperature is the major determinant of the timing of sleep onset and/or duration (37,130).

However, from its slow decrease, beginning prior to sleep onset, central body temperature regularly shows a further stepwise drop associated with sleep onset (and/or descent toward SWS). This phenomenon has been evident in data of numerous studies for many years. It therefore seems quite probable, as proposed earlier (20,24), that the sleep onset of the major sleep period repeatedly phase resets the circadian body temperature rhythm.

A second main factor frequently implicated in the control of timing of sleep onset and termination is the duration of prior wakefulness. It has been assumed (16,49) that some hypnogenic factor (or factors) is accumulated in wakefulness, builds up to a threshold inducing sleep, and is then dissipated within sleep. Such conceptualizations go back at least to the series of "hypnotoxin" studies of Legendre and Piéron [cf. (103)] and are not without subjective and objective support. The latter comes from many sources, including those indicating the existence of sleep-inducing factors (such as neuroactive peptides) that are increased by prior wakefulness (100), and studies of recuperative sleep following deprivation both by classical visual analysis (11,41,87), and more recently by automatic analysis of the sleep EEG (17).

Studies of humans in time-free environments have proven that endogenous circadian biorhythmicity of sleep exists [cf. (7)] and that its natural period is 24.5 to 25.5 hr with a mean of some 25.0 hr (130), rather than the entrained one of 24.0 hr. There is a general tendency for diurnal animals like humans to have a period rather longer than 24.0 hr and for nocturnal animals to have one shorter than 24.0 hr. The reason for this consistent pattern remains obscure. Another finding from such studies is the frequent dissociation of the circadian sleep/wake pattern from that of body temperature, a phenomenon first described by Aschoff and Wever (6) as "internal desynchronization" to emphasize biorhythm uncoupling. It is difficult to resolve the concept that the circadian central body temperature rhythm is the major determinant of the circadian sleep/wake rhythm with the phenomenon of internal desynchronization.

Slow wave sleep (SWS) in the entrained state shows a typical circadian distribution. Of course, there are major ontogenetic changes in the nature and amount of SWS with a dramatic decrease being observed across the decades. (In infants and young children, SWS represents some 40% to 50% of night sleep, in adolescents 15% to 20%, in middle age 10% to 15%, and in old age 0% to 5%. It has been shown that this decrease is attributable to simple amplitude reduction of individual delta waves, as the number of delta waves detected by baseline crossing analysis remains essentially constant.) In all ages, however, the expression of SWS is normally most pronounced in the first portion of the night. Insofar as sleep per se shows a circadian periodicity, the inherent SWS therefore also exhibits circadian periodicity. In adults its acrophase (moment of greatest probability of occurrence) is within the first two sleep cycles (more precisely, approximately 1 hr after sleep onset). It is evident that its rhythmicity is largely, if not wholly, sleep dependent, as the acrophase appears more or less totally dependent on the time of sleep onset. Delaying sleep (with inherent sleep deprivation) simply increases the amount of SWS and changes the clock hour of the acrophase; as soon as sleep onset occurs, the descent toward SWS returns.

Rapid eye movement (REM) sleep in recent years has been increasingly considered from a circadian, as well as an ultradian, perspective. It had long been known in adults that there is a general trend for progressive increase in REM period length from the first to the third REM periods and that REM sleep occurrence is most probable in the latter one-third of the night. Studies of extended night sleep into the later morning hours (54,123) and studies of naps at different times of the day (67,91,126) have suggested that the probability of REM sleep occurrence remains high throughout the morning period well after the usual hour of morning awakening. Endo et al. (46) in particular have been interested in determining the circadian acrophase of REM sleep probability.

Further analysis of the 15-hr extended sleep data of Gagnon and De Koninck (55) (with sleep onset at midnight and termination at 1500 hr) has shown that the circadian REM acrophase occurs at approximately 0730 hr and REM sleep is well sustained (i.e., more than 90% of its maximum level) until 1100 hr in subjects who normally awaken at approximately 0730 hr (27). The 0730-hr REM acrophase, located between the two SWS peaks, is illustrated in Fig. 1 (27). The increased propensity for REM sleep appears closely related to the rising slope in the circadian central body temperature cycle (38).

Sleep "depth" can also be considered from a circadian viewpoint. In general, sleepers have fewest body movements and are most difficult to arouse fully in the first third of the night, and especially in SWS. Few studies, however, have differentiated circadian time-of-day (or rather time-of-night) effects from state effects. Nevertheless, many studies have shown that SWS awakenings have the greatest propensity to cause nocturnal sleep inertia or full-blown confusional "sleep drunkenness" (19,58). SWS arousals, compared with awakenings at other times of the night, lead to greater impairment of simple reaction time (116), choice reaction time (50), retrograde and anterograde memory problems (15,58,121), and the brain's response to stimuli tested by the evoked potential technique (18,113). In all of these studies REM sleep arousals produced comparably lesser deficits than did SWS awakenings.

Early threshold studies performed to decide whether SWS or REM sleep is deeper have shown that the results depend on the nature of the awakening stimulus. Using nonsignificant stimuli, subjects can be extremely difficult to awaken from REM sleep, and in such studies arousal threshold was often highest in this state (132). However, using significant stimuli such as the sleeper's name, it has been repeatedly found that subjects are hardest to awaken in SWS in the first third of the night and awaken quite easily from REM sleep (104), even in infancy (107). The circadian acrophase of "sleep depth" (awakening threshold) therefore depends on the nature of the stimuli.

Morning awakening also, of course, shows a circadian timing. In many

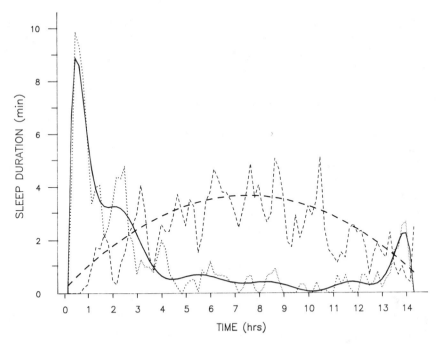

FIG. 1. Evolution of circasemidian pattern of slow wave sleep (SWS) and circadian pattern of REM sleep across 14 hr of extended sleep with sleep onset at midnight. The graphs represent raw data plus best-fitted curves by orthogonal polynomial regression. For SWS the best-fitted curve was expressed as a function of the 14th power with the first peak around 0.5 hr after sleep onset and the second peak 13.5 hr later. For REM sleep the best fit was a function of the second power with the acrophase of the circadian REM distribution at 0730 hr, i.e., around the usual time of awakening. (···) SWS observed; (—) SWS fitted; (---) REM observed; (– –) REM fitted. (From ref. 27.)

industrial societies it is the need to awaken (usually induced by an alarm clock) to go to work that repeatedly times final morning awakening. Without an alarm many individuals would continue to sleep for a period dependent on the time of sleep onset, degree of accumulated sleep deprivation, presence of other environmental stimuli (e.g., sunlight, noise), and other factors such as conditioned time of awakening. However, in all stabilized conditions without artificial alarms, including the disentrained environment, termination of the major sleep period does attain circadian predictability, although the factors that determine this timing are not yet fully elucidated. Rising central body temperature and circadian REM acrophase are two implicated factors. At other times of the day, and particularly in the early morning hours, awakening is extremely infrequent. Some have referred to this as the "forbidden wake-up zone" (68) or "verboten-to-wake zone" (133).

Overall *arousal level* may also be said to exhibit a circadian pattern. Subjective alertness, objective minimal sleep tendency (by the multiple sleep latency test or its variants), central body temperature, and performance on tests not heavily memory loaded all tend to cycle together in parallel fashion around the 24 hr, usually peaking in the early to midevening and being minimal in the early morning hours. Memory functions, however, show a curious and as yet unexplained independent peak in the morning hours (8,79). Superimposed on this overall circadian fluctuation is a circasemidian phenomenon described below. The overall slow circadian fluctuation in arousal can be schematized by 24-hr clock diagrams, as in Broughton (20).

The evening peak of relative inability to sleep is associated with heightened body temperature and subjective alertness. It is a very robust phenomenon seen even in the extreme sleep fragmentation inherent in an ultrashort sleep schedule consisting of repeated 7 min for sleep and 13 min for wake (i.e., a 20-min "day"). At this time sleep is so infrequent that Lavie (82) has called it a "forbidden zone" for sleep (see Chapter 6).

Circasemidian Rhythms

The proposal that human sleep expresses a circasemidian or, as Kronauer (76) would prefer, a hemicircadian endogenous biorhythm with a major sleep period that typically occurs at night and a second minor period in the midafternoon was first made more than a decade ago by the author (20). It was further proposed that the afternoon sleep tendency reflected pressure for SWS as part of an approximately 12-hr bimodal SWS biorhythm. Although there already existed a considerable preceding literature on human midafternoon decreases in performance associated with an increase in subjective sleepiness, this proposal came at a time when sleep itself was just beginning to be widely considered a biorhythm, and it was certainly the first proposal of an endogenous circasemidian sleep/wake biorhythm. Subsequent research by numerous groups has both confirmed the presence of such a rhythm in sleep tendency and provided some degree of support for the SWS rhythm. These findings will be briefly summarized.

Evidence for an afternoon increase in sleep tendency is supported by the following.

1. The usual timing of the last nap given up in growth and development is in the afternoon, with sleep generally becoming monophasic as the child begins to go to school (71) (see Chapter 3).

2. The siesta, seen especially in cultures that typically restrict their nighttime sleep by eating late, is essentially always in the midafternoon (22) (see Chapter 12).

3. Afternoon napping in adults is much more frequent than generally appreciated. Naps taken by adults, either as replacement naps following reduced night sleep or as appetitive naps (48), are almost always at this time. The timing of these naps relative to nocturnal sleep is midday, approximately

12 hrs from midnocturnal sleep (44,45) (see Chapter 9). After retirement, when social demands for sustained wakefulness across the day are substantially weakened, this predilection for afternoon sleep remains intact.

4. It is the moment of the so-called "post-lunch dip" in performance ability first described by Blake (14). Blake's data, however, show that it occurs whether or not food is in fact taken; and no comparable decrease is induced by food intake at breakfast or supper [cf. (61)]. It is therefore related to time of day rather than to food intake (see Chapter 9). This afternoon transient decrease in performance can, moreover, be potentiated by sleep loss (3,43).

5. Sleepiness measured objectively as a decrease in sleep latency using the multiple sleep latency test (MSLT), maintenance of wakefulness test (MWT), and their variants shows a transitory increase at this time of day (see Chapter 4). MSLTs, in fact, show almost as rapid a sleep onset in the afternoon as in the middle of the night, as is illustrated in Fig. 2, taken from Richardson et al. (110). The 2/day increase in sleep tendency remains apparent in altered sleep/wake schedules, as in shift-work studies (2) and even in extreme ultrashort sleep paradigms like the 20-min day of Lavie and colleagues (81,83) (see Chapter 6).

6. Significant social events related to sleepiness show a similar pattern. Midafternoon decreases in work performance and/or an increase in accidents are well documented (13,33,53,60,134). Automobile traffic accidents attrib-

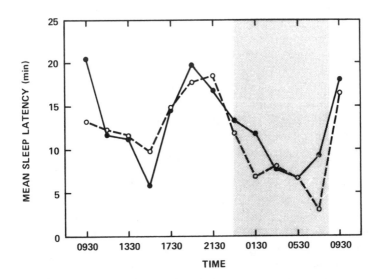

FIG. 2. Mean sleep latency on MSLT in the two populations of normal adults. (O) *n* = 8 young subjects; (●) *n* = 10 old subjects. Note the very short sleep latencies (rapid transition from wakefulness to sleep) during the overnight sleep period 0130–0730 hr and again in the midafternoon (trough at 1530 hr). This circasemidian pattern of sleepiness is interdigitated with similar circa 12-hr periods of highest alertness in the morning and again in the evening. (From ref. 110.)

uted to sleep or sleepiness show a circasemidian distribution identical to that of sleep tendency with a first peak in the middle of the night and a second in the midafternoon (84,109). Even deaths from all causes show a secondary midafternoon peak after a nocturnal one (120).

7. Finally, in pathology, the midafternoon period is the privileged time for sleep episodes in narcolepsy, whether studied by continuous in-lab monitoring (12) or by ambulant monitoring in the home environment (29,30) (see Chapter 13). This holds true also for other conditions such as idiopathic hypersomnia and sleep apnea.

The endogenous biorhythmic nature of this circasemidian sleep tendency is supported by the ubiquity of the phenomenon across all ages, in both normals and patients with clinical sleep disorders. It is provable, however, only by recordings in time-free environments facilitating disentrainment. Unfortunately, in almost all such early studies the subjects were asked to avoid napping. Repeated personal suggestions throughout a number of years to the late Elliot Weitzman to permit his subjects recorded in time-free conditions to sleep *ad libitum* were not implemented, because of other priorities.

Zulley and Campbell (136) published data on disentrainment conditions in which subjects either acted against instruction not to nap or were allowed to nap. It is evident from their results that an endogenous circasemidian sleep rhythm indeed exists. Sleep regularly splits into a major sleep period and a minor sleep period (nap). The moment of highest probability of the major sleep period coincides approximately with that of the circadian nadir in central body temperature and that of the nap with highest temperature levels (see Chapter 7).

These results are lawful and replicable. Strogatz (122) averaged across 359 time-free studies emanating from 15 protocols done in four countries (Germany, France, England, and the United States) and confirmed the ubiquity of the results. The fact that this powerful second peak in sleep propensity is related to a temperature maximum further weakens the argument that increased sleep onset probability necessarily reflects low central body temperature.

That pressure for SWS is involved in the afternoon increase in sleep tendency is also supported by several studies. Afternoon naps, compared with those in the morning or evening hours, have long been known to contain high amounts of SWS (47,66,67,127). If sleep is extended beyond 12 hr, a second increase in SWS appears in the afternoon. This phenomenon was noted in passing by Weitzman and colleagues (128) and was first systematically studied by Gagnon and De Koninck (54). These authors found that the mean interval from the onset of the first SWS pulse to that of the much later one was 12.4 hr. The robust nature of this secondary major peak in deep nonrapid eye movement (NREM) sleep is shown by its persistence even in the ultrashort sleep schedule of Lavie and Scherson (83). In their data, although stage 1 sleep showed an essentially equal distribution throughout the 12-hr daytime period tested, there was a very robust transitory increase in stage 2 sleep (the deepest NREM sleep available in the 7-min nap period) that peaked at 1400 to 1600 hr for all three experimental conditions of a normal night of sleep, after sleep deprivation, and after selective REM deprivation.

This transient afternoon SWS increase is also seen in pathology. It has been documented for daytime sleep in narcoleptics whether recorded in the lab (12) or by ambulant home monitoring (29,30). In the latter study, the average interval between the onset of daytime and preceding nighttime SWS was essentially identical to the results of extended sleep in normals (i.e., approximately 13 hr).

SWS has generally been assumed to directly reflect the amount of prior wakefulness (127) or perhaps that of wakefulness plus REM sleep (64). This mechanism is insufficient to explain the increase in SWS in afternoon versus evening naps and incompatible with the results of shifted extended sleep by Gagnon et al. (55). In the latter study, subjects with extended sleep were required to delay sleep onset by 4 hr (from 2400 to 0400 hr). On the first day of the shift, three major SWS peaks occurred: one in the first 3 hr after sleep onset, the second in the midafternoon at the time of the previous secondary increase, and the third some 12.5 hr after the first. By the third day of shifted extended sleep, the middle peak had disappeared and only the initial and 12.5-hr subsequent peaks were maintained. There was a negative, rather than positive, correlation between the magnitude of the later SWS peak and the amount of intercalated wakefulness (sleep period fragmentation), stage 1 sleep, or wakefulness plus REM sleep (27). The delayed SWS pulse was therefore greatest in magnitude when preceding extended sleep was deepest and least fragmented.

The results indicate that sleep onset had phase-set a bimodal SWS rhythm with a period of approximately 12.5 hr and that for at least 1 day the influence of the preceding entrained rhythm persisted as the middle peak in the midafternoon. Both sleep-dependent and, less strikingly, circadian mechanisms were therefore involved in the delayed SWS of extended sleep, a phenomenon that is not explicable by wakefulness after sleep onset.

Results presented by Campbell and Zulley (Chapter 7), moreover, further substantiate the existence of the circasemidian SWS biorhythm. Employing four different time-free disentrained conditions these authors report that sleep regularly becomes bimodal with the major and minor sleep periods being approximately 180° out of phase and that both contain SWS shortly after sleep onset.

Although SWS shows the circasemidian rhythm, SWS pressure may not underlie it. There are alternative explanations for the circasemidian rhythm. It may represent a rhythm of sleep propensity alone (as measured, for example, by sleep latency). The findings of Dijk et al. (42) support such a mechanism. Alternatively, it may represent a fundamental two-per-day rhythm of *wakefulness* rather than of sleep. This seems most plausible from at least two perspectives. First, wakefulness has hierarchical state priority. Much evidence exists that sleep normally can be suppressed for quite long periods of time (i.e., across several days) by heightened arousal. It appears that sleep is only permitted when the waking arousal level decreases below a (yet unspecified) threshold and disinhibits the active mechanisms for sleep production. Prolonged sleep, by comparison, never suppresses wakefulness to a similar extent, except perhaps in rare cases of neurological pathology. Second, the essential functions of wakefulness (eating, defense of self and territory, propagation of species) are well documented and clear, whereas the more postponable ones of sleep remain uncertain or unknown. In short, it makes greater biological sense that the brain be programmed for two daily prolonged periods of wakefulness than for two of sleep.

The circasemidian rhythm is not restricted to humans. The animal literature contains, however, more frequent reference to activity cycles than to sleep per se. It has long been known that many animals show two activity peaks, usually referred to as "splitting" (4,62,105,106). The paper by Aschoff (4), for example, contains references to split activity peaks in a number of lower species.

It is evident that the endogenous circasemidian sleep rhythm is almost as strong as the circadian rhythm and is certainly stronger than the waking ultradian rhythms described in the next section. Whether the phenomenon should be considered as an independent and separate approximately two-per-day rhythm (and therefore, more appropriately, an ultradian rhythm of unusually long period) or as an inherent tendency for the circadian sleep rhythm to express itself in a bimodal fashion remains uncertain (5,23). There is evidence that the phenomenon may involve interaction between the two suprachiasmatic nuclei, as splitting may be abolished by lesioning one of them (102). Above all, it should be noted that this powerful rhythm is not predicted by any of the published chronobiological sleep/wake models other than my brief earlier proposal (20), and a recent three-oscillator model of Kronauer (76).

Intermediate Ultradian Rhythms

There is evidence for sleep/wake biorhythms with periods falling between the circa 12-hr (circasemidian) and the circa 1.5-hr [basic rest/activity cycle (BRAC)] rates. Although certainly less well studied, these rhythms are reasonably well documented in humans with further evidence of their existence being present in lower animals. They tend to cluster around the periods of approximately 6 and 3 hr.

Circa 6- to 8-hr (3–4/day) rhythms have been noted for humans in ultrashort sleep/wake studies of "sleepability" (85), which show a prominent peak in the spectral analysis of the averaged data for 3.6 cycles per day. Similarly, normal humans studied under disentrained conditions of 60 hr of bed rest showed an average sleep/wake cycle of 6.0 hr but with a high standard deviation of 2.7 hr (35). Zulley and Wever (137) have noted that sleep onset times in free-running subjects "were not randomly distributed" but showed ultradian peaks with a mean peak-to-peak interval of 5.7 hr. In rats, Honma and Honma (63) found a 5- to 6-hr rhythmicity in activity levels and central body temperature under continuous light (LL) conditions. Rusak (111,112) has noted that with the loss of circadian rhythmicity from suprachiasmatic lesions, hamsters under LL conditions develop short inactivity periods (apparent sleep) occurring with an 8-hr (and 12-hr) periodicity.

Strong evidence exists for a circa 3- to 4-hr (6–8/day) sleep/wake rhythm. Infants fed by on-demand schedules display a striking sleep/wake rhythm at a period of 3 to 4 hr (71,94). Nakagawa (97) found a similar periodicity of sleep probability in adult subjects recorded for 24 hr under continuous bed-rest conditions. Using the ultrashort sleep schedule paradigm, Lavie and co-workers (83,85) found increases in sleepability with this periodicity in two studies, particularly during sleep deprivation conditions. Similar cyclic increases in sleepiness as indexed by augmented daytime (8 per day) parietal EEG spectral power in several frequency bands (alpha, theta, overall power) were noted by Manseau and Broughton (89,90). Zulley (135) subsequently described an approximately 4-hr rhythmicity of sleep itself in the disentrained state. Extended sleep data (27) show some evidence for increases in SWS every 3 to 4 hr, and this periodicity of SWS has also been found in 24-hr recordings of narcoleptics under bed-rest conditions (De Koninck and Billiard, personal communication). The phenomenon does not appear to be restricted to humans. A 3- to 4-hr sleep/wake rhythm has also been described in rats with a corresponding rhythmic pulsation of growth hormone timed with the periods of sleep (69).

Further evidence for intermediate-range ultradian sleep/wake rhythms may be found in the landmark volume edited by Schulz and Lavie (115). It is important to note that these periodicities are repeatedly evident in raw data, as well as in analyses using spectral techniques. Unfortunately, the latter can introduce totally artifactual harmonics of a dominant periodicity such as the circadian one (23,32).

BRAC-Rate Ultradian Rhythms

The NREM/REM sleep cycle within sleep with its inherent physiological, endocrine, and neurochemical fluctuations represents the best-documented ultradian sleep rhythm. The SWS component alone may be considered a dampened ultradian rhythm across the night, as first proposed by Lubin et al. (88). Of course, a large number of studies have documented REM periodicity, although it should be noted that the period of this rhythm is much less stable than generally believed (125).

The issue of whether the ultradian NREM/REM rhythm is sleep dependent or independent is very important for adequate modeling but remains controversial. Studies collapsing wake time, such as those in humans by Moses et al. (95,96) and in the cat by Ursin et al. (124), support its being sleep dependent. On the other hand, studies in humans who sleep much of the daytime such as narcoleptics (40) suggest that the subsequent nocturnal REM sleep cycle can be predicted by the timing of daytime REM sleep. Findings like the latter support the existence of an ultradian rhythm throughout 24 hr, manifested as alternations in NREM/REM states during sleep and in somnolence/alertness during wakefulness in a manner compatible with the BRAC hypothesis of Kleitman (70–72).

It has long been evident that NREM and REM sleep are mutually inhibitory, although at times so-called intermediate sleep states occur that combine features of both. Increased pressure for deep NREM (slow wave) sleep at the start of the night appears to inhibit REM sleep, increasing REM latency and shortening the initial REM period duration. In childhood, when maximum SWS pressure is present, the first REM period is often skipped, and the REM latency greatly increased. Conversely, this mutual inhibitory interaction also appears present in instances of early REM periods or actual sleep-onset REM periods—phenomena that can also be considered as a phase advance of the circadian REM acrophase. They tend to occur in situations in which the circadian SWS acrophase is weakened or lacking. Consider the two prototype conditions, narcolepsy and endogenous depression, in which sleep-onset or early REM periods are frequently encountered. In narcoleptics, SWS tends to be equally distributed across the thirds of the night (31), whereas in depressives, SWS is greatly reduced in amount and often essentially absent (59).

Suppression of REM sleep expression by increased pressure for SWS is also evident in the circasemidian sleep rhythm. As well as containing more SWS than do naps at other times of the day, midafternoon naps contain less REM sleep. This holds true for normals (67), as well as for ambulant recordings of daytime sleep in narcoleptics (29,30). The phenomenon is also observed in the structured naps inherent in MSLTs, which, in narcoleptics, show significantly higher probability of REM sleep in the 1000-hr naps and of SWS in afternoon naps (1). Indeed this suppression of REM sleep by SWS has been documented even within individual NREM/REM cycles across the night (73).

A daytime ultradian variation in sleepiness (and sleep) with a periodicity similar to nocturnal NREM/REM cycle, predicted by Kleitman's BRAC hypothesis, is now well documented. Its further prediction that these daytime fluctuations are in phase with nocturnal NREM/REM cycles, however, is far less well substantiated. Indeed, the evidence is conflicting. Ultrashort sleep schedule studies have found a trend toward decreased sleepability (higher alertness) in phase with prior REM periods (85). Results from a vigilance task have suggested poorer performance with fewer detections and more false positives (indicating lowered alertness) in phase with prior REM periods (23). Determining which is correct is an essential step to more adequate modeling. Furthermore, a number of waking ultradian rhythms at approximately the NREM/REM cycle period exist that are not related to alertness and appear totally independent of the putative BRAC. These interesting latter findings are irrelevant to our present purposes and are reviewed elsewhere by Lavie (81).

A number of approaches have documented ultradian rhythms in daytime alertness at approximately the period of the within-sleep NREM/REM cycle. Electroencephalographic (EEG) measures have long been used in this regard. Early support came from the studies of Kripke (74). Our laboratories have investigated changes in EEG frequency bands at both frontal and parietal electrode sites over both hemispheres. Subjects were in a partial isolation chamber for 8 hr and lacked any performance demands that could influence endogenous rhythmicity. We documented approximately 14-per-day (72–120 min) EEG increases in both overall EEG content (2.5–20 Hz) and in frontal theta (significant only on the right), with both hemispheres synchronizing in phase (89,90). Increased frontal theta and overall increase in the EEG power are, in fact, sensitive indicators of drowsiness. Similar results have been reported by Okawa et al. (98) using retrospective analysis of prolonged recordings.

Numerous other approaches substantiate this ultradian rhythm of alertness/drowsiness. Performance tests sensitive to fluctuations of alertness, such as a visual vigilance task, exhibit rhythmic fluctuations at this rate (34,99). Pupillometry, a sensitive measure of alertness (86), shows 90- to 120-min oscillations in awake normals (80) and awake narcoleptics (108). Increases in ability to sleep recurred with 90- to 120-min periodicity in the ultrashort sleep schedules of Lavie and Scherson (83).

The ultradian rhythm is also expressed in pathology. Ultradian cyclic appearance of sleep attacks has been described in narcolepsy for sleep attacks both beginning in REM sleep (101,114) and in NREM sleep (9). Petit mal absences, which are facilitated by drowsiness, can be shown in children with frequent attacks to have an ultradian daytime distribution (23).

The intensity of the ultradian fluctuations appears to depend on the overall level of arousal (23). Like the expression of the circasemidian rhythm, it is enhanced in all situations that lead to dearousal such as boredom, sleep deprivation, or sleep pathologies characterized by excessive daytime sleepiness. Conversely, it is masked by sustained high arousal levels. Indeed, if subjects are extremely motivated and stressed, the ultradian rhythms may remain totally undetectable for long periods of testing, as in the study by Kripke et al. (75).

There is evidence that the magnitude of the ultradian alertness rhythm may relate to personality and other measures. In the studies of Manseau and Broughton (89,90), only two of some 64 psychological-behavioral factors correlated with the magnitude of the waking ultradian alertness rhythm. These were introversion (Eysenckian scale) and subjective poor quality sleep. It would therefore appear that good quality night sleep leads to the ability to maintain more even levels of daytime alertness and that the latter may be associated with the personality trait of greater emotional stability.

MODELS OF SLEEP/WAKE BIORHYTHMS

Circadian Models

Several models exist to explain the circadian distribution of sleep.

A single-oscillator two-process model proposed by Borbély (16) consists of a process S, similar to the classical concept of a hypnotoxin, in combination with a circadian oscillator, process C. Process S is considered to build up in wakefulness, decline exponentially in sleep, and be physiologically reflected in the spectral power of EEG slow wave activities. Process C is a rhythmic circadian variation in sleep propensity paralleling central body temperature fluctuations. The probability of both sleep onset and sleep duration are reasonably predicted by the model. (The separate within-sleep NREM/REM sleep alternation is assumed in this model to be owing to an unrelated sleep-dependent ultradian oscillation resulting from reciprocal interaction of the sleep states). Daan et al. (39) further elaborated the model adding high and low thresholds for sleep onset and termination and found that the model allowed computer simulation of several main features of sleep timing. These included internal desynchronization in the absence of time cues, sleep fragmentation during continuous bed rest, and circadian phase dependence of sleep duration, all of which were predicted for three different experimental manipulations (isolation from time cues, recovery from sleep deprivation, and shift work). The model has also been applied with some success to the sleep of depressed patients (78).

A two-oscillator model, based on mutually coupled van der Pol oscillators, was proposed by Kronauer et al. (77) initially to explain data obtained in temporal isolation. One oscillator (X) is considered a strong oscillator that dominates central body temperature. The weaker Y oscillator regulates the sleep/wake state. These oscillators are analogous to Wever's (130) earlier Types I and II oscillators. The oscillators differ in their period range and intrinsic period stability. Their attributes and interactions are expressed by differential equations. The model does not accommodate the homeostatic aspects of sleep. Nevertheless, numerous close predictions concerning the circadian temporal aspects of sleep are provided by the model, including the phenomenon of "phase trapping." Indeed, Gander et al. (56) applied simulated zeitgebers to the model to specifically investigate six different types of synchronization with the model predictions expressed as entrainment maps. These predictions were then tested by existing experimental data (57).

A three-oscillator circadian model, which involves separate oscillators for central body temperature, wake onset, and sleep onset, was proposed by Kawato et al. (68). The functional interrelationships are expressed by differential equations. Unlike the model of Kronauer et al. (77), in which both oscillator phase and amplitude variables are stipulated, the oscillator states are described only by phase. The model reasonably simulates sleep duration as a function of sleep onset time and in irregular free-running patterns, and also reproduces the circadian forbidden wake-up zone.

All of these models have reasonable predictive validity for the circadian features of a monophasic sleep pattern in which no napping occurs. None predicts, however, either the circasemidian or the ultradian sleep/wake patterns described earlier.

Circasemidian Models

My initial proposal of the existence of an endogenous two-per-day biorhythm of sleep discussed three main aspects (20). The first was the existence of a two-per-day decrease in vigilance facilitating sleep onset, the initial peak normally beginning in the evening and the daytime peak in the afternoon. Second, it was considered that the circasemidian sleep propensity rhythm might be an integral (super) harmonic component of the basic circadian rhythm. Third, it was postulated that there existed an approximately 12-hr "bipolar" rhythm of SWS, whose first pole was set by sleep onset.

Evidence for the existence of a circasemidian sleep propensity was subsequently strengthened by the results from the ultrashort sleep/wake schedule of Lavie and Scherson (83). Lavie later added this afternoon peak to the major nocturnal peak in his schematic representation of 24-hr variations in the probability of wake-to-sleep transitions (81). Recently, Kronauer (76) extended his two-oscillator model, splitting his weak Y oscillator into two reciprocal inhibitory subcomponents (Y_1 and Y_2) to incorporate the circasemidian rhythm.

Ultradian Models

For many years the essentially rhythmic nature of the NREM/REM sleep alternation has been recognized and analyzed by various techniques. As we have seen, some authors have considered this cyclicity to be a sleep-dependent phenomenon, others as being sleep independent. Kleitman (70,71) proposed his BRAC hypothesis, which is not an actual model in the strict sense. It postulated the existence of an ultradian rhythm, which persists around the 24-hr period, expressed during sleep periods as the NREM/REM cycle and in wakefulness as oscillations in alertness and movement at essentially the same periodicity. As we have also seen, there is now very solid evidence for ultradian variations in waking alertness at approximately the nocturnal NREM/REM sleep periodicity, although proof of it being in phase with nocturnal cycles remains inconclusive. Evidence for the predicted parallel periodicity in movement is essentially nonexistent.

The within-sleep NREM/REM cycle was formally modeled by McCarley and Hobson (92) from a structural and mathematical perspective. They employed Lotka-Volterra equations to express the interaction between mutually interactive REM-on and REM-off neural populations. This reciprocal-interaction model showed only neutral stability, its long-term behavior being totally set by the initial conditions. An update of the model has more recently been presented by McCarley and Massaquoi (93), in which physiological limits on neuronal firing rates were added and the mathematical functions better describe neuronal activities at low discharge rates. This resulted in a model that displays limit cycle stability and is highly predictive of the timing of REM periods in both entrained and free-running human data. The model also provides phase-response curve predictions of relevance to pharmacological effects on REM sleep and to REM cycle patterns in depression. The model, however, is essentially a sleep-dependent one and does not predict waking ultradian fluctuations in sleepiness/alertness variations.

Modeling of REM sleep cycle has also been done mathematically by assuming the ultradian rhythm to be an autonomous self-sustained oscillator using two mutually coupled second-order equations or an equivalent fourth-order equation (129–131). This approach has been used by Wever to characterize sleep-dependent REM cyclicity.

Kobayashi et al. (73) considered the single NREM/REM sequence to be a basic sleep unit that is normally repeated four or five times per night in humans. They analyzed the interactions between the two components across the sleep period. This approach has helped to clarify those variables affecting the range of sleep-cycle duration, as well as the interaction between the sleep states within the sleep cycle. It involves an algorithm expressing a closed-loop relationship between length and content of the ultradian NREM/REM cycles. The approach characterizes the negative interaction between the SWS portion of NREM and REM sleep, including the existence of an occasional fully skipped REM period in the first sleep cycle.

It is notable that most, if not all, ultradian models consider the NREM/REM cycle (usually conceived as the REM cycle) to be rhythmic in nature. This assumption is highly questionable. Even the most casual perusal of raw data will show that the NREM/REM sequence is highly asymmetrical. Although descent into deeper NREM stages is essentially always slow and progressive, the onset of REM sleep plotted on the same time base is much more abrupt. A sleep-stage histogram, or even spectral analysis of EEG content across time, therefore, resembles a sawtooth pattern more so than a sine wave. In short, the data strongly suggest that REM sleep is a (more or less) periodic, rather than a rhythmic, event. Models might therefore profitably be pursued that represent REM sleep as an essentially periodic all-or-none event that is increasingly inhibited by greater pressure for SWS. They would much more adequately characterize data of human sleep patterns than do models that assume a sinusoidal rhythmicity.

Combined Circadian-Circasemidian-Ultradian Models

A number of factors, not the least of which is insufficient data, explain the relative paucity of attempts to conceptually integrate the multiple temporal orders of sleep/wake biorhythms.

This author's first postulate (20) was that humans generally tend to exhibit biological rhythms that may be considered as harmonics of the basic circadian (cosmic solar) day. These include both the 48-hr bicircadian (36) subharmonic and the approximately 12-hr (circasemidian), 6-, 3-, and 1.5-hr superharmonic rhythms. Assuming that these apparently privileged periodicities occur in the raw data, a more plausible current hypothesis is that there exists an evolutionary natural selection for those periodicities that show fixed integer ratios to each other and to the basic 24-hr solar day (23). Briefly stated, although oscillators with all conceivable periodicities might have appeared at some point in evolution, there would be survival value (more economic use of available energy resources) to maintain rhythms that can be consistently phase related to each other and to the major external zeitgeber (viz., the solar day). It is also evident that the approximately 12.5-hr circasemidian rhythm is close to the mean tidal period and that the circa 90-min periodicity is close to that of Schuler's constant for the gyrocompass (118). This raises the interesting speculation that a number of prominent biorhythmicities other than the circadian one may also parallel geophysical environmental fluctuations.

Wever (129–131) combined his mathematical models of circadian and ultradian rhythms with a sleep/wake threshold level that creates curves that reasonably mimic human circadian and ultradian data. They do not, however, predict or explain the powerful circasemidian component, nor, as the ultradian rhythm is considered sleep dependent, do they explain the waking ultradian rhythm in sleepiness/alertness.

This overview of attempts to model human sleep/wake states leads to the conclusion that no comprehensive adequate model currently exists and, moreover, that lack of knowledge on a number of important issues impedes fuller modeling of these rhythms. These issues include:

1. To what extent, and how, do body temperature, duration of prior wakefulness, and underlying endogenous circadian, circasemidian, and ultradian sleep/wake rhythms *interact* to define the probability of sleep onset, maintenance, and termination?

2. To what extent, and how, do personal choice of time of going to bed, as well as various sleep-enhancing (quiet environment, optimal room temperature, comfortable bed and clothing) and sleep-impeding factors, combine with the underlying chronobiological sleep/wake and central body temperature biorhythm factors to lead to resultant sleep/wake structure?

3. Is the circasemidian sleep propensity simply inherent to the circadian sleep rhythm? Or does an independent circa 12-hr sleep rhythm (based or not on a 12-hr SWS rhythm) exist? Or is, as proposed here, the fundamental rhythm one of wakefulness?

4. Does the BRAC hypothesis hold true with respect to its prediction that waking oscillations in alertness/sleepiness should be in phase with the within-sleep REM periodicity? Or is the REM cycle completely sleep dependent?

5. To what extent are the three basic biological states (wakefulness, NREM sleep, REM sleep) mutually inhibitory and necessarily exclusive? Conversely, do sleep and waking states simply passively fill in when the other is missing? Moreover, to what extent and under what conditions can the states coexist, such as in the so-called intermediate sleep showing some simultaneous features of NREM and REM sleep, and in so-called double consciousness with simultaneous perception of dream activity and the environment (21)? Indeed it seems possible, and there is some evidence (10) that some subcomponents of the brain can be physiologically asleep while others are awake.

SCHEMA OF A TYPICAL ENTRAINED 24-HR DAY

Admittedly, then, the lawful chronobiological relationships between the various sleep/wake rhythms, the role and limits of voluntary selection of time of sleep onset, and other major factors affecting sleep/wake structure are not yet fully known. Nevertheless, as stated in the introduction, it would seem worthwhile to attempt to characterize a human adult's typical well-entrained 24-hr day. Such an attempt could highlight issues requiring further clarification and might even give explicatory insights into phenomena still not adequately explained.

The schema provided in Fig. 3 summarizes a number of important variables across 27 hr, beginning 1 hr prior to sleep onset. Sleep onset is arbitrarily set at midnight and morning awakening at 0745 hr. The sleep period is shaded, and increased levels of arousal are plotted up. The circasemidian

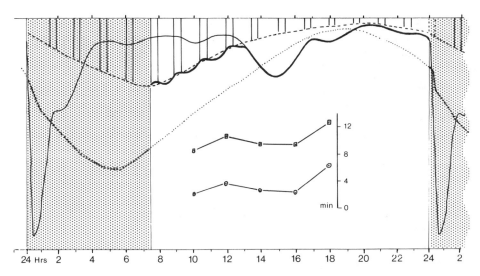

FIG. 3. Schematic representation of the entrained adult evolution of sleep and alertness across the 24 hr beginning 1 hr prior to sleep onset. The overnight sleep period is stippled. The circasemidian pressure for SWS and the circadian modulation of periodic REM sleep pressure (*vertical bars*) are actual data (up to 1500 hr) from studies of extended sleep (27). Also shown are the circadian curve of central body temperature (dotted and darkened during night sleep) and MSLT data from 21 controls (*above*) and matched untreated narcoleptics (*below*) (28). Waking alertness level is the area below the darkened line in the daytime period. See text for detailed description.

peaking of pressure for SWS with superimposed 4-hr variations represents actual lab data (27) up to 1500 hr. REM sleep pressure is shown as periodic pulses [vertical bars whose circasemidian envelope with peak at time of awakening represent actual data from Broughton (24) up to 1500 hr]. The circadian variation of central body temperature with its nadir at approximately 0530 hr is derived from several published sources. Level of waking arousal is represented as the white (awake area) below the dark line following pressure mainly for REM sleep in the morning and SWS in the afternoon. Actual lab MSLT data (27) for 21 controls (above) and 21 matched narcoleptics (below) are plotted in the appropriate location within wakefulness. Also important (but not shown) is the circasemidian facilitation of sleep onset proposed by Broughton (20) and Lavie (81).

For the purpose of discussion this schematic representation could equally have been begun at time of morning awakening or been double plotted. It makes a number of assumptions.

1. The time of going to bed (and resultant actual sleep onset time) is assumed to be voluntarily selected at a period of escalating subjective sleepiness, which is related to a combination of sufficient prior wakefulness and of decreasing central body temperature, combined with a "gate" of in-

creased facilitation of transition into sleep. It is known that in the entrained condition central body temperature does in fact normally start to decrease some time before sleep onset and that in the free-running condition the major sleep period is normally selected near the temperature minimum (whereas the main nap, if taken, is not).

2. Sleep onset (or perhaps the physiological mechanisms of descent toward SWS) causes a further decrease in central body temperature and appears to phase-reset the circadian temperature oscillation (20,24).

3. Sleep onset also determines the circadian acrophase of the first SWS peak and therefore also the phase of the secondary peak some 12.5 hr later. Maximum pressure for SWS is shown to occur mainly in the first 3 hr of sleep. This period of time corresponds to Winfree's forbidden wake-up zone (68,133). It can be visually interpolated as being at approximately 0100 to 0200 hr. The second major peak of SWS, which is manifested in a voluntary nap or is experienced as marked pressure for sleep, if remaining awake, is shown at approximately 1500 hr (i.e., some 12.5 hr later).

4. Pressure for SWS diminishes during the night in a more or less exponential fashion, modulates the intensity of REM sleep, and may also phase-reset the ultradian REM periodicity. Decreased pressure for SWS early in the night (either from overall SWS reduction, as in depression, or from more even SWS distribution, as in narcolepsy) results in a precocious first REM period. Conversely, increased SWS pressure at the start of the night (as in childhood or recovery from sleep deprivation) retards REM latency and may indeed totally suppress expression of the initial REM period.

5. The period between the two circasemidian peaks of pressure for SWS is associated with minimal suppression of the ultradian REM propensity, and the circadian REM acrophase occurs at this time. During the overnight sleep period, this is expressed as increasing amounts of REM sleep toward the end of the night. If sleep is extended into midmorning or if voluntary naps are taken at that time, high amounts of REM sleep are still encountered.

If the subject is awake throughout the morning, high intensity of the underlying ultradian facilitation of REM sleep may explain two further interesting phenomena at this time of day. The first is the fact that daytime ultradian oscillations of alertness/sleepiness are much greater than later in the day, as indicated in Fig. 3 and shown by the ultrashort sleep schedule approach (85). The second is the previously noted finding that daytime memory functions are optimal at this time (8,52,53,79). Positive correlations between REM sleep and memory mechanisms have, in fact, been known for some time (51,119). Indeed, recall of material presented prior to REM naps has been shown to be superior to NREM naps (117). The waking expression of the effects of increased pressure for REM versus NREM sleep have already been documented for subjective sleepiness, objective sleepiness, and cerebral evoked potentials (25,26). It is notable that individual episodes of

REM sleep or pressure for REM sleep lead to relative increases in vigilance during sleep periods and relative decreases during wake periods of the day.

6. During the afternoon period, the second sleep gate of the circasemidian rhythm and a transient increase in pressure for SWS would appear to be the basis for both the flattening of ultradian oscillations in alertness/sleepiness in awake subjects (by suppressing REM propensity) and for the postlunch dip in performance efficiency (from subjects trying to suppress this transient period of augmented sleepiness). Of course, it would also directly explain the high frequency of SWS at this time (relative to others), if a nap (siesta) is taken.

7. In the early evening hours, there is temporal coincidence of three factors minimizing sleepiness and optimizing alertness. These include low pressure for SWS (whose circasemidian peaks occurred earlier in the day), minimum pressure for REM sleep (whose acrophase was in the morning), and highest central body temperature. This coincidence would appear to explain why it is at this time of day that subjects more or less universally find it most difficult to fall asleep, as demonstrated by MSLTs throughout the 24-hr day and ultrashort sleep schedules. It is, in fact, Lavie's forbidden zone of sleep.

8. The evolution of the waking arousal levels is the inverse of those of sleep pressure, as sleep and wakefulness are mutually competitive and essentially mutually exclusive. Following the curve of waking arousal indicates high levels of ultradian BRAC-rate oscillations of alertness in the morning when periodic pressure for REM sleep continues to be high. There is a late-morning transient peak of alertness (and lowered sleep probability) around noon, as the circadian facilitation of REM sleep is waning, and prior to the midafternoon increase in sleep pressure. Finally, as mentioned, arousal typically reaches its maximum in the early evening. These fluctuations, which are derived by subtracting sleep pressure during the daytime, and which are based on extended sleep data (27), are followed precisely by independent observations of MSLT data averaged for large numbers of untreated narcoleptics and controls (28).

9. Somewhat later in the evening, the central body temperature begins to decrease, sleepiness related to accumulated wakefulness continues to increase, and the major circadian gate is approached. The decision to go to bed in order to sleep is then made and the cycle is resumed.

Finally, the magnitudes of the circadian, circasemidian, and ultradian sleep-facilitating rhythms reflect overall pressure for sleep. Sleep pressure will be increased by such factors as prior sleep loss, sedative drugs, and certain sleep pathologies. Each of these would enhance the expression of all sleep-facilitating rhythms. Conversely, factors optimizing the daily recuperative value of sleep for maximum alertness (in particular, obtaining sufficient amounts of good quality sleep at regular hours) would minimize the daytime ultradian and circasemidian expressions of sleepiness.

Whether the optimum efficiency of human higher nervous functions is best served by such a single nocturnal monophasic sleep pattern, by night "anchor" sleep with a single well-timed afternoon nap, or by polyphasic napping (catnapping) at the most propitious moments, remains to be determined. The nature and timing of work demands, of course, will further interact to give differing optimal schedules for different circumstances.

ACKNOWLEDGMENT

Personal research cited in this chapter was supported by the Medical Research Council of Canada, of which the author is a Career Investigator Awardee.

REFERENCES

1. Aguirre, M., and Broughton, R. (1987): Complex event-related potentials (P300 and CNV) and MSLT in the assessment of excessive daytime sleepiness in narcolepsy-cataplexy. *Electroencephalogr. Clin. Neurophysiol.*, 67:298–316.
2. Åkerstedt, C., and Gillberg, M. (1982): Experimentally displaced sleep: Effects on sleepiness. *Electroencephalogr. Clin. Neurophysiol.*, 54:220–226.
3. Angus, R. G., and Heslegrave, R. J. (1985): Effects of sleep loss on sustained cognitive performance during a command and control stimulation. *Behav. Res. Meth. Instr. Comput.*, 17:55–67.
4. Aschoff, J. (1966): Circadian activity pattern with two peaks. *Ecology*, 47:657–662.
5. Aschoff, J., and Gerkema, M. (1985): On diversity and uniformity of ultradian rhythms. In: *Ultradian Rhythms in Physiology and Behavior*, edited by H. Schulz and P. Lavie, pp. 321–334. Springer-Verlag, Berlin/New York.
6. Aschoff, J., and Wever, R. (1962): Spontanperiodik des Menschen bei Ausschluss aller Zeitgeber. *Naturwissenschaften (Berlin)* 49:337–342.
7. Aschoff, J., and Wever, R. (1982): The circadian system in man. In: *Handbook of Behavioral Neurobiology: Biological Rhythms, Vol. 2,* edited by J. Aschoff, pp. 311–332. Plenum Press, New York.
8. Baddeley, A. D., Hatter, J., Scott, D., and Snashall, A. (1970): Memory and time of day. *Br. J. Psychol.*, 22:605–609.
9. Baldy-Moulinier, M., Arguner, A., and Besset, A. (1976): Ultradian and circadian rhythms in sleep and wakefulness. In: *Narcolepsy*, edited by C. Guilleminault, P. Passouant, and W. C. Dement, pp. 485–498. Spectrum, New York.
10. Ball, N. J., Amlander, C. J., Shafferty, J. P., and Opp, M. R. (1988): Asynchronous eye-closure and unihemispheric quiet sleep of birds. In: *Sleep '86,* edited by W. Koella, pp. 151–153. Fischer-Verlag, Stuttgart.
11. Berger, R. J., and Oswald, I. (1962): Effect of sleep deprivation on behaviour, subsequent sleep and dreaming. *J. Ment. Sci.*, 108:457–465.
12. Billiard, M., Quera Salva, M., De Koninck, J., Besset, A., Touchon, J., and Cadhilac, J. (1986): Daytime characteristics and their relationships with night sleep in the narcoleptic patient. *Sleep*, 9:167–174.
13. Bjerner, B., Holm, A., and Swenson, A. (1955): Diurnal variations in mental performance: A study of three-shift workers. *Br. J. Ind. Med.*, 12:103–130.
14. Blake, M. J. F. (1967): Time of day effects on performance in a range of tasks. *Psychosom. Sci.*, 9:349–350.
15. Bonnet, M. H. (1983): Memory for events occurring during arousal from sleep. *Psychophysiology*, 20:81–87.

16. Borbély, A. A. (1982): A two process model of sleep regulation. *Hum. Neurobiol.*, 1:195–204.
17. Borbély, A. A., Baumann, F., Brandeis, D., Stauch, I., and Lehmann, D. (1981): Sleep deprivation effect on sleep stages and EEG power density in man. *Electroencephalogr. Clin. Neurophysiol.*, 51:483–493.
18. Broughton, R. J. (1968): Sleep disorders: Disorders of arousal? *Science*, 150:1070–1078.
19. Broughton, R. J. (1973): Confusional sleep disorders: Interrelationship with memory consolidation and retrieval in sleep. In: *A Triune Concept of the Brain and Behavior*, edited by T. J. Boag and D. Campbell, pp. 115–127. University of Toronto Press, Toronto.
20. Broughton, R. J. (1975): Biorhythmic variations in consciousness and psychological functions. *Can. Psychol. Rev.*, 16:217–230.
21. Broughton, R. J. (1982): Human consciousness and sleep/wake rhythms: A review and some neuropsychological considerations. *J. Clin. Neuropsychol.*, 4:193–218.
22. Broughton, R. J. (1983): The siesta: Social or biological phenomenon? *Sleep Res.*, 12:28.
23. Broughton, R. J. (1985): Three central issues concerning ultradian rhythms. In: *Ultradian Rhythms in Physiology and Behavior*, edited by H. Schulz and P. Lavie, pp. 217–233. Springer-Verlag, Berlin/Heidelberg/New York/Tokyo.
24. Broughton, R. J. (1988): The circasemidian sleep rhythm and its relationship to the circadian and ultradian sleep-wake rhythms. In: *Sleep '86*, edited by W. P. Koella, F. Obal, H. Schulz, and P. Visser, pp. 41–43. Fischer-Verlag, Stuttgart.
25. Broughton, R. J., and Aguirre, M. (1985): Evidence for qualitatively different types of excessive daytime sleepiness. In: *Sleep '84*, edited by W. P. Koella, E. Ruther, and H. Schulz, pp. 86–87. Fischer-Verlag, Stuttgart.
26. Broughton, R. J., and Aguirre, M. (1987): Differences between REM and NREM sleepiness measured by event-related potentials (P300, CNV), MSLT and subjective estimate in narcolepsy-cataplexy. *Electroencephalogr. Clin. Neurophysiol.*, 67:317–326.
27. Broughton, R. J., De Koninck, J., Gagnon, P., Dunham, W., and Stampi, C. (1988): Chronobiological aspect of SWS and REM sleep in extended night sleep of normals. *Sleep Res.*, 17:361.
28. Broughton, R. J., Duchesne, P., Dunham, W., Aguirre, M., Newman, J., and Lutley, K. (1988): A single nap is as accurate as the MSLT in diagnosing EDS in narcolepsy-cataplexy. *Sleep Res.*, 17:150.
29. Broughton, R. J., Dunham, W., Rivers, M., Lutley, K., and Duchesne, P. 24-hour ambulatory monitoring of sleep-wake patterns in narcoleptics and controls. *Electroencephalogr. Clin. Neurophysiol. (in press)*.
30. Broughton, R. J., Dunham, W., Suwalski, W., Lutley, K., and Roberts, J. (1986): Ambulant 24-hour sleep-wake recordings in narcolepsy-cataplexy. *Sleep Res.*, 15:109.
31. Broughton, R. J., and Mamelak, M. (1980): Effects of gamma-hydroxybutyrate on sleep/waking patterns in narcolepsy-cataplexy. *Can. J. Neurol. Sci.*, 7:23–31.
32. Broughton, R. J., Stampi, C., Romano, S., Cirignotta, F., Baruzzi, A., and Lugaresi, E. (1985): Do waking ultradian rhythms exist for petit mal absences? A case report. In: *Biological Rhythms and Epilepsy*, edited by A. Martins da Silva, C. Binnie, and H. Meinardi, pp. 95–105. Raven Press, New York.
33. Browne, R. C. (1949): The day and night performance of teleprinter switchboard operators. *Occup. Psychol.*, 23:121–126.
34. Busby, K., and Broughton, R. J. (1983): Waking ultradian rhythms in hyperkinetic and normal children. *J. Abnorm. Child Psychol.*, 11:431–442.
35. Campbell, S. S. (1983): Human sleep patterns under conditions of disentrainment. In: *Sleep 1982*, edited by W. P. Koella, pp. 212–215. Karger, Basel.
36. Chouvet, G., Mouret, J., Koindet, J., Siffre, M., and Jouvet, M. (1974): Periodicité bicircadienne du cycle veille-sommeil dans des conditions hors du temps: Etude polygraphique. *Electroencephalogr. Clin. Neurophysiol.*, 37:367–380.
37. Czeisler, C. A., Weitzman, E. D., Moore-Ede, M. C., Zimmerman, J. C., and Kronauer, R. S. (1980): Human sleep, its duration and organization depend on its circadian phase. *Science*, 210:1264–1267.
38. Czeisler, C. A., Zimmerman, J. C., Ronda, J. M., Moore-Ede, M. C., and Weitzman, E. D. (1980): Timing of REM sleep is coupled to the circadian rhythm of body temperature in man. *Sleep*, 2:329–346.

39. Daan, S., Beersma, D. G. M., and Borbély, A. A. (1984): The timing of human sleep: A recovery process gated by a circadian pacemaker. *Am. J. Physiol.,* 246(Regulatory, Integrative Comp. Physiol. 12):R161–R178.

40. De Koninck, J., Quera Salva, M., Beseet, A., and Billiard, M. (1986): Are REM cycle narcoleptic patients governed by an ultradian rhythm? *Sleep,* 9:162–166.

41. Dement, W. C. (1960): The effect of dream deprivation. *Science,* 131:1705–1707.

42. Dijk, D. J., Beersma, D. G. M., and Daan, S. (1987): EEG power density during naps; reflections of an hourglass measuring the duration of prior wakefulness. *J. Biol. Rhythms,* 3:207–219.

43. Dinges, D. F., Orne, M. T., and Orne, E. C. (1984): Sleepiness during sleep deprivation: The effects of performance demands and circadian phase. *Sleep Res.,* 13:189.

44. Dinges, D. F., Orne, M. T., Orne, E. C., and Evans, F. J. (1980): Voluntary self-control of sleep to facilitate quasi-continuous performance. U.S. Army Medical Research and Development Command, Fort Detrick, Frederick, MD (NTIS No. AD-A102264).

45. Dinges, D. F., Orne, M. T., Orne, E. C., and Evans, F. J. (1981): Behavioral patterns in habitual nappers. *Sleep Res.,* 10:126.

46. Endo, S., Kobayashi, T., Yamamoto, T., Fukuda, H., Sasaki, M., and Ohta, T. (1981): Persistence of the circadian rhythms of REM sleep: A variety of experimental manipulations of the sleep-wake cycle. *Sleep,* 4:319–328.

47. Evans, F. J., Cooke, M. R., Cohen, D. H., Orne, E. C., and Orne, M. T. (1977): Appetitive and replacement naps: EEG and behavior. *Science,* 197:687–689.

48. Evans, F. J., and Orne, M. T. (1975): Recovery from fatigue. U.S. Army Medical Research and Performance Command, Fort Detrick, Frederick, MD (NTIS No. A100347).

49. Feinberg, I. (1974): Changes in sleep cycles with age. *J. Psychiatr. Res.,* 10:283–306.

50. Feltin, M., and Broughton, R. J. (1968): Differential effects of arousal from slow wave sleep and REM sleep. *Psychophysiology,* 5:231.

51. Fishbein, W., and Gutwein, B. M. (1977): Paradoxical sleep and memory storage processes. *Behav. Biol.,* 19:425–464.

52. Folkard, S., Knauth, P., Monk, T. H., and Rutenfranz, J. (1976): The effect of memory load on the circadian variation in performance efficiency under a rapidly rotating shift system. *Ergonomics,* 19:479–488.

53. Folkard, S., Monk, T. H., and Lobban, M. C. (1978): Short and long-term adjustment of circadian rhythms in 'permanent' night nurses. *Ergonomics,* 21:785–799.

54. Gagnon, P., and De Koninck, J. (1984): Reappearance of EEG slow waves in extended sleep. *Electroencephalogr. Clin. Neurophysiol.,* 58:155–157.

55. Gagnon, P., De Koninck, J., and Broughton, R. J. (1985): Reappearance of electroencephalographic slow waves in extended sleep with delayed bedtime. *Sleep,* 8:118–128.

56. Gander, P. H., Kronauer, R. E., Czeisler, C. A., and Moore-Ede, M. C. (1984): Simulating the action of zeitgebers on a coupled two-oscillator model of the human circadian system. *Am. J. Physiol.,* 247 (Regulatory Integrative Comp. Physiol. 12):R418–R426.

57. Gander, P. H., Kronauer, R. E., Czeisler, C. A., and Moore-Ede, M. C. (1984): Modelling the action of zeitgebers on the human circadian system: Comparisons of simulations and data. *Am. J. Physiol.,* 247(Regulatory Integrative Comp. Physiol. 16):R427–R444.

58. Gastaut, H., and Broughton, R. J. (1965): A clinical and polygraphic study of episodic phenomena during sleep. Academic address (Sakel lecture). *Recent Adv. Biol. Psychiatr.,* 7:197–221.

59. Gillin, J. C., Duncan, W., Pettigrew, K. D., Frankel, B. L., and Snyder, F. (1979): Successful separation of depressed, normal and insomniac subjects by EEG sleep data. *Arch. Gen. Psychiatry,* 36:85–90.

60. Hildebrandt, G., Rohmert, W., and Rutenfranz, J. (1974): 12 and 24 h rhythms in error frequency of locomotive drivers and the influence of tiredness. *Int. J. Chronobiol.,* 2:175–180.

61. Hockey, G. R. J., and Colquhoun, W. P. (1972): Diurnal variation in human performance: A review. In: *Aspects of Human Efficiency,* edited by W. P. Colquhoun, pp. 1–23. The English University Press, London.

62. Hoffman, K. (1971): Splitting of the circadian rhythm as a function of light intensity. In: *Biochronometry,* edited by M. Menaker, pp. 130–148. Natural Academy of Science, Washington, DC.

63. Honma, K., and Honma, S. (1985): Ultradian rhythms in locomotor activity, deep body temperature and plasma corticosterone levels in rats: Two different origins? In: *Ultradian Rhythms in Physiology and Behavior,* edited by H. Schulz and P. Lavie, pp. 77–94. Springer-Verlag, Berlin/New York.
64. Horne, J. (1986): Organization and regulation of sleep: Various models. Presented at the 8th European Congress of Sleep Research, Szeged, Hungary, Sept. 1–5.
65. Horne, J. A., and Ostberg, O. (1977): Individual differences in human circadian rhythms. *Biol. Psychol.,* 5:179–190.
66. Hume, K. I., and Mills, J. N. (1977): Rhythms of REM and slow-wave sleep in subjects living on abnormal time schedules. *Waking Sleeping,* 1:291–296.
67. Karacan, I., Finley, W. W., Williams, R. L., and Hursch, C. J. (1970): Changes in stage I-REM and stage 4 sleep during naps. *Biol. Psychiatry,* 2:261–265.
68. Kawato, M., Fujita, K., Suzuki, R., and Winfree, A. T. (1982): A three-oscillator model of the human circadian system controlling core temperature rhythm and the sleep-wake cycle. *J. Theor. Biol.,* 98:369–392.
69. Kimura, F., Praputpittaya, C., Mitsugi, N., Hashimoto, R., and Suzuki, R. (1985): Relationship between ultradian rhythms of the sleep-wakefulness cycle and growth hormone and corticosterone secretion in rats. In: *Ultradian Rhythms and Physiology and Behavior,* edited by H. Schulz and P. Lavie, pp. 61–76. Springer-Verlag, Berlin/New York.
70. Kleitman, N. (1961): The nature of dreaming. In: *The Nature of Sleep,* edited by G. E. W. Wolstenholme and M. O. O'Connor, pp. 349–364. Churchill, London.
71. Kleitman, N. (1963): *Sleep and Wakefulness.* University of Chicago Press, Chicago.
72. Kleitman, N. (1982): Basic rest-activity cycle—22 years later. *Sleep,* 5:311–317.
73. Kobayashi, T., Tsuji, Y., and Endo, S. (1985): Sleep cycles as a basic unit of sleep. In: *Ultradian Rhythms in Physiology and Behavior,* edited by H. Schulz and P. Lavie, pp. 260–269. Springer-Verlag, Berlin/New York.
74. Kripke, D. F. (1972): An ultradian biological rhythm associated with perceptual deprivation and REM sleep. *Psychosom. Med.,* 34:221–234.
75. Kripke, D. F., Mullaney, D. J., and Fleck, P. A. (1985): Ultradian rhythms during sustained performance. In: *Ultradian Rhythms in Physiology and Behavior,* edited by H. Schulz and P. Lavie, pp. 200–216. Springer-Verlag, Berlin/New York.
76. Kronauer, R. E. (1987): Temporal subdivision of the circadian cycle. *Lect. Math. Life Sci.,* 19:63–120.
77. Kronauer, R. E., Czeisler, C. A., Pilato, S. F., Moore-Ede, M. C., and Weitzman, E. D. (1982): Mathematical model of the human circadian system with two interacting oscillators. *Am. J. Physiol.,* 242(Regulatory Integrative Comp. Physiol. 11):R3–R17.
78. Kupfer, D. G., Ulrich, R. F., Coble, P. A., et al. (1984): Application of automated REM and slow wave sleep analysis. II. Testing the assumptions of the two-process model of sleep regulation in normal and depressed subjects. *Psychiatr. Res.,* 13:335–343.
79. Laird, D. A. (1925): Relative performance of college students as conditioned by time of day and day of week. *J. Exp. Psychol.,* 8:50–63.
80. Lavie, P. (1979): Ultradian rhythm in alertness—a pupillometric study. *Biol. Psychol.,* 9:49–62.
81. Lavie, P. (1985): Ultradian rhythms: Gates of sleep and wakefulness. In: *Ultradian Rhythms in Physiology and Behavior,* edited by H. Schulz and P. Lavie, pp. 148–164. Springer-Verlag, Berlin/New York.
82. Lavie, P. (1986): Ultrashort sleep-waking schedule. III. Gates and "forbidden zones" for sleep. *Electroencephalogr. Clin. Neurophysiol.,* 63:414–425.
83. Lavie, P., and Scherson, A. (1981): Ultrashort sleep-waking cycle. I. Evidence of ultradian rhythmicity in "sleepability." *Electroencephalogr. Clin. Neurophysiol.,* 52:163–174.
84. Lavie, P., Wollman, M., and Pollack, J. (1986): Frequency of sleep related traffic accidents and hour of day. *Sleep Res.,* 15:275.
85. Lavie, P., and Zomer, J. (1984): Ultrashort sleep-waking schedule. II. Relationship between ultradian rhythms in sleepability and the REM-NREM cycles and effects of the circadian phase. *Electroencephalogr. Clin. Neurophsyiol.,* 57:35–42.
86. Lowenstein, O., and Loewenfeld, J. (1964): The sleep-wake cycle and pupillary activity. *Ann. NY Acad. Sci.,* 17:142–256.
87. Lubin, A., Moses, J. M., Johnson, L. C., and Naitoh, P. (1974): The recuperative effects

of REM and stage 4 sleep on human performance after complete sleep loss. *Psychophysiology,* 11:133.

88. Lubin, A., Nate, C., Naitoh, P., and Martin, W. B. (1973): EEG delta activity during sleep as a damped ultradian rhythm. *Psychophysiology,* 10:27–35.

89. Manseau, C., and Broughton, R. J. (1983): Ultradian variations in human daytime EEGs: A preliminary report. In: *Sleep 1982,* edited by W. Koella, pp. 196–198. Karger, Basel.

90. Manseau, C., and Broughton, R. J. (1984): Bilaterally synchronous ultradian EEG rhythms in adult humans. *Psychophysiology,* 21:265–273.

91. Maron, L., Rechtschaffen, A., and Wolpert, E. A. (1964): Sleep cycling during napping. *Arch. Gen. Psychiatry,* 11:503–508.

92. McCarley, R. W., and Hobson, J. A. (1975): Neuronal excitability modulation over the sleep cycle: A structural and mathematical model. *Science,* 189:58–60.

93. McCarley, R. W., and Massaquoi, S. G. (1986): A limit cycle mathematical model of the REM sleep oscillator system. *Am. J. Physiol.,* 251(Regulatory Integrative Comp. Physiol. 20):R1011–R1029.

94. Meier-Koll, A., Hall, U., Hellwig, U., Kott, G., and Meier-Koll, U. (1978): A biological oscillator system and the development of sleep-waking behavior during early infancy. *Chronobiologia,* 6:301–308.

95. Moses, J., Lubin, A., Johnson, L. C., and Naitoh, P. (1977): Rapid eye movement cycle is a sleep dependent rhythm. *Nature,* 265:360–361.

96. Moses, J., Naitoh, P., and Johnson, L. C. (1978): The REM cycle in altered sleep/wake schedules. *Psychophysiology,* 15:207–211.

97. Nakagawa, Y. (1980): Continuous observations of EEG patterns at night and in daytme of normal subjects under restrained conditions. I. Quiescent state when lying down. *Electroencephalogr. Clin. Neurophysiol.,* 49:524–537.

98. Okawa, M., Matousek, M., and Petersen, I. (1984): Spontaneous vigilance fluctuations in the daytime. *Psychophysiology,* 21:207–211.

99. Orr, W. C., Hoffman, H. G., and Hegge, F. W. (1974): Ultradian rhythms in extended performance. *Aerospace Med.,* 45:995–1000.

100. Pappenheimer, J. R., Miller, T. B., and Goodrich, C. A. (1967): Sleep-promoting effects of cerebrospinal fluid from sleep-deprived goats. *Proc. Natl. Acad. Sci. USA,* 58:513–518.

101. Passouant, P., Halberg, F., Genicot, R., Popoviciu, L., and Baldy-Moulinier, M. (1969): La periodicité des accès narcoleptiques et le rythme ultradian du sommeil rapide. *Rev. Neurol. (Paris),* 121:155–164.

102. Pickard, G. E., and Turek, F. E. (1983): The suprachiasmatic nuclei: Two circadian clocks? *Brain Res.,* 268:201–210.

103. Piéron, H. (1913): *Le Problème Physiologique du Sommeil.* Paris, Masson.

104. Pisano, M., Rosadini, G., Rossi, G. F., and Zattoni, J. (1966): Relation between threshold arousal and EEG patterns during sleep in man. *Physiol. Behav.,* 1:55–58.

105. Pittendrigh, C. S. (1960): Circadian rhythms and the circadian organization of living systems. *Cold Spring Harbor Symp. Quant. Biol.,* 25:159–184.

106. Pittendrigh, C. S., and Daan, S. (1976): A functional analysis of circadian pacemakers in nocturnal rodents. V. Pacemaker structure: A clock for all seasons. *J. Comp. Physiol.,* 106:333–355.

107. Poitras, R., Thorkildsen, A., Gagnon, M. A., and Naiman, J. (1973): Auditory discrimination during REM and non-REM sleep in women before and after delivery. *Can. Psychiatr. Assoc. J.,* 18:519–525.

108. Pressman, M., Spielman, A., and Kroczyn, A. (1980): Pupillometry in normals and narcoleptics throughout the course of a day. *Sleep Res.,* 9:218.

109. Prokop, O., and Prokop, L. (1955): Ermudung und einschlafen am Steuer [Fatigue and falling asleep while driving]. *Deut. Z. Gericht. Med.,* 44:343–355.

110. Richardson, G. S., Carskadon, M. A., Orav, W. C., and Dement, W. C. (1982): Circadian variation of sleep tendency in elderly and young adult subjects. *Sleep,* 5(suppl. 2):S82–S94.

111. Rusak, B. (1977): The role of the suprachiasmatic nuclei in the generation of circadian rhythms in the golden hamster. *Mesocricetus auratus. J. Comp. Physiol.,* 118:145–164.

112. Rusak, B. (1982): Physiological models of the rodent circadian system. In: *Vertebrate Circadian Systems: Structure and Physiology*, edited by J. Aschoff, S. Daan, and G. A. Groos, pp. 62–74. Springer-Verlag, Berlin/New York.

113. Saier, J., Regis, H., Mano, T., and Gastaut, H. (1968): Potentiels évoqués visuels pendant les differents phases du sommeil chez l'homme: étude de la réponse visuelle évoquée après le reveil. In: *The Abnormalities of Sleep in Man*, edited by H. Gastaut, E. Lugaresi, G. Berti Ceroni, and G. Coccagna, pp. 55–66. Aulo Gaggi, Bologna.

114. Schulz, H. (1985): Ultradian rhythms in the nycthemeron of narcoleptic patients and normal subjects. In: *Ultradian Rhythms in Physiology and Behavior*, edited by H. Schulz and P. Lavie, pp. 165–185. Springer-Verlag, Berlin/New York.

115. Schulz, H., and Lavie, P. (1985): *Ultradian Rhythms in Biology and Behavior*, Springer-Verlag, Berlin/New York.

116. Scott, J., and Snyder, F. (1968): "Critical reactivity" (Piéron) after abrupt awakenings in relation to EEG stages of sleep. *Psychophysiology*, 4:370.

117. Scrima, L. (1982): Isolated REM sleep facilitates recall of complex associative information. *Psychophysiology*, 19:252–259.

118. Shapiro, A. (1970): Comments on the 90-minute sleep-dream cycle. In: *Sleeping and Dreaming*, edited by E. Hartmann. Little, Brown, Boston.

119. Smith, C. T. (1985): Sleep states and learning: A review of the animal literature. *Neurosci. Biobehav. Rev.*, 9:157–168.

120. Smolensky, M., Halberg, F., and Sargent, F. (1972): Chronobiology of the life sequence. In: *Advances in Climatic Physiology*, edited by S. Ito, K. Ogata, and Y. Yoshimura, pp. 515–516. Igatu Shoin, Tokyo.

121. Stones, M. J. (1977): Memory performance after arousal from different sleep stages. *Br. J. Psychol.*, 68:177–181.

122. Strogatz, S. H. (1986): *The Mathematical Structure of the Human Sleep-Wake Cycle. Lecture Notes in Mathematics No. 69.* Springer-Verlag, Berlin.

123. Taub, J. M., Hollingsworth, H. H., and Bruce, N. S. (1983): Effects on the polysomnogram and waking electrocorticogram of ad-libitum extended-delayed sleep. *Int. J. Neurosci.*, 9:173–178.

124. Ursin, R., Moses, J., Naitoh, P., and Johnson, L. C. (1983): REM-NREM cycle in the cat may be sleep dependent. *Sleep*, 6:1–9.

125. Webb, W. B. (1982): Sleep and biological rhythms. In: *Biological Rhythms, Sleep and Performance*, edited by W. B. Webb, pp. 87–110. Wiley, Chichester/New York.

126. Webb, W. B., and Agnew, H. W., Jr. (1967): Sleep cycling within twenty-four hour periods. *J. Exp. Psychol.*, 74:158–160.

127. Webb, W. B., and Agnew, H. W., Jr. (1971): Stage 4 sleep: Influence of time course variables. *Science*, 174:1354–1356.

128. Weitzman, E. D., Czeisler, C. A., Zimmerman, J. C., and Ronda, J. M. (1980): Timing of REM and stages 3 and 4 sleep during temporal isolation in man. *Sleep*, 2:391–408.

129. Wever, R. A. (1975): The circadian multi-oscillator system in man. *Int. J. Chronobiol.*, 3:19–55.

130. Wever, R. A. (1979): *The Circadian System in Man*. Springer-Verlag, Berlin/New York.

131. Wever, R. A. (1985): Modes of interaction between ultradian and circadian rhythms toward a mathematical model of sleep. In: *Ultradian Rhythms in Physiology and Behavior*, edited by H. Schulz and P. Lavie, pp. 309–317. Springer-Verlag, Berlin/New York.

132. Williams, H. L., Hammack, J. T., Daly, R. L., Dement, W. C., and Lubin, A. (1964): Responses to auditory stimulation, sleep loss and EEG stages of sleep. *Electroencephalogr. Clin. Neurophysiol.*, 16:206–279.

133. Winfree, A. T. (1982): Circadian timing of sleepiness in man and woman. *Am. J. Physiol.*, 243(Regulatory Integrative Comp. Physiol. 1):R193–R204.

134. Wojtczak-Jaroszowa, J., and Pawlowska-Skyba, K. (1967): Praca nocna i zmianowa: I. Dobowe wahania sprawnosci a wydajnosc pracy [Work at night and shiftwork: I. Day and night oscillations of working capacity and the work efficiency]. *Medycyna Pracy*, 18:1–10.

135. Zulley, J. (1988): The four-hour sleep wake cycle. *Sleep Res.*, 17:403.

136. Zulley, J., and Campbell, S. (1985): Napping behavior during "spontaneous internal desynchronization": Sleep remains in synchrony with temperature. *Hum. Neurobiol.*, 4:123–126.
137. Zulley, J., and Wever, R. (1982): Interaction between the sleep-wake cycle and the rhythm of rectal temperature. In: *Vertebrate Circadian Systems Structure and Physiology*, edited by J. Aschoff, S. Daan, and G. Groos, pp. 253–261. Springer-Verlag, Berlin/New York.

Sleep and Alertness: Chronobiological,
Behavioral, and Medical Aspects of
Napping, edited by
D. F. Dinges and R. J. Broughton.
Raven Press, Ltd., New York © 1989.

6

To Nap, Perchance to Sleep—Ultradian Aspects of Napping

Peretz Lavie

Sleep Laboratory, Faculty of Medicine, Technion-Israel Institute of Technology,
Haifa 32 000, Israel

The term *nap* is defined in *Webster's Dictionary* as a "short sleep." But in fact the term nap is used in the sleep-related literature as a neologism for different types of sleep that share only a single attribute: being shorter than the habitual nocturnal sleep period. We refer to a brief morning recovery sleep after a sleepless night as a nap. The term is also used to describe a prematurely aborted nocturnal sleep period, and a sleep episode occurring regularly at midafternoon, which supplements a full-length nocturnal sleep period. Still more confusing, involuntary sleep episodes of patients suffering from excessive daytime somnolence, such as narcoleptics and sleep apneic patients, are also termed naps.

In spite of their common attribute of being shorter than normal sleep episodes, the aforementioned naps differ greatly in respect to the dynamics of their underlying regulatory mechanisms. This leads to important differences in sleep depth and sleep stage content (5). Therefore, the term nap cannot be used indiscriminately to denote just any short sleep. It should be qualified by descriptors identifying the location of the short sleep episode on the circadian time scale; its distance from previous and, in some cases, future nocturnal sleep periods; and whether it replaces or supplements a nocturnal sleep period. In this chapter, use of the term without any descriptors will be reserved for brief periods of sleep occurring during the habitual waking period that alternates with nocturnal sleep periods.

It is difficult to ascertain how much the confusion in terminology has forestalled research on the dynamics and mechanisms of napping. Not only has napping been scarcely investigated, but also in isolation studies that researched the circadian organization of sleep/wake cycles—studies in which napping could have been optimally studied—napping often was explicitly prohibited. In cases in which brief sleep periods nevertheless occurred, these were typically treated as noise in the data and excluded from the analysis. Thus, at least implicitly, it was assumed that napping is an insignificant phenomenon that is not important to the understanding of sleep/wake cyclicity. The elimination of naps in the isolation studies has led to theoretical models of sleep/wake regulation that ignored napping as an inherent component of the sleep/wake cycle. Reanalysis of some of the data obtained in isolation has indeed brought into question the arbitrary decision to disregard napping (see Chapter 7).

The present chapter investigates the argument that during the wake portion of the 24-hr cycle, there are spontaneous variations in the level of sleepiness, which at certain times can be overtly manifested as napping behavior. It is suggested that both the amplitude and the temporal organization of the fluctuations in arousal are important aspects of the circadian sleep/wake regulation. It is further argued that both aspects constitute stable individual characteristics.

ULTRASHORT SLEEP/WAKE SCHEDULES

There is ample evidence that the level of arousal during the habitual waking day is neither stable nor homogeneous but varies in a periodic or quasiperiodic manner, with a prominent midafternoon dip in arousal. Broughton (1) and then Webb (24) first proposed that the midday dip in arousal may be an endogenously rather than a culturally determined habit. A variety of methods have been applied to investigate the possible existence of spontaneous fluctuations in arousal across the span of the waking day. Frequent or continuous electrophysiological recordings of electroencephalographic (EEG) activity, autonomic activity, gross body motility, and perceptual motor performance have been employed (9). Although many of these studies presented convincing evidence of the existence of ultradian variations in arousal, centered on 1.5 hr, they accounted for only a relatively small proportion of the variance, and their relevance to sleep/wake behavior remained unclear.

In recent years, therefore, we have used a different approach to investigate the variations in arousal during the day. Utilizing an ultrashort sleep/wake schedule, our laboratory has investigated subjects' ability to fall asleep at different times of the day under a variety of experimental conditions (11,16,19). The use of an ultrashort sleep/wake cycle is not new. Carskadon and Dement (2) put their subjects on a 90-min day for five 24-hr periods. Similarly, Weitzman et al. (26) instructed their subjects to fall asleep for 60 min every 120 min, whereas Webb and Agnew (25) studied subjects living on a schedule of 3 hr of rest, 6 hr of activity for 6 days (see Chapter 8.)

There are, however, two inherent problems in these earlier studies of ultradian schedules that limit their direct application to the question of sleep/

wake regulation and napping. First, their sampling intervals were too long to allow an accurate description of the fluctuations in arousal during the day. Second, the relatively long periods allocated for sleep could result in sleep accumulation, which might affect subsequent sleep episodes and distort the 24-hr cycle of arousal. These studies, therefore, could not contribute to our understanding of the minute details of the daily fluctuations in arousal. Their most important contributions were in the areas of the circadian sleep/wake organization and the relationship between nap content and pre- and postnap arousal levels.

To overcome these difficulties, we utilized a much faster 20-min sleep/wake cycle consisting of either 5 min sleep/15 min wake, or 7 min sleep/13 min wake. Such schedules allow only a minimal accumulation of sleep, thus permitting a more accurate description of the diurnal variations in arousal. The basic procedure involved the following: Every 20 min subjects were instructed to lie in a bed and attempt to fall asleep in a darkened, sound-attenuated bedroom. Electrophysiological recordings were carried out during the 5-min or 7-min sleep attempts to determine sleep stages. At the end of each trial, whether asleep or awake, the subjects were requested to leave the bedroom. In the middle of the 13-min or 15-min scheduled wake periods, during which they were under supervision, psychomotor tests were conducted. Approximately equal-sized meals of light snacks and soft drinks were available every 2 hr throughout the experimental regimen. The subjects were not isolated from time cues and knew the time of day.

The ability to fall asleep during such schedules was investigated under the following experimental conditions: after waking from a normal night's sleep, at the end of a normal working day, after partial sleep deprivation, and after selective rapid eye movement (REM) sleep deprivation.

In some of the studies, a second experimental condition was added (i.e., resisting sleep). We then investigated the temporal structure of subjects' abilities to resist sleep during the 7-min periods on the schedule. The specific instructions were to lie in bed with eyes closed and try to resist sleep. Electrophysiological recordings were performed during the 7-min resisting-sleep trials as in the attempting-sleep condition. At the end of the 7-min trials, whether asleep or awake, subjects were taken outside the bedroom and subjected to psychomotor tests. To increase motivation, monetary rewards were promised to the subjects who performed the best in each condition (i.e., to subjects obtaining the largest amounts of sleep in the attempting-sleep condition and to the subjects with the least amount of sleep in the resisting-sleep condition). The resisting-sleep paradigm was investigated after 24 hr or 28 hr of sleep deprivation and after a normal day (11). It was also studied under several drug treatments (13) and in combination with interjected 2-hr/day periods (17).

Sleepiness was investigated under both experimental conditions to determine whether the temporal structure of the ability to fall asleep is different from that of resisting sleep. A similar temporal structure would support the interpretation that the 24-hr variations in sleepiness are regulated by an active modulation of underlying hypnogenic processes. A different temporal structure, on the other hand, would weaken this interpretation and suggest instead the possible influence of factors related to the subject's acquired sleep/wake habits. We now review the results of these studies of attempting sleep and resisting sleep during ultrashort sleep/wake schedules.

BASAL STRUCTURE OF SLEEPINESS

The structure of sleepiness obtained by the ultrashort sleep/wake schedule under basal conditions of normal nocturnal sleep agrees closely with the results obtained by other methods. The lower panel of Fig. 1 depicts the mean 24-hr sleep-stage histogram of four subjects who were investigated from 7 A.M. to 7 A.M. under the attempting-sleep paradigm after an 8-hr sleep period. Sleepiness shows a clear bimodal distribution with midafternoon and nocturnal peaks. The peaks are characterized by a short latency to stage-2 sleep and even some slow wave sleep (stages 3 and 4) during the night period. In addition to the two sleepiness peaks, there was evidence of ultradian variations in sleep stage 1. In an earlier study, conducted on a larger group of subjects, Lavie and Scherson (16) showed that these episodic occurrences of stage-1 sleep tended to appear in clusters approximately 90 min apart. These ultradian variations were not stationary, however. They were most prominent during the morning to noon period but were masked by the mid-afternoon sleepiness peak and the evening increase in arousal, and again reappeared later in the evening.

As expected, the ability to fall asleep increased toward the night hours. From the first nocturnal trial containing several minutes of stage-1 sleep, sleepiness gradually increased, reaching a peak within four to five trials or 80 to 100 min. Then there was a subsequent nadir followed by a second increase. Thus, nocturnal sleepiness does not increase in a smooth monotonic manner but in a series of steps superimposed on the ascending slope. The nonmonotonic increase in sleepiness is further demonstrated in Fig. 2, which depicts individual histograms of total sleep for the four subjects comprising the bottom panel of Fig. 1. The same stepwise increase in nocturnal sleepiness was seen in the data of eight subjects who were tested on both the attempting- and resisting-sleep conditions, from 7 P.M. to 7 P.M., after a normal working day.

The basic structure of sleepiness did not change significantly when the ultradian sleep/wake cycle was lengthened to 15 min sleep/30 min wake (Lavie, unpublished results). In this study, four 17-year-old subjects were tested under the attempting-sleep condition from 9 A.M. to 9 A.M. after an 8-hr nocturnal sleep period. Their data also show a stepwise increase in nocturnal sleepiness similar to that observed with the 7/13-min regimen.

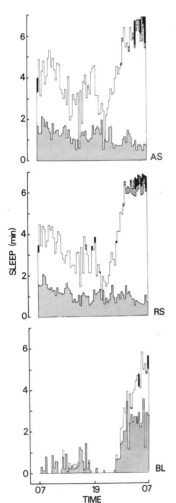

FIG. 1. Mean sleep histograms of four subjects who were tested under three different conditions: attempting sleep after a normal night sleep (**BL**), attempting sleep after 1 night of sleep deprivation (**AS**), and resisting sleep after 1 night of sleep deprivation (**RS**). (■) REM; (◩) stage 3; (□) stage 2; (▦) stage 1.

Thus, it can be concluded that under normal sleep/wake conditions, the nocturnal increase in sleepiness is not a smooth process but is comprised of two components: a slow, near linear component and a superimposed, fast ultradian component. Hence, sleep can be more easily initiated at certain "gates" that are spaced approximately 1.5 to 2 hr apart along the ascending slope of sleepiness. Interestingly, this structure agrees with the subjective experience of evening sleep spells that are temporarily relieved by brief catnaps, and with the subjective experience of missing the sleep gate, i.e., the subjective introspection of some sleep disorder patients that sleep must come at a certain precise time and, if missed, can cause difficulties initiating sleep.

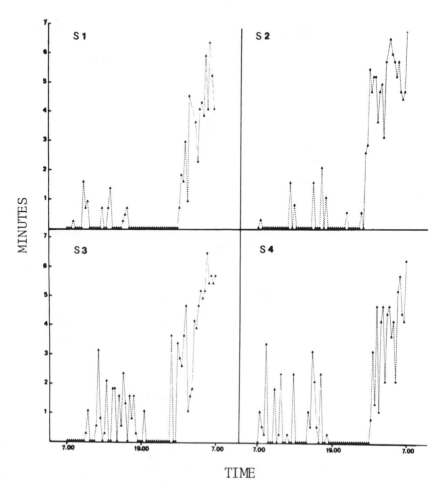

FIG. 2. Individual sleep histograms for the baseline condition of the four subjects whose mean data are presented in Fig. 1.

STRUCTURE OF SLEEPINESS FOLLOWING SLEEP DEPRIVATION

The structure of sleepiness revealed by the ultradian sleep/wake schedule under conditions of sleep deprivation is consistent with the sleepiness structure observed under normal nonsleep-deprived conditions. Sleep deprivation also uncovers several phenomena, however, that were not evident during normal conditions. The upper panels of Fig. 1 present the 24-hr mean sleep histograms of the same four subjects whose baseline data are depicted in the lower panel, after 1 night of sleep deprivation, for both the attempting- and resisting-sleep paradigms (AS and RS, respectively). The variations in sleepiness are clearly more pronounced after sleep loss. Under both conditions, the subjects showed initial high levels of sleepiness that probably reflect an immediate compensatory phenomenon to the sleep deprivation; sleepiness then decreased, reaching a nadir between 1200 and 1500 hr, with subsequent midafternoon peaks (circa 1500–1900 hr). The midafternoon peaks are much more prominent than in the baseline studies. In three afternoon trials in the resisting-sleep condition, there were brief periods of REM sleep; in two afternoon trials in the attempting-sleep condition, there were brief periods of stages 3 and 4 sleep. In both conditions, the midafternoon peaks are followed by a pronounced decrease in sleepiness, reaching a nadir at precisely the same time, between 2000 and 2100 hr. This period of decreased sleepiness was followed by an abrupt increase. After a single trial during which subjects were able to fall asleep within 3 min, on each of the subsequent trials they could easily initiate sleep. The nocturnal crest in sleepiness included sleep stages 3 and 4 and REM sleep.

It should be emphasized that the change in instructions from attempting- to resisting-sleep did not alter the results. There were neither significant differences between the amounts of sleep in the two conditions nor any major changes in the temporal structure of sleepiness. There was, however, one important difference between the two experimental conditions. The midafternoon peak and the early evening nadir in sleepiness were more pronounced in the resisting- than in the attempting-sleep condition. This is exemplified in Fig. 3, which presents the mean difference between total sleep in the RS and AS conditions for 8 subjects tested from 0700 to 0700 hr. The timing of the midafternoon sleepiness peak is indicated by the dotted area [for more details, see (11)]. As discussed below, this interesting finding may be interpreted as providing further support for the claim that the variations in sleepiness are actively generated by endogenous switching on and off of hypnogenic mechanisms.

A second finding, not directly related to the structure of sleepiness, was that some aspects of the perceptual motor task performed during the 13-min intervening wake episodes were significantly better in the attempting-sleep condition. This finding was interpreted within the framework of the multiple resource theory of human information processing and response limitations.

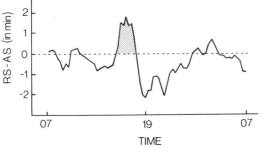

FIG. 3. Mean difference of total sleep (n = 8) for the resisting- (RS) and attempting-sleep (AS) conditions. *Dotted area* indicates the time of midafternoon sleepiness peak.

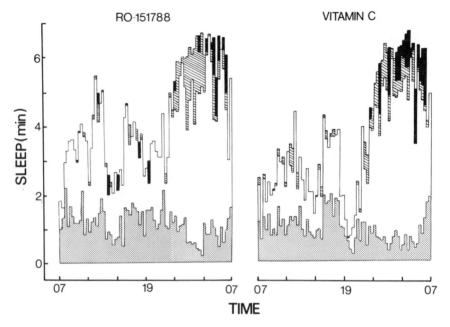

FIG. 4. Mean sleep histograms (*n* = 6) for resisting-sleep condition after 1 night of sleep deprivation under repeated vitamin C and RO-151788 treatments. (■) REM; (▨) stage 3; (□) stage 2; (▦) stage 1. (From ref. 15.)

It indicates that subjects investing effort in resisting sleep deployed their energetic resources to such a degree that it decreased performance level (14).

The 24-hr structure of sleepiness observed under sleep deprivation was also minimally affected by drug treatment (13). In the experiment investigating drug effects, six subjects were tested on the 7/13 resisting-sleep schedule for 24 hr after a night of sleep deprivation, under two treatment conditions: 100 mg vitamin C (as placebo) and 30 mg of the benzodiazepine receptor antagonist RO-151788. Treatment was given in a double-blind crossover design every 4 hr, starting at 7:00 A.M. Figure 4 presents the 24-hr sleep histograms for the two treatment conditions. Although 30 mg of RO-151788 significantly increased sleepiness, particularly during the early evening hours, it did not alter its overall temporal structure. Under both conditions, sleepiness followed the same pattern observed in the previous experiments—an initial early morning increase followed by midafternoon and nocturnal peaks that were separated by an early evening nadir.

PRIMARY AND SECONDARY SLEEP GATES

The consistent occurrence of the midafternoon peak in sleepiness and the distinctiveness of the nocturnal increase in sleepiness, particularly under the sleep deprivation condition, suggest the existence of primary and secondary sleep gates along the circadian cycle. At two distinct time zones, there is a switching on of hypnogenic mechanisms that greatly facilitates the transition from wake to sleep. The length of prior wakefulness appears to change the intensity but not the timing of these gates. Whereas under normal conditions the ascending slope of sleepiness appears to be composed of several gates with increasing sleep propensity, under sleep deprivation the nocturnal sleep period was initiated in a single trial. This was virtually an all-or-none phenomenon. After that critical trial, there was a rapid sleep onset in every subsequent trial. Conceivably, this critical gating trial coincided with the first nocturnal gate under the baseline condition. Unfortunately, we do not have enough data on the same subjects, tested under both baseline and sleep deprivation conditions, to test this hypothesis experimentally.

The augmented midafternoon peak, under the resisting-sleep condition, provides further support for the claim that the sleepiness gates are generated by activation of hypnogenic mechanisms. Subjects could more easily resist sleep during time periods outside the sleep gates. During the sleep gates they not only were unable to resist sleep but showed a compensatory increase in sleepiness. This was evident both in the augmented midafternoon peak and in the abruptness of the nocturnal gates. In all experiments involving the two experimental conditions, when the sleep histograms were synchronized to the primary sleep gate (i.e., the nocturnal sleep gate), the increase in sleepiness was more abrupt and pronounced under the resisting-sleep condition. This is exemplified in Fig. 5, which presents the average histograms for total sleep of all subjects participating in the three experiments reported by Lavie (11), synchronized to the nocturnal sleep gate.

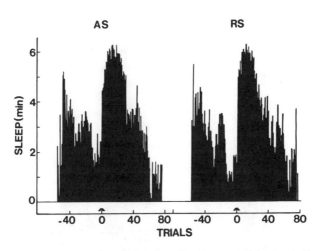

FIG. 5. The mean histogram of total sleep for all subjects participating in the attempting-(AS) and resisting-sleep (RS) conditions (n = 22), synchronized to the individual sleep gate (trial 0).

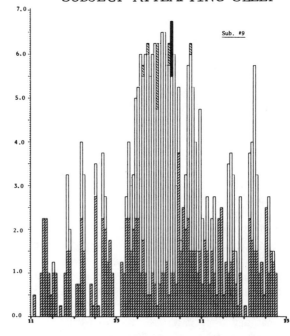

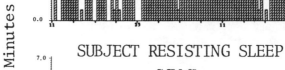

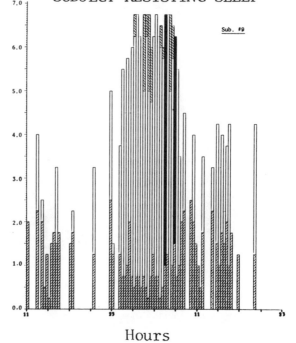

The changes in amplitude of sleepiness agree with the conclusions reached by Strogatz (22) regarding the interaction of prior wakefulness and circadian phase on sleep onset times. Based on his analysis of sleep/wake behavior of internally desynchronized free-running subjects, he showed that although the rate of sleep onset depends on both factors, prolonged wakefulness increases the rate of sleep onset at particular phases of the circadian cycle.

Consistency of the Primary Sleep Gate

The data reported by Lavie (11) revealed another important feature of the primary sleep gate (i.e., its intrasubject stability). Operationally, the primary sleep gate was defined as the first trial after 1900 hr during which there was at least 50% sleep, followed by at least five of six trials meeting the same criteria. Defined thusly, a sleep gate could be identified in 20 of the 22 subjects who were investigated under the attempting- and resisting-sleep conditions. The timing of the primary gate was remarkably consistent for each subject and was not influenced by the contrasting experimental instructions. Figure 6 presents the sleep histograms of one subject tested after one night of sleep deprivation for 36 hr, from 1100 hr (on day 1) to 2300 hr (on day 2). The sleep gate opened at approximately midnight in both conditions. Figure 7 presents the scatter diagram of the primary (i.e., nocturnal) sleep gate times in the two conditions (RS and AS). The correlation coefficient between the times in the two conditions was 0.72 ($p < 0.01$), clearly demonstrating the great intrasubject stability of the gate.

Further results from our laboratory demonstrated that the nocturnal sleep gate remains stable even when 2-hr recuperative naps were interjected at different times during the 7/13 resisting-sleep schedule (17). In that study, nine subjects were tested twice on the 7/13 resisting-sleep paradigm, after one night of total sleep deprivation. In contrast to the other experiments, here the ultradian sleep/wake cycle was interrupted at either 1500 hr or 1900 hr, when subjects were allowed uninterrupted 2-hr sleep periods. At the end of these periods (at 1713 and 2113 hr, respectively) the ultradian schedule was resumed until 0400 hr. Subjects did not know in advance the timing of the naps nor their planned length.

Although the general pattern of sleepiness was in good agreement with our previous results, the 2-hr sleep periods had differential effects on the postsleep sleepiness levels. Comparison of the sleepiness levels during the first 2 hr after the 2-hr interjected naps revealed that the early sleep period

FIG. 6. Sleep histograms of one subject for the attempting- and resisting-sleep conditions. Note the close similarity in the occurrence of the nocturnal sleep gate. (■) REM; (▨) stage 1; (□) stage 2; (▧) stages 3 and 4.

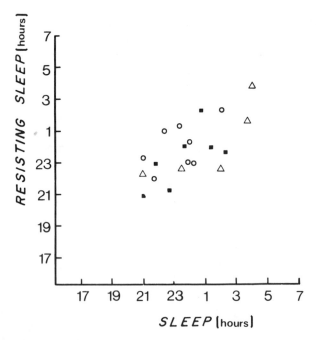

FIG. 7. Scatter diagram of the timing of the nocturnal sleep gate in the attempting- and resisting-sleep conditions in three different experiments. (△) experiment 1; (○) experiment 2; (■) experiment 3. (From ref. 11.)

was much more efficient (i.e., lowered postnap sleepiness) than the late nap period.

Despite the differential effects of the interjected sleep (nap) periods on subsequent sleepiness levels, there was no change in the timing of the nocturnal sleep gate. In five of the nine subjects, a distinct sleep gate, as previously defined, could be identified for both experimental conditions of early and late sleep periods. The mean times of the gates following each 2-hr nap were very close—2340 and 2400 hr for the early and late sleep periods, respectively. As in earlier studies, this similarity was clearly reflected in the correlation coefficient between the times of the gates ($r = 0.68$, $p < 0.025$).

Consistency of the Secondary Sleep Gate

Further analysis of the data reported by Lavie (11) revealed that there was also great consistency in the occurrence of the secondary midafternoon sleep gate. The timing of the midafternoon peak in sleepiness appeared to be locked to the timing of the primary gate, preceding it by 7 to 8 hr. Thus, subjects who had late nocturnal gates also had late midafternoon sleepiness peaks. This relationship is demonstrated in Fig. 8, which shows the 24-hr sleep histograms of 11 subjects who had a late nocturnal sleep gate (all after midnight) and 9 subjects whose sleep gate appeared before 11:00 P.M. Since there was no difference between the resisting- and attempting-sleep conditions, the data of the two conditions were pooled.

For subjects who showed an early primary (nocturnal) gate, the midafternoon peak occurred at 1500 hr, 2 hr and 20 min earlier than the time of the midafternoon peak in subjects who showed a late primary gate. Similarly, there were comparable differences with respect to the timings of the morning and evening nadirs in sleepiness. These occurred at 1200 hr and 2000 hr and at 1500 hr and 2220 hr, in the early and late groups, respectively. Interestingly, subjects with late gates had lower levels of diurnal sleepiness and considerably fewer episodes of REM sleep.

Thus, it can be concluded that the timing of the primary and secondary sleep gates is a stable individual characteristic. Most probably this underlying structure of sleepiness is the basis of the differentiation commonly made between evening and morning circadian types.

Other Evidence for the Bimodality of the Sleep Gates

In recent years, more data coming from different and seemingly unrelated studies have provided support for the bimodality of the sleep initiating zones, or sleep gates. Some of these findings are summarized in histograms of Fig. 9 (graphs a, b, and c). Histogram 9a is taken from Strogatz (22). It displays the distribution of self-selected bedtimes during internal desynchronization of 15 free-running normal young adults. The phase of the temperature cycle was adjusted so that phase 0 corresponds to the temperature minimum. The histogram shows two peaks in the circadian cycle: a major peak near phase 0, which corresponds to the timing of the primary gate at approximately 0600 to 0700 hr and a secondary peak near circadian phase 9 to 10, which corresponds to the midday sleepiness peak at 1500 to 1600 hr. It is significant that Strogatz noted that the secondary peak was not owing to naps in the usual sense, since some of the sleep periods initiated at that time lasted for more than 20 hr. This justified the term secondary sleep gate. As noted before, Zulley and Campbell (28) reported similar findings using subjectively designated naps of free-running subjects who were also internally desynchronized (see Chapter 7).

Histogram 9b depicts the distribution of bedtimes selected by a patient with a non-24-hr sleep/wake cycle during a 4-year period (27). This clearly resembles the distribution of the sleep onset times of the internally desynchronized subjects. This patient, in spite of his apparently chaotic sleep/wake organization, showed a bimodal distribution of selected bedtimes, a primary peak at approximately 0200 to 0300 hr, and a secondary peak at midafternoon.

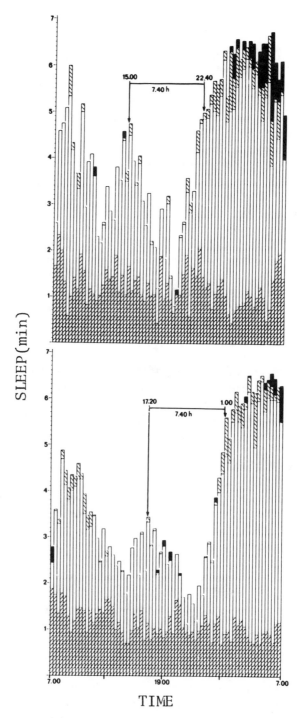

FIG. 8. Mean sleep histograms of subjects who had early (**top**) or late (**bottom**) nocturnal sleep gates. (■) REM; (▨) stage 1; (☐) stage 2; (▧) stages 3 and 4.

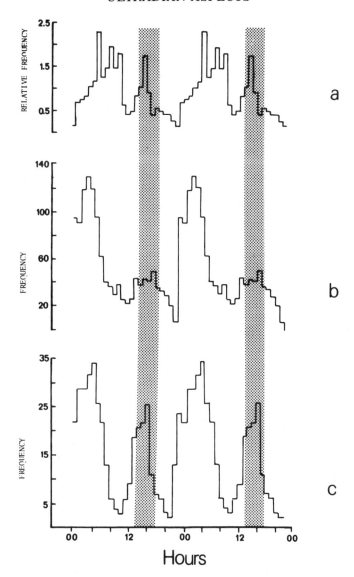

FIG. 9. **a:** Frequency distribution of self-selected sleep times of 15 free-running internally desynchronized subjects. (From ref. 22.) **b:** Frequency distribution of self-selected bedtimes during a 4-year period of a patient with a hypernycthemeral sleep/wake cycle. **c:** Hourly distribution of traffic accidents in Israel between 1978 and 1984, caused by falling asleep while driving.

Histogram 9c provides still further evidence for the bimodality of the sleepiness cycle based on a different sort of data (18). It shows the hourly distribution of traffic accidents in Israel caused by falling asleep while driving for the years 1978 to 1985. The timing of the peaks in accident occurrence corresponds to the primary and secondary peaks of sleepiness seen in graphs a and b of Fig. 9. Similar data had been previously presented by Hildebrandt (6).

"FORBIDDEN ZONE" FOR SLEEP

In addition to the prominence of the sleep gates, the ultradian sleep/wake cycles revealed that the secondary and primary gates of sleepiness are separated by a distinct zone of decreased sleepiness or increased arousal, which we termed the "forbidden zone" for sleep. This was particularly impressive under the sleep-deprivation conditions. In spite of the fact that in the sleep deprivation paradigm, subjects accumulated a considerable amount of sleep loss, which was only partially compensated for by the sleep achieved during the attempting- and resisting-sleep trials, there was a pronounced evening decrease in sleepiness. This was even more impressive in the study that lasted for 35 hr after 28 hr of sleep deprivation. In spite of the accumulation of 58 hr of almost total sleep deprivation, there was a clear decrease in sleepiness during the last 6 hr of the experiment, under both resisting- and attempting-sleep conditions (see Fig. 6).

Supporting evidence for the existence of a forbidden zone for sleep during the early evening hours has been provided by other studies as well. Strogatz's (22) reanalysis of sleep/wake cycles of internally desynchronized subjects (Fig. 9a) showed what he called "wake maintenance" zones during which subjects rarely initiated a sleep period. The evening wake maintenance zone occurs just before the entrained bedtime, coinciding with the forbidden zone for sleep. Interestingly, their morning wake maintenance zone occurs at approximately 1000 to 1100 hr, which also coincides with a local minimum in sleep propensity observed under the ultradian sleep/wake cycle paradigm.

The sleep/wake data of the patient suffering from hypernycthemeral day (Fig. 9b) also support the existence of an evening forbidden zone for sleep. They show a pronounced nadir in the frequency of sleep onsets at 2100 to 2300 hr. The patient started a sleep period at 2300 hr only six times in 4 years, whereas he started a sleep period 95 times 1 hr later, at midnight. Thus, the data of a free-running subject in the normal environment are in excellent agreement with the data of internally desynchronized free-running subjects in isolation. The traffic accident data also show a similar phenomenon. Sleep-related traffic accidents were least likely to occur at 1000 and 2100 hr. As was shown before, the timing of the forbidden zone was related to the timing of the two sleep gates. It occurred at 2000 hr in subjects with an early sleep gate, and 2 hr later, at approximately 2200 to 2300 hr, in subjects with a late gate. Consequently, in both groups the forbidden zone preceded the nocturnal sleep gate by approximately 2 to 3 hr.

WINDING THE REM CLOCK

The fact that REM sleep episodes frequently occurred in the ultradian sleep/wake cycle made it possible to investigate the temporal relation between the primary sleep gate and REM propensity (12).

Like the first REM sleep latency (approximately 90 min) in an uninterrupted nocturnal sleep period, in each of the three experiments investigating sleepiness for 24 hr, there was a minimum REM latency of at least 160 min (wake plus sleep) from the primary sleep gate (under both attempting- and resisting-sleep conditions). The mean latency was remarkably close in both experimental conditions—295 and 292 min in the resisting- and attempting-sleep conditions, respectively. Subtracting the minutes of intervening wake episodes revealed that the amounts of accumulated nonrapid eye movement (NREM) sleep were identical to those in uninterrupted nocturnal sleep periods (i.e., 91 and 93 min).

In contrast to the first REM periods, which appeared to be dependent on the accumulation of a critical amount of NREM sleep, the regulation of subsequent REM periods was sleep independent. Once the first REM sleep appeared, the subsequent REM episodes appeared approximately every 90 min, irrespective of the amount of intervening NREM sleep. Overall, the mean inter-REM interval of wake plus sleep time was 86.5 min, which is very close to the inter-REM interval in uninterrupted nocturnal sleep. In the ultradian studies, however, only 14 min of this interval were accounted for by NREM sleep.

These findings suggest that once the REM oscillator was activated, it continued to function during short periods of waking, maintaining an average cycle of approximately 1.5 hr. The conclusion that the REM oscillator operates during waking was also supported by the finding that a relatively large number of REM episodes tended to occur in adjacent trials or in trials separated by a single trial containing NREM sleep. This tendency was also observed in our previous studies, which did not involve sleep deprivation (16,19). Most probably, as in fragmented REM periods during uninterrupted sleep, these reflected the same REM period.

SLEEPY AND ALERT SUBJECTS

In addition to the consistency in the occurrence of the sleep gates and the overall structure of sleepiness demonstrated by Lavie (11), there was also evidence for a considerable within-subjects consistency in the total amount of sleep. In fact, based on their obtained amounts of sleep in the attempting- and resisting-sleep paradigms, subjects could be divided into two types: sleepy and alert. Sleepy subjects could easily fall asleep when instructed to do so and also fell asleep when instructed to remain awake. Alert subjects, on the other hand, could not easily fall asleep but could easily resist sleep.

In the three experiments reported by Lavie (11), the differences between the most sleepy and least sleepy subjects for total sleep and stage-2 sleep ranged from 189% to 406%. The correlation coefficient calculated across all subjects between total sleep in the resisting- and attempting-sleep conditions was $r = 0.42$ ($p < 0.05$). Significant coefficients were also obtained for stage-2 sleep ($r = 0.42$, $p < 0.05$), stages-3 and -4 sleep ($r = 0.43$, $p < 0.05$) and REM sleep ($r = 0.40$, $p < 0.06$).

The consistency was considerably higher when the stability was investigated within experimental conditions. In the study which investigated the effects of interjected 2-hr sleep periods on the structure and level of sleepiness, subjects performed the resisting-sleep paradigm twice: once under a condition of an early (1500–1700 hr) interjected sleep period and a second time under a condition of a late sleep period (1900–2100 hr). There was a highly significant correlation between the minutes of total sleep in the ultradian sleep paradigm in the two experimental conditions ($r = 0.87$, $p < 0.002$).

Thus, it can be concluded that the diurnal level of sleepiness constitutes a stable individual characteristic, which, together with the temporal structure of the gates, establishes individual ultradian and circadian sleepiness signatures.

THEORETICAL IMPLICATIONS

The wealth of experimental data accumulated on the ultradian sleep/wake schedule, together with the results of the reanalysis of the sleep/wake cycles of internally desynchronized subjects, sheds new light on the 24-hr organization of arousal levels. Instead of focusing on variations in the loosely defined arousal level, the emphasis has been shifted to sleepiness or, more specifically, to the probability of initiating a sleep period at different times during the 24 hr [$p(w \rightarrow s)$]. The results have uniformly shown that there are two distinct gates during the 24-hr cycle: a nocturnal primary gate and a midafternoon secondary gate. During these gates, the transition from wake to sleep is greatly facilitated—most probably by the endogenous activation of neural hypnogenic mechanisms.

The observed structure of sleepiness links findings on ultradian variations in arousal, evident in a variety of variables, with the regulation of the circadian sleep/wake cycle. The two sleep gates can be seen as two privileged phases of a more general ultradian cyclic process in sleepiness, which is barely detected at other times. The capricious nature of the ultradian cycles in sleepiness can be explained by their instability and dependency on prior sleep loss. In fact, there is evidence that in some pathological conditions the ultradian cyclicity surfaces prominently. Both Passouant and colleagues (21) and Volk and colleagues (23) showed that narcoleptic sleep attacks tend to occur cyclically during the day with periodicities within the ultradian range. These may reflect pathological exacerbation of the normally low amplitude ultradian cyclicity in sleepiness.

The ultradian sleep/wake cycle also provides evidence linking the infrastructure of sleep to the gating mechanism that facilitates the transition from wake to sleep. Once the primary sleep gate opens, it activates the REM-generating mechanism, which starts accumulating NREM time. The first REM appears after a critical amount of NREM sleep has been accumulated. Once the first REM appears, however, subsequent REM periods tend to recur periodically, irrespective of the amount of intervening NREM sleep. Although the primary sleep gate switches on the REM-generating mechanism, the phase of the sleep cycle on awakening from sleep influences the phase of the ultradian variations in arousal (19). Awakening from REM sleep is associated with higher levels of arousal as measured by the attempting-sleep paradigm, as compared with awakening from NREM sleep. The differential effects were shown to persist for some 30 min after the awakening. There is, therefore, a close functional relationship between the ultradian cycle during the waking and sleeping portions of the 24-hr cycle. Elsewhere, I discussed the notion that both the sleep and wake rhythms complement each other by providing multiple potential transition points from one state to another (10).

The consistent bimodality of the sleepiness structure with the intervening forbidden zone is not easily handled using current models of sleep/wake regulation. Daan and colleagues (4) proposed an elaborate model that combined homeostatic principles and a single circadian oscillator. They assumed that a sleep-promoting substance S builds up during active waking and subsequently decreases during sleep. Sleep onset is triggered when S approaches an upper threshold, and waking occurs when S reaches a lower threshold. These thresholds, it was suggested, were controlled by a circadian oscillator. It is difficult to reconcile the 24-hr pattern of sleepiness observed in the ultradian studies with a linear or exponential accumulation of a sleep-promoting factor. Even if we assume that the sleep gate is abruptly triggered when the sleep-promoting factor reaches an upper threshold, the forbidden zone for sleep, which occurs just before the gate, cannot be accounted for by such a process.

Similarly, the two-oscillator model of Kronauer et al. (8) cannot account for the secondary sleep gate and the forbidden zone for sleep. Previously, I mistakenly attributed to Strogatz and Kronauer observations of wake-maintenance zones in internally desynchronized subjects as conclusions derived from the two-oscillator model (11). Actually, these were purely empirical findings. After a detailed examination of all the prevailing models of sleep/wake behavior, Strogatz (22) summarized his dissertation by saying that "the existence of the nap phase, the evening wake-maintenance zone and anomalous circadian phase during entrainment, are all beyond the reach of the current models."

PRACTICAL IMPLICATIONS

The division of the waking portion of the 24-hr cycle into sleep gates and forbidden zones for sleep also has important practical implications. As mentioned before, the subjective experience of missing the right moment for sleep is common among many patients complaining of sleep disorders. In these patients, delaying or advancing bedtime may cause difficulties initiating sleep. It is possible, therefore, that for some individuals the timing of the primary sleep gate is rather rigid, and the time of sleep should be carefully maintained from night to night. It is also possible that unusually prolonged forbidden zones for sleep or misalignment of the forbidden zone with respect to the sleep gate are at the base of some cases of sleep onset insomnia. This may be particularly true in patients suffering from phase-delay sleep onset (3). Thus, information about the individual structure of sleepiness throughout the day may be very valuable in explaining sleep difficulties. It also should be determined if tolerance to shift work and the disruptive effects of jet lag are influenced by this structure.

Another possible implication concerns the placement of recuperative naps. In fact, the midafternoon sleep period, which is so prevalent in many cultures (see Chapter 12), may be seen as an adaptation to the secondary sleep gate. Speculations about the endogenous origin of the so-called "post-lunch dip" were made as early as 1963 by Kleitman (7), who thought that it may reflect a pronounced rest phase of the basic rest activity cycle. Broughton (1) suggested that it may be a part of a 12-hr sleepiness cycle (see Chapter 5). The present data suggest that the two gates are approximately 8 hr apart, but this may be different from the interval between the midpoints of the two sleepiness zones themselves (see Chapter 9).

What are the operational consequences of the temporary decrease in arousal? Preliminary results from our laboratory indeed demonstrate that there are important differences between recuperative naps placed during the forbidden zone and at the secondary gate for sleep. The latter had higher sleep efficiency indices, more stages 3 and 4 sleep, and were more efficient in reducing immediate postnap sleepiness levels. Thus, sleep management designs (20) aiming at prolonging continuous performance by interjecting brief recuperative naps should take the structure of sleepiness during the day into consideration. On the other hand, operators who continuously work around the clock should be warned about the imminent midafternoon decrease in arousal. Many still attribute this feeling of sleepiness to the after-effects of a heavy meal. Workers in occupations requiring constantly high levels of vigilance, such as professional drivers, air controllers, and pilots should be re-educated regarding the diurnal structure of sleepiness and its determinants. For such people, a special emphasis should be put on the augmentation of the diurnal sleepiness structure under sleep deprivation conditions.

ACKNOWLEDGMENTS

Some of the studies summarized in this chapter were supported by the U.S. Army Institute for the Behavioral and Social Sciences, through its Liaison Office in London under contract No. DAJA-83-C-0047. The invaluable help of the staff of the Technion Sleep Lab is greatly appreciated.

REFERENCES

1. Broughton, R. (1975): Biorhythmic variations in consciousness and psychological functions. *Can. Psychol. Rev.*, 16:217–239.
2. Carskadon, M. A., and Dement, W. C. (1980): Distribution of REM sleep on a 90-minute sleep-wake schedule. *Sleep*, 2:309–317.
3. Czeisler, C. A., Richardson, G. S., Coleman, R. M., Zimmerman, J. C., Moore-Ede, M. C., Dement, W. C., and Weitzman, E. D. (1981): Chronotherapy: Resetting the circadian clocks of patients with delayed sleep phase insomnia. *Sleep*, 4:1–21.
4. Daan, S., Beersma, D. G. M., and Borbély, A. A. (1984): Timing of human sleep: Recovery process gated by a circadian pacemaker. *Am. J. Physiol.*, 246:R161–R178.
5. Evans, F. J., Cook, M. P., Cohen, H. D., Orne, E. C., and Orne, M. T. (1977): Appetitive and replacement naps: EEG and behavior. *Science*, 197:687–689.
6. Hildebrandt, G. (1975): Outline of chronohygiene. *Chronobiology*, 3:113–127.
7. Kleitman, N. (1963): *Sleep and Wakefulness.* University of Chicago Press, Chicago.
8. Kronauer, R. E., Czeisler, C. A., Pilato, S. F., Moore-Ede, M. C., and Weitzman, E. D. (1982): Mathematical model of the human circadian system with two interacting oscillators. *Am. J. Physiol.*, 242:R3–R17.
9. Lavie, P. (1982): Ultradian rhythms in human sleep and wakefulness. In: *Biological Rhythms, Sleep and Performance*, edited by W. B. Webb, pp. 239–272. Wiley, Chichester.
10. Lavie, P. (1985): Ultradian rhythms: Gates of sleep and wakefulness. In: *Ultradian Rhythms in Physiology and Behavior*, edited by H. Schulz and P. Lavie, pp. 110–124. Springer-Verlag, Heidelberg.
11. Lavie, P. (1986): Ultrashort sleep-waking schedule. III. Gates and "forbidden zones" for sleep. *Electroencephalogr. Clin. Neurophysiol.*, 63:414–425.
12. Lavie, P. (1987): Ultrashort sleep-wake cycle: Timing of REM sleep—evidence for sleep dependent and sleep independent components. *Sleep*, 10:62–68.
13. Lavie, P. (1987): RO 15-1788 decreases hypnotic effects of sleep deprivation. *Life Sci.* 41:227–233.
14. Lavie, P., Gopher, D., and Wollman, M. (1987): Thirty-six hour correspondence between performance and sleepiness cycles. *Psychophysiology*, 24:430–438.
15. Lavie, P., Peled, R., Wollman, M., Zomer, J., and Tzischinsky, O. (1987): Agonistic effects of RO 15-1788. *Neuropsychobiology*, 17:72–76.
16. Lavie, P., and Scherson, A. (1981): Ultrashort sleep-waking schedule. I. Evidence of

ultradian rhythmicity in "sleepability." *Electroencephalogr. Clin. Neurophysiol.*, 52:163–174.

17. Lavie, P., and Veler, B. Timing of naps: Effects on post nap sleepiness levels. *Work and Stress (in press)*.
18. Lavie, P., Wollman, M., and Pollack, I. (1986): Frequency of sleep related traffic accidents and hour of day. *Sleep Res.*, 15:282.
19. Lavie, P., and Zomer, J. (1984): Ultrashort sleep-waking schedule. II. Relationship between ultradian rhythms in sleepability and the REM-NONREM cycles and effects of the circadian phase. *Electroencephalogr. Clin. Neurophysiol.*, 57:35–42.
20. Naitoh, P. (1982): Chronobiological approach for optimizing human performance. In: *Rhythmic Aspects of Behavior*, edited by F. M. Brown and R. C. Graeber, pp. 41–103. Lawrence Erlbaum, Hillsdale NJ.
21. Passouant, P., Halberg, F., Genicot, R., Popoviciu, L., and Baldy-Moulinier, M. (1969): La périodicité des accès narcoleptiques et le rhytme ultradien du sommeil rapide. *Rev. Neurol. (Paris)*, 121:155–164.
22. Strogatz, S. H. (1986): The mathematical structure of the human sleep-wake cycle. In: *Lecture Notes in Biomathematics Vol. 69*, edited by S. Levin, pp. 1–239. Springer, Berlin.
23. Volk, S., Simon, O., Schultz, H., Hansert, E., and Wilde-Frenz, J. (1984): The structure of wakefulness and its relationship to daytime sleep in narcoleptic patients. *Electroencephalogr. Clin. Neurophysiol.*, 57:119–128.
24. Webb, W. B. (1978): Sleep and naps. *Specul. Sci. Tech.*, 1:313–318.
25. Webb, W. B., and Agnew, W. H. (1975): Sleep efficiency for sleep-wake cycles of varied length. *Psychophysiology*, 12:637–641.
26. Weitzman, E. D., Nogeire, C., Perlow, M., et al. (1974): Effects of a prolonged 3-hour sleep-wake cycle on sleep stages, plasma cortisol, growth hormone, and body temperature in man. *J. Clin. Endocrinol. Metab.*, 38:1018–1030.
27. Wollman, M., and Lavie, P. (1985): Hypernycthemeral sleep-wake cycle: Some hidden regularities. *Sleep*, 9:324–334.
28. Zulley, J., and Campbell, S. S. (1985): Napping behavior during "spontaneous internal desynchronization": Sleep remains in synchrony with body temperature. *Hum. Neurobiol.*, 4:123–126.

*Sleep and Alertness: Chronobiological,
Behavioral, and Medical Aspects of
Napping*, edited by
D. F. Dinges and R. J. Broughton.
Raven Press, Ltd., New York © 1989.

7

Napping in Time-Free Environments

*Scott S. Campbell and †Juergen Zulley

*Institute for Circadian Physiology Boston, Massachusetts 02215, USA,
and †Max-Planck Institute for Psychiatry, Munich, Federal Republic of Germany*

After a necessarily brief historical orientation, we describe some of the
ways in which naps are studied in time-free environments. We then examine
the data obtained from such studies, in an effort to address several questions
concerning the nature of naps. In the final section of the chapter, we relate
these findings to the hypothesis that napping is an expression of part of the
biological system mediating the rhythmic occurrence of human sleep.

HISTORICAL PERSPECTIVE

The study of napping behavior in human chronobiology has neither an
extensive nor long history. The first study to actually report objective data
relative to spontaneous napping in a time-free environment was published
only 20 years ago (23). In that study, the duration of naps was the only
parameter reported, and then only in passing, in the discussion section of
the paper. Since the time of that report, only a handful of studies have
examined spontaneous sleep within the circadian day.

The reasons for the paucity of data in this area may be traced to the original
aims of time-free studies and the resulting conditions under which subjects
were maintained. Such studies were initially undertaken in an effort to eliminate the possible masking influences imposed on putative endogenous
rhythms by social and environmental time cues. Subjects were isolated from
diurnal variations in light and temperature, as well as from the entraining
influences of cultural habit, such as work schedules and the demands of
other human beings.

At the same time, however, virtually all studies employing time-free environments specifically prohibited napping by requesting that subjects lead
a "regular" life, with three meals taken in normal sequence, and no daytime

TABLE 1. *Descriptions of studies examining napping in the time-free environment*

Study	Approach	N	Placement of naps	Duration of naps	Rate of occurrence	Structure addressed
Schaefer et al. (23)	Permitted naps	2	Afternoon	121–212 min	1/day	No
Webb and Agnew (28)	Permitted naps	14	Variable	Not stated	Not stated	No
Nakagawa (21)	Encouraged naps	20	Late morning/ afternoon	92 min	3.5/day	Yes
Weitzman et al. (30)[a]	Encouraged naps	3	Temperature maximum	<half of overall mean	1/day (n = 2) 1/3 days (n = 1)	No
Campbell (7)	Encouraged naps	9	Late morning/ afternoon	115 min	2.2/day	Yes[b]
Campbell and Zulley (9)	Encouraged naps	9	Temperature maximum	102 min	~1/day	Yes
Zulley and Campbell (34,35)	Prohibited naps	6	Temperature maximum	144 min	23% of cycles	No

[a] Abstract.
[b] Campbell and Zulley (10).

sleep. Such instructions were imposed probably for two reasons. First, common experience indicated that the typical sleep/wake system of adult humans was organized monophasically. Instructions not to nap, therefore, were given in an effort to study normal patterns of sleep and wakefulness. In addition, from the standpoint of circadian research, naps were frequently viewed as sources of experimental variance, having the potential to distort the course of other rhythms of interest, such as body temperature, renal excretion, or performance (26).

The imposition of such instructions made it highly likely that compliant, well-motivated subjects would exhibit monophasic, circadian sleep patterns. In effect, cultural demands were replaced by experimental commands or demand characteristics (22). Thus, the original goal of recording biological rhythms in the absence of environmental and social constraints was only partially realized. The alternation of sleep and wakefulness in time-free environments remained, by and large, under the influence of external controls.

Nevertheless, several studies have been published in which at least some features of napping have been described. These are listed in Table 1, along with their general findings with regard to the nature of naps. Also shown in Table 1 is the approach taken by each study to extreme napping behavior. A more detailed description of these strategies is the topic of the next section.

RESEARCH APPROACHES

To be effective, the strategies used to study napping (i.e., sleep episodes other than major nocturnal sleep) in time-free environments must reduce the constraints imposed on sleep by experimental demands. Such approaches can be divided into three general categories, based on the degree to which behavioral controls on the initiation of sleep are a feature of the environment.

Prohibited Naps

Although sleep/wake behavior is clearly influenced in the time-free environment by instructions against napping, total compliance has not always been achieved. In an early paper describing studies in the time-free environment, Aschoff (1) recounted his own experience as a subject:

> On day 8, I got up after only 3 hours of sleep. Shortly after breakfast I wrote in my diary: "Something must be wrong. I feel as if I am on dogwatch." I went to bed again and started the day anew after three more hours of sleep.

The recognition that additional sleep episodes might be, at times, unavoidable led to the practice of requiring subjects to signal the beginning and end of subjectively perceived naps, even though they continued to be prohibited (31).

Presumably, as a result of the original circadian orientation of the studies, these signaled naps were not considered as sleep in subsequent analyses of the data. Yet the reanalysis of such data has proved to be a useful source of information regarding the chronobiology of napping (34,35). Because subjects are required to specifically designate naps, findings from this approach permit the examination of relationships between the perception of what a nap is, and corresponding behavioral and physiological aspects of napping. As will be seen, such analyses allow, on the one hand, inferences to be made regarding biological determinants of why a nap is viewed as such, and on the other hand, what influence this subjective perception may have on the subsequent sleep episode.

Permitted Naps

In a few studies, subjects have been informed that other-than-nocturnal sleep would be allowed during their time in isolation. In one of the first electrographic studies of sleep and waking in the time-free environment, Webb and Agnew (27) gave their subjects the following instructions:

> During isolation we ask that you "loosen" your sleep-waking process and let it follow a course of least resistance. On the one hand, it is possible to take a set of maintaining your typical biphasic [*sic*] sleep and waking pattern. On the other hand, it is possible that you attempt some other pattern, such as a sleep-wake-nap-wake-sleep pattern. We hope that you will do neither. Rather we hope that you will sleep whenever you find this an acceptable response.

The advantage of such instructions for the study of napping in the time-

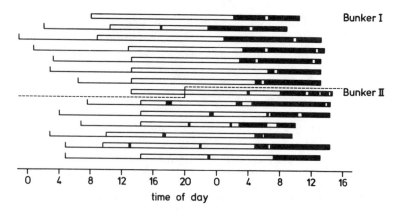

FIG. 1. Episodes of sleep (*solid bars*) and wakefulness (*open bars*) for a single subject living in isolation for 2 weeks. During the first week (Bunker I), the subject was requested to structure her days and to sleep only during her subjective night. During the second week (Bunker II), the subject was instructed to eat and sleep when inclined to do so.

free environment is obvious. Subjects feel no compulsion to structure their days in response to external demands and are more likely, therefore, to express a closer approximation of physiological sleep tendency. The degree to which such instructions can influence the occurrence of sleep is illustrated in Fig. 1. Shown are episodes of sleep and wakefulness exhibited by a subject during a 2-week stay in the time-free environment (Campbell and Zulley, unpublished data). During both weeks, the subject (an artist) was permitted to continue her usual daily activities, which consisted primarily of drawing and painting, exercise, reading, and writing. Prior to the start of the first week (Bunker I), the subject was instructed to organize her days by eating three meals in normal sequence and by sleeping only during her subjective night. With one exception, the subject complied with these instructions and exhibited a typical monophasic sleep pattern.

At the beginning of week 2 (Bunker II), instructions to the subject were changed. She was simply told to eat and sleep when inclined to do so. The result was a more than twofold increase in the number of sleep episodes taken during the week. The subject napped daily, and on three occasions, napped twice a day.

This example notwithstanding, behavioral controls on sleep are not completely removed simply by the removal of experimental restriction on sleep. Consider an individual who agrees to live in the time-free environment because (s)he feels the time would be well-spent studying for upcoming final exams or writing a term paper. This subject may well employ self-imposed behavioral controls, such as drinking a cup of coffee or transiently increasing activity, to overcome periodic episodes of drowsiness, thereby influencing the extent to which napping behavior is observed. As Mills and co-workers have pointed out (9,11,18), even under these permissive conditions there may be a strong tendency for some individuals to continue their usual patterns of behavior. Habits developed throughout many years may be difficult to change simply by the removal of time cues, if the opportunity to continue habitual behaviors remains a feature of the environment.

Encouraged Naps

The persistence of habitual behaviors in time-free conditions can be used to advantage, of course, if the behavior in question is habitual napping. The removal of time cues and cultural demands from the environments of habitual nappers probably leads to an even greater likelihood that their apparent predisposition for napping will be behaviorally expressed. Several authors have employed this approach to examine the chronobiology of napping (23, 30).

For subjects who are not in the habit of napping, such behavior can be encouraged by reducing the number of behavioral options in the environment that may be incompatible with the initiation of sleep episodes. One such experimental approach has been referred to by us as "disentrainment" (6,7). The most radical version of this design entails the elimination of experimental instructions relative to when and when not to sleep, the minimization of behavioral alternatives to sleep (e.g., reading, writing, engaging in conversation, listening to music), and enforced bed rest (7,10,21). A slightly less drastic form of disentrainment permits free movement around the isolation unit but continues to prohibit virtually all behavioral options to do anything other than sleep (9).

It can be argued that such basal static conditions constitute the best environment in which to study the physiological propensity for sleep. Simply from lack of viable alternatives, subjects studied under these conditions are more likely to express a biological sleep tendency in the unequivocal form of napping behavior or polycyclic sleep tendencies.

On the other hand, it may be argued that sleep patterns exhibited under these conditions are abnormal, reflecting a physiological response to boredom rather than natural sleep tendency. An added disadvantage of employing the disentrainment design is the relatively short time during which subjects can be studied under such conditions; 3 to 4 consecutive days is probably the maximum interval that subjects would tolerate. Thus, parameters such as reliability of phase relationships and stability of sleep/wake patterning over time may not be adequately examined using this strategy.

Pathological Naps

Finally, in addition to the approaches mentioned above, it may be of value to examine the napping behavior in time-free environments of individuals with sleep pathology characterized by daytime sleepiness. Narcoleptics may be of particular interest in this regard, since there is some evidence that daytime sleep attacks in this disorder may be periodic in nature, perhaps reflecting an imbalance between circadian and ultradian components of the biological system mediating the timing of sleep (16,20). To our knowledge, however, there are no published reports of sleep patterns of such populations under time-free conditions.

CHRONOBIOLOGY OF NAPPING

Implicit in the rationale behind studying napping in a time-free environment is the assumption that napping represents a component of the endogenously mediated sleep/wake system of adult humans. This assumption, in turn, suggests certain hypotheses regarding the nature of naps: (a) naps should maintain some characteristic pattern of occurrence with a given time frame, (b) the occurrence of naps should maintain a stable phase relationship with other components of the human circadian system, such as core body temperature, and (c) the duration and infrastructure of such sleep episodes should be governed by the same circadian principles governing the length and architecture of typical nocturnal sleep. The following sections will examine these hypotheses by addressing several basic questions regarding the nature of naps under time-free conditions.

When Do Naps Occur?

There is good agreement across studies that the initiation of naps in the time-free environment does not occur randomly. Depending on the conditions under which they are studied and the manner in which napping is defined, such sleep episodes may show one, or several, preferred times of occurrence within the circadian day.

Consistently, naps show a tendency to cluster around a phase position corresponding to the time of maximum values in the circadian course of body core temperature (see Fig. 2). Indeed, this phase of the circadian day has been referred to as the "nap domain" (30). This designation was based on findings from studies of two elderly habitual nappers who lived in time-free environments for 14 and 18 days, and a third subject who was not a habitual napper but was strongly encouraged to nap, during 31 days under time-free conditions. The sleep/wake profiles of all three subjects were characterized by the presence of relatively short sleep episodes (i.e., less than half the mean duration of all sleep episodes taken during the free-run) occurring "approximately 180° before mid-low temperature, close to the maximum temperature."

That a preferred phase position for the occurrence of naps exists around

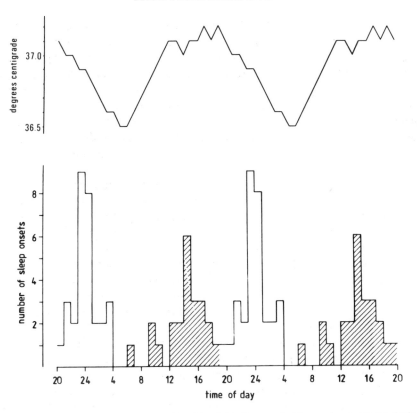

FIG. 2. Distribution of sleep onsets (double plotted) recorded from nine subjects during 72-hr disentrainment period (9). *Hatched area* shows sleep episodes designated as naps in subsequent analyses (see text). Above the distribution is the course of body core temperature (rectal) averaged for the nine subjects.

the temperature maximum is supported by studies involving nonhabitual nappers as well. For example, six subjects who signaled having taken naps, contrary to experimental instructions, initiated naps between 7 and 14 hr after the temperature minimum, an interval corresponding to the temperature maximum (35). (It should be noted that under these conditions, in which subjects designated naps, they sometimes called sleep episodes that occurred at the temperature minimum "naps." These sleep episodes were of a length generally associated with major sleep. Thus, major sleep episodes may be subjectively perceived as naps as well. The reasons for such perceptions are discussed in a later section.)

In most cases, the maximum in body temperature corresponds to a time approximately halfway between successive major sleep episodes, as shown in Fig. 2. This figure shows the distribution of sleep onsets recorded from nine young adults during 72 hr in a disentrained environment (9). In this study, subjects were permitted free movement around the isolation unit but were allowed virtually no behavioral alternatives other than to sleep. With respect to successive onsets of major, nocturnal sleep episodes, naps showed a tendency to begin approximately 60% of the way through the cycle. Specifically, the average onset of naps was 14.6 hr after the onset of the preceding nocturnal sleep and 9.9 hr prior to the onset of the subsequent nocturnal sleep episode. Viewed in a different way, this places naps, on average, approximately halfway (48.7%) through the waking episode separating successive nocturnal sleep periods (i.e., between the termination of one and the onset of the next major sleep episode).

Although naps show a strong tendency to cluster around the temperature maximum, and thus the middle of the typical waking period, their occurrence is not restricted to that phase of the circadian day. Under certain experimental conditions, a relatively robust propensity for napping in the late morning has been reported. Nakagawa (21), for example, reported that subjects confined to bed for 12 hr during the day, following a full night's sleep, spent a greater proportion of time asleep between 0800 and 1100 hr than during any other 3-hr time block (63.2%). For the group, the tendency to initiate sleep at approximately 1030 hr was as strong as that observed about 4 hr later, at the expected nap phase.

In a similar study, in which subjects were confined to bed for 60 consecutive hr (7), more sleep episodes were initiated between 1000 and 1200 hr (17% of total) than during any other 2-hr interval within the 24-hr day. The two additional peak times for sleep onset were between 1500 and 1700 hr and between 0200 and 0300 hr. Overall, 42% of all sleep recorded during the 60 hr was initiated during these intervals, despite the fact that they comprise only 21% of the 24-hr day.

It is unclear how napping at this phase relates to the circadian course of body temperature. This variable was not recorded in either of the studies mentioned. However, there is no evidence to suggest that individual differences in nap domains (i.e., temperature acrophase) are responsible for the occurrence of two peaks within the circadian day. To the contrary, in the study just described (7), sleep episodes initiated between 1000 and 1200 hr were contributed by eight of the nine subjects studied. Furthermore, more than half of these late morning naps were followed several hours later by an additional nap, suggesting that the bimodal distribution of nap times is not simply a reflection of differential placement of naps from one day to the next.

It has been suggested that the two preferred phase positions for the occurrence of naps may reflect an ultradian rhythm in sleep and wakefulness, in the 4- to 6-hr range (7,9,21). Such polyphasic organization of rest and activity is widespread across the animal kingdom (2). There is no reason, therefore, why humans should not exhibit such patterns as well, given an environment in which the expression of these patterns is not masked by behavioral options that may be incompatible with the initiation of sleep episodes.

On the other hand, the environments in which this bimodal distribution in napping is observed can clearly be considered unusual. The question remains whether such sleep/wake patterns reflect natural sleep tendency or whether they should be considered behavioral responses to the extremely static, basal condition characterizing the environments in which they occur.

How Long Do Naps Last?

It is clear from the previous discussion that, for the most part, the timing of naps is tied to the circadian rhythm of body core temperature. Likewise, the relationship between the durations of sleep episodes and the circadian course of body core temperature is well established (12,33,36). In general, the longest sleep episodes are initiated several hours prior to the temperature minimum, with decreasing durations as onset times approach the temperature maximum. In light of this relationship and that between the initiation of naps and temperature phase, it is no surprise that naps are typically short, relative to major nocturnal sleep episodes.

As with the placement of naps within the circadian day, there is general agreement across studies with regard to the duration of such sleep episodes, as shown in Table 1. As a general rule, naps taken in time-free environments continue for longer durations than do naps under entrained conditions (see Chapter 9). For the 71 naps recorded during 12 hr of bed rest, Nakagawa (21) reported a mean duration of 92.3 min. Similar values (115.8 min) were reported for day phase (0600–1800 hr) sleep episodes recorded during 60 hr of bed rest (7) and for naps occurring in coincidence with temperature maxima during 72 hr of disentrainment, without bed rest (102 min) (7).

In the earliest study to employ continuous polygraphic recording in the time-free environment, Schaefer and co-workers (23) reported average nap durations of 212 and 121 min, for two young adults recorded for 8 consecutive days. Both subjects also napped during a 4-day control period immediately prior to the start of the time-free condition, with average nap durations of 103 and 117 min. Thus, one subject lengthened his average nap duration by a factor of 2 in the time-free environment, whereas the other showed no change.

Weitzman et al. (30) referred to a similar lengthening of nap duration for two elderly subjects studied in isolation. Instead of a consistent pattern of short nap/long sleep observed under entrained conditions, the total daily sleep of these subjects was divided into "2 episodes of varying length" during the time-free portion of the study. Nevertheless, as mentioned earlier, sleep episodes occurring within the nap domain consistently lasted for less than half the mean sleep length for the entire experimental period.

Because of the relationship between circadian phase and sleep duration, it might be expected that the late morning naps observed under bed rest conditions would have a longer average duration than those taken in the afternoon. Alternatively, given the well-established relationship between sleep duration and prior wakefulness one could predict the converse—that afternoon naps would be longer than those occurring closer to the termination of preceding nocturnal sleep episodes. The limited available data suggest a substantial influence of time-of-day, but little effect of prior wakefulness on nap duration.

In the disentrainment study described earlier (7), naps that occurred between 1000 and 1200 hr were significantly longer than those initiated between 1500 and 1700 hr (130.2 versus 96 min), despite the fact that the early naps were preceded by significantly less prior wakefulness than were the afternoon naps (95 versus 172 min). The overall correlation coefficient between prior wakefulness and nap duration was +0.28.

An interesting additional influence on nap duration must be considered under conditions in which subjects have been instructed to signal their intentions to nap. In such cases, the subjective perception of an imminent sleep episode as a nap appears to be associated with the subsequent duration of the episode. This point is illustrated in Fig. 3. The average durations of sleep episodes, as a function of circadian phase of onset, are shown for six subjects who signaled naps during their stays in the time-free environment (34,35). When compared with major sleep episodes taken at the same circadian phase, subjectively designated naps were, on average, 57% shorter.

Thus, simply by deciding that an upcoming sleep episode will be a nap,

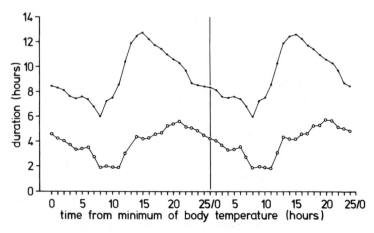

FIG. 3. Average durations of sleep episodes, as a function of circadian phase of onset, for six subjects who signaled naps during their stay in the time-free environment (34,35). The designation of a sleep episode as a nap was associated with a 57% reduction of subsequent sleep period duration. (●) major sleep; (○) nap sleep.

an individual effectively places an upper limit on the duration of that episode. Whether this is primarily a result of the behavioral set accompanying such a designation, or the consequence of physiological factors (e.g., effects of prior wakefulness), is unclear. However, we have all had the experience of lying down for a nap, frequently in an illuminated room, fully clothed, and on the bed rather than in the bed. These conditions are clearly not conducive to extended sleep and very likely play a role in the restriction of nap duration.

How Are Naps Structured?

The common view of naps is that they are short and involve lighter sleep than nighttime sleep episodes. We have just shown that, although naps taken in the time-free environment are typically longer than those taken under entrained conditions, they are nevertheless characterized by shorter durations than major nocturnal sleep.

The very limited data that address the structure of naps in the time-free environment suggest that the common perception of naps as lighter sleep episodes is also generally correct. This is illustrated in Table 2, which compares several structural parameters of normal nocturnal sleep (32) with those of daytime sleep episodes derived from three different studies of spontaneous napping (9,10,21). In all three of these studies, naps were characterized by larger proportions of stages 1 and 2 sleep and less slow wave sleep (SWS), when compared with normative data on nocturnal sleep episodes.

TABLE 2. *Mean values of sleep-stage parameter recorded from spontaneous daytime sleep episodes in time-free environments, relative to normal nocturnal sleep values*

Study	Source of data	Sleep stage percentages					REM latency
		1	2	3	4	REM	
Williams et al. (32)[a]	Normal nocturnal sleep (ages 20–29)	4.2	45.5	6.2	14.6	28.0	94.2
Nakagawa (21)	12-hr bed rest	15.4	51.0	7.5	4.3	21.7	Not reported
Campbell (7)	Disentrainment (bed rest)	2.8	60.8	4.9	11.0	17.5	35.4
Campbell and Zulley (9)	Disentrainment (no bed rest)	6.2	58.7	7.5	7.4	14.5	46.1
Campbell (8)	1400–1700-hr naps ($n = 13$)	7.5	53.8	9.8	10.6	13.3	
	All other naps ($n = 13$)	4.8	63.6	5.2	4.7	15.7	

[a] Normative nocturnal sleep values for healthy young adults.

Perhaps as a result of this reduction in SWS (3,4,8), average latency to rapid eye movement (REM) sleep in naps in temporal isolation was substantially shortened relative to typical nocturnal sleep. Indeed, sleep onset REM periods were a common feature of naps taken during bed rest (8,10,21). The average REM latency for naps recorded in the studies shown in Table 2 was about 45 min—approximately half that of normal nocturnal sleep.

Although generally accurate, the profile of naps as lighter sleep episodes than nocturnal sleep requires an important qualification. A closer inspection of the composition of individual naps taken during 72 hr of disentrainment (8) indicates that this view may be slightly misleading. There appears to be a subgroup of naps that tends to cluster around a specific phase position in the circadian day and more closely resembles typical nocturnal sleep in its basic structure. Shown at the bottom of Table 2 is a comparison of sleep-stage parameters between two groupings of naps that were separated on the basis of their placement within the nap distribution (i.e., their times of onset). Naps in the middle of the distribution (see Fig. 2 for the distribution of naps used in these analyses), with onsets between 1400 and 1700 hr, showed a mean SWS percentage of more than two times the average SWS for all other naps. Nakagawa (21) likewise found a higher percentage of SWS in naps with onsets between 1400 and 1700 hr than in naps initiated during any other 4-hr block during the day. In neither of these studies was duration of prior wakefulness a significant factor in the differential placement of SWS. (Interestingly, 1400 to 1700 hr is the time of day that naps are most likely to occur in healthy young adults—see Chapter 9.)

These findings strongly support Broughton's (5; Chapter 5) hypothesis that there exists a circasemidian rhythm in the occurrence of SWS, with a specific relatively well-delineated phase position within the circadian day during which SWS is likely to approach typical nocturnal percentages. The findings further suggest the existence of essentially two types of naps, at least with respect to structure. One group corresponds to the common perception of naps as lighter, more fragile sleep episodes. A second group of naps, however, located centrally within the nap distribution, tends to look very much like miniatures of major nocturnal sleep.

How Common Are Naps?

To a certain extent, the frequency with which napping occurs in the time-free environment depends on how often it is allowed to occur. Under the most permissive conditions [i.e., disentrainment with bed rest (7)], daytime sleep episodes (onsets between 0700 and 1900 hr) comprised 36% of all sleep recorded. All nine subjects in this study exhibited some degree of napping behavior, with an average of 2.2 daytime sleep episodes per 24 hr. Nakagawa (21) reported a similar frequency of occurrence for 20 subjects studied during

12 hr of bed rest. Nakagawa's subjects reported that they were periodically "overcome by an uncontrollable desire to sleep," despite explicit instructions to the contrary. Given these basal static conditions, then, napping seems to be an unavoidable component of the sleep/wake system of all subjects.

The degree to which bed rest specifically contributed to the occurrence of naps can be seen by comparing these results with those of a study that employed a disentrainment design, without confinement to bed (9). Under these conditions, all nine subjects continued to exhibit daily napping behavior, although these sleep episodes comprised a substantially smaller proportion of total sleep time (TST) than in the bed-rest condition.

However, as behavioral alternatives to sleep become a more prominent feature of the environment, individual differences in the propensity to nap become more evident. We recently completed a study in which 10 subjects lived for 2 weeks in the time-free environment, with instructions to eat and sleep when inclined to do so (Campbell and Zulley, unpublished data). Whereas 7 of the 10 subjects exhibited substantial napping, 3 subjects exhibited no napping during their time in isolation. Likewise, an examination of the individual sleep patterns of subjects studied by Webb and Agnew (28) (Fig. 2), under similar conditions, reveals substantial interindividual variability in the frequency with which napping occurred. It seems clear from these examples that, just as there are long and short sleepers (28,29), and morning and evening types ("larks" and "owls") (15), there are individual differences in the predisposition to nap (see Chapter 9). There are nappers and nonnappers. Perhaps this point is most strongly emphasized by the finding that even under conditions in which subjects were encouraged to continue their normal activities and specifically instructed to avoid napping, a certain proportion of these individuals found it impossible to comply (34).

Why Are Naps Perceived as Such?

From the examination of Fig. 3, it can be concluded that subjectively perceived naps, as well as major sleep episodes, may occur at any time, although preferred phase positions for sleep clearly exist. These preferred phase positions are not mutually exclusive but rather overlap to a certain extent (34,35). That is, subjectively perceived naps may occur around the temperature minimum and subjectively perceived major sleep episodes may occur around the maximum. What conditions, then, make a subject decide that a sleep episode is a nap rather than a major sleep?

Although it is clear that the duration of sleep episodes is an effective discriminator, it is only so in retrospect. Since subjects signaled their intention to nap prior to the sleep episode, different factors must have been involved in their decisions. Daily logs kept by subjects during isolation reveal

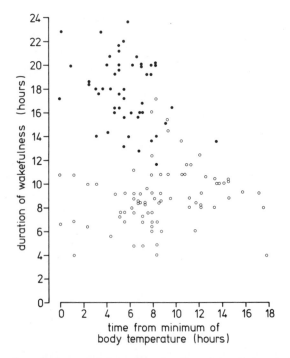

FIG. 4. Distribution of prior wakefulness, as a function of duration and circadian phase, recorded from six subjects who signaled naps during their time in isolation (34,35). (O) waking episodes preceding subjectively perceived naps; (●) those preceding major sleep. On average, naps were preceded by waking episodes of approximately half the duration of those prior to major sleep.

that the decision to call an impending sleep episode a nap was based on the subject's perception that not enough time had elapsed since the previous awakening for the episode to be considered a full day.

The data support the validity of this perception. Figure 4 shows the distribution of prior wakefulness, as a function of duration and phase, recorded from six subjects who signaled naps during their time in isolation (34). Sleep episodes designated as naps were preceded by significantly shorter episodes of wakefulness (8.8 hr) than were major sleep episodes (17.9 hr). There was virtually no overlap in the distributions of prior waking for the two groups of sleep episodes.

NAPPING AS A BIOLOGICAL RHYTHM

Of the experimental paradigms used to study napping, the time-free environment is probably the best suited to address the question of whether napping behavior is the expression of part of an endogenously mediated tendency for the recurrence of sleep throughout the nycthemeron. Within the general framework of time-free environments, the degree to which the putative biological tendency for sleep is expressed as napping behavior is strongly influenced by the number of behavioral controls extant in the experimental environment. Yet, even under conditions in which subjects are encouraged to continue their normal daily activities, and in which it is expressly prohibited, more than half of all subjects find it impossible to entirely suppress napping behavior. The very strength of this tendency suggests that napping is the response to some physiological predisposition to rest periodically throughout the day.

It was suggested throughout the beginning of this chapter that if the propensity to nap is, indeed, a component of the endogenous rhythm of sleep and waking, nap sleep should conform to the same rules that govern the placement, duration, and internal organization of major nocturnal sleep. Taken together, the studies reviewed in the previous sections provide convincing evidence in support of this hypothesis.

The placement of naps, like that of major sleep episodes recorded in time-free environments, is not random, but rather is closely related to the circadian oscillation of body core temperature. Likewise, the duration of naps is determined primarily by their placement within the circadian day, a relationship well-established for major nocturnal sleep as well. With regard to structure, the strong circadian influence is largely responsible for the mode of appearance of REM sleep in nocturnal sleep episodes, which is evident in naps, both in percentage and latency values. In addition, the degree to which SWS occurs within naps appears to be substantially determined by circadian factors.

The hypothesis that napping is the expression of an endogenous propensity for sleep raises additional questions regarding the nature of the biological system mediating sleep and waking. For example, what kind of rhythmic system includes one component which, in normal adults, is most conspicuous by its frequent lack of occurrence? If napping is part of a biologically mediated rhythmic system, what are the consequences of its chronic absence?

With regard to the first question, it is clear that the human sleep/wake system is a flexible one (see Chapter 8). Both the onset and termination of major nocturnal sleep are routinely shifted in response to the demands of daily life. The duration of major sleep episodes can be dramatically and chronically curtailed, and shift work requires the frequent rotation of major sleep to virtually any placement within the 24-hr day. Indeed, the very fact that the circadian component of the sleep/wake rhythm is forever entrained to a 24-hr periodicity is clear testimony to its flexible nature.

As with any instinct, the sleep/wake system interacts with and is responsive to the environmental conditions under which it functions. The absence of naps in daily life, therefore, may be a more accurate reflection of the environment than of the rhythm [cf. (24)]. Simply, the expression of this component of the sleep/wake rhythm may be effectively masked by social and occupational demands. This view is supported by the observation that certain populations report an increased incidence in napping behavior, in response to lifestyles that are less structured than the usual "8-to-5" work environment (17,25,26; see also Chapters 9 and 10). Results of studies using the various time-free environments cited throughout this chapter also reflect the strong influence imposed on the occurrence of naps by behavioral controls.

Although the sleep/wake system is flexible, it is generally not manipulated without cost. For example, both chronic restriction of nocturnal sleep and sleep period time displacement (e.g., shift work) are accompanied by decrements in performance, alterations in mood, and changes in sleep itself. If napping is an integral component of the sleep/wake system, the exclusion of naps from our daily schedules might be expected to result in similar decrements. Stated differently, the inclusion of napping in our daily regimens might be expected to augment such measures. Few studies have adequately addressed the effect of napping on nonsleep-deprived subjects (see Chapter 9). It is, therefore, difficult to estimate the consequences of its chronic absence. There is, however, limited evidence to suggest that napping may indeed be effective in enhancing performance over nonnap levels in normal rested adults (13,14).

Such findings raise interesting questions. Do our nonnapping societies impose an artificial ceiling on performance potential? Could we function better by adapting our behavior to reflect more closely underlying biological sleep tendency? Is the siesta a natural, yet generally ignored, aid to performance? Common experience indicates that most people perform reasonably well without benefit of regular napping. However, that we can get along adequately without naps should not be interpreted to mean that we would not function more efficiently with them.

ACKNOWLEDGMENTS

We gratefully acknowledge the support of the Max-Planck Society and Grant AG05131. Special thanks are extended to Dr. Hartmut Schulz, who sponsored S.S.C.'s Visiting Scientist Fellowship at the Max-Planck Institute for Psychiatry, Munich, and who contributed a great deal to the content of this chapter through numerous stimulating discussions.

REFERENCES

1. Aschoff, J. (1965): Circadian rhythms in man. *Science*, 148:1427–1432.
2. Aschoff, J. (1966): Circadian activity pattern with two peaks. *Ecology*, 47(4):657–662.

3. Borbély, A. A. (1984): Sleep regulation: Outline of a model and its implications for depression. In: *Sleep Mechanisms*, edited by A. Borbély and J.-L. Valatx, pp. 272–284. Springer-Verlag, Berlin.
4. Borbély, A. A., and Wirz-Justice, A. (1982): Sleep, sleep deprivation and depression. *Hum. Neurobiol.*, 1:205–210.
5. Broughton, R. (1975): Biorhythmic variations in consciousness and psychological functions. *Can. Psychol. Rev.*, 16:217–239.
6. Campbell, S. S. (1983): Human sleep patterns under conditions of "disentrainment." In: *Sleep '82: Proceedings of the 6th European Congress on Sleep Research*, edited by W. P. Koella, pp. 212–215. Karger, Basel.
7. Campbell, S. S. (1984): Duration and placement of sleep in a "disentrained" environment. *Psychophysiology*, 21:106–113.
8. Campbell, S. S. (1988): Bed rest induces depressed sleep patterns in healthy young adults. In: *Sleep '86: Proceedings of the 8th European Congress on Sleep Research*, edited by W. P. Koella, pp. 223–226. Fischer, New York.
9. Campbell, S. S., and Zulley, J. (1985): Ultradian components of human sleep/wake patterns during disentrainment. In: *Ultradian Rhythms in Physiology and Behavior*, edited by H. Schulz and P. Lavie, pp. 234–255. Springer-Verlag, Berlin.
10. Campbell, S. S., and Zulley, J. (1987): Induction of depressive-like sleep patterns in normal subjects. In: *Chronobiology and Neuropsychiatric Disorders*, edited by A. Halaris, pp. 117–132. Elsevier, New York.
11. Conroy, R. T., and Mills, J. N. (1970): *Human Circadian Rhythms*. Churchill, London.
12. Czeisler, C. A., Weitzman, E. D., Moore-Ede, M. C., Zimmerman, J. C., and Knauer, R. S. (1980): Human sleep: Its duration and organization depend on its circadian phase. *Science*, 210:1264–1267.
13. Godbout, R., and Montplaisir, J. (1986): The performance of normal subjects on days with and days without naps. *Sleep Res.*, 15:71.
14. Godbout, R., and Montplaisir, J. (1986): Effects of napping on performance of normal non-sleep-deprived subjects. Paper presented at the Eighth European Congress of Sleep Research, Szeged, Hungary.
15. Horne, J., and Ostberg, O. (1977): Individual differences in circadian rhythms. *Biol. Psychol.*, 5:179–190.
16. Kripke, D. F. (1976): Biological rhythm disturbances might cause narcolepsy. In: *Narcolepsy: Advances in Sleep Research, Vol. 3*, edited by C. Guilleminault, W. C. Dement, and P. Passouant, pp. 475–484. Spectrum, New York.
17. Lewis, H., and Masterton, J. (1957): Sleep and wakefulness in the Arctic. *Lancet*, 1:1262–1266.
18. Mills, J. N. (1973): Transmission processes between clock and manifestations. In: *Biological Aspects of Circadian Rhythms*, edited by J. N. Mills, pp. 28–74. Plenum Press, New York.
19. Mills, J. N., Minors, D. S., and Waterhouse, J. M. (1974): The circadian rhythms of human subjects without timepieces or indication of the alternation of day and night. *J. Physiol.*, 240:567–594.
20. Mosko, S. S., Holowach, J. B., and Sassin, J. F. (1983): the 24-hour rhythm of core temperature in narcolepsy. *Sleep*, 6:137–146.
21. Nakagawa, Y. (1980): Continuous observation of EEG patterns at night and in daytime of normal subjects under restrained conditions. I. Quiescent state when lying down. *Electroencephalogr. Clin. Neurophysiol.*, 49:524–537.
22. Orne, M. T. (1962): On the social psychology of the psychological experiment: With particular reference to demand characteristics and their implications. *Am. Psychol.* 17:776–783.
23. Schaefer, K. E., Clegg, B. R., Carey, C. R., Dougherty, J. H., and Weybrew, B. B. (1967): Effect of isolation in a constant environment on periodicity of physiological functions and performance levels. *Aerospace Med.*, 38:1002–1018.
24. Soldatos, C. R., Madianos, M. G., and Vlachonikolis, I. G. (1983): Early afternoon napping: A fading Greek habit. In: *Sleep '82: Proceedings of the 6th European Congress on Sleep Research*, edited by W. P. Koella, pp. 202–205. Karger, Basel.
25. Tune, G. (1968): Sleep and wakefulness in normal human adults. *Br. Med. J.*, 2:269–271.

26. Webb, W. B. (1978): Sleep and naps. *Specul. Sci. Technol.*, 1:313–318.
27. Webb, W. B., and Agnew, H. W. (1970): Sleep stage characteristics of long and short sleepers. *Science*, 168:146–147.
28. Webb, W. B., and Agnew, H. W. (1974): Sleep and waking in a time-free environment. *Aerospace Med.*, 45:617–622.
29. Webb, W. B., and Friel, J. (1971): Sleep stage and personality characteristics of "natural" long and short sleepers. *Science*, 171:587–588.
30. Weitzman, E. D., Kronauer, R., Fookson, J., Moline, M., Zimmerman, J., and Ronda, J. (1982): Endogenous oscillators control nap and sleep during non-entrainment. *Sleep Res.*, 11:221.
31. Wever, R. A. (1979): *The Circadian System of Man: Results of Experiments Under Temporal Isolation.* Springer, New York.
32. Williams, R. L., Karacan, I., and Hursch, C. J. (1974): *Electroencephalography (EEG) of Human Sleep: Clinical Applications.* Wiley and Sons, New York.
33. Zulley, J. (1976): Schlaf und Temperatur unter freilaufenden Bedingungen. In: *Bericht uber den 30. Kongres der dtch. Ges. fur Psych.*, edited by W. H. Tack, pp. 398–399. Hogrefe, Gottingen.
34. Zulley, J., and Campbell, S. S. (1985): Napping behavior during "spontaneous internal desynchronization": Sleep remains in synchrony with body temperature. *Hum. Neurobiol.*, 4:123–126.
35. Zulley, J., and Campbell, S. S. (1985): The coupling of sleep-wake patterns with the rhythms of body temperature. In: *Sleep '84: Proceedings of the 7th European Congress on Sleep Research*, edited by W. P. Koella, E. Ruther, and H. Schulz, pp. 81–85. Fischer, New York.
36. Zulley, J., Wever, R., and Aschoff, J. (1981): The dependence of onset and duration of sleep on the circadian rhythm of rectal temperature. *Pflugers Arch.*, 391:314–318.

Sleep and Alertness: Chronobiological, Behavioral, and Medical Aspects of Napping, edited by D. F. Dinges and R. J. Broughton. Raven Press, Ltd., New York © 1989.

8

Ultrashort Sleep/Wake Patterns and Sustained Performance

Claudio Stampi

Human Neurosciences Research Unit, University of Ottawa, Ottawa, Ontario, Canada K1H 8M5

In those situations in which more or less continuous and prolonged performance demands are experienced for days and weeks, the habitual monophasic nocturnal sleep pattern can rarely, if ever, be accomplished. The need to sleep and recuperate performance effectiveness may sometimes force the individual to adopt ultrashort sleep/wake patterns (i.e., schedules with multiple and repeated naps).

In this chapter, it is postulated that humans may have an endogenous, although usually masked, ability to adapt to polyphasic sleep patterns, and that this tendency is clearly revealed in certain experimental situations, as well as during prolonged sustained tasks in which individuals are free to choose an alternative pattern to monophasic sleep. Findings that strongly suggest a propensity for polyphasic sleep are discussed, the limited literature on ultrashort sleep is critically reviewed, a field study of ultrashort sleep/wake patterns during continuous work (CW) is presented, and factors relevant to implementing such schedules are discussed. Because of the plasticity of the human sleep system, systematic and prolonged use of ultrashort sleep/wake schedules is proposed as a feasible and promising solution to maintain high levels of efficiency during continuous work situations.

SLEEP LOGISTICS DURING SUSTAINED TASKS

In industrialized societies, there is an increasing number of occupations that may require a person to do prolonged periods of CW (several days or weeks). Some of these sustained operations (SUSOPs) scenarios also involve performance of essential services and crucial high-responsibility tasks. The best examples of SUSOPs are undoubtedly those experienced by military and defense personnel during combat or an emergency (see Chapter 11). However, a number of other groups may occasionally be exposed to relatively extended periods of CW. These groups include fire fighters; police officers; health care providers; rail, sea, and air transportation personnel; nuclear, petrochemical and steel industry workers; expeditionary groups (e.g., astronauts, support crews); and rescue and emergency personnel dealing with man-made (e.g., hazardous materials catastrophes) and natural disasters (e.g., earthquake, forest fire, volcanic eruption).

A particular characteristic of most CW scenarios is that the demand for activity, performance, or simply being on watch is either prolonged or comes at very short or unpredictable intervals, which prevent the person from sleeping in the habitual monophasic 6- to 8-hr nocturnal manner. On the other hand, because of the highly specific or skilled nature of the tasks involved, the demands on the individual cannot be met easily by sharing the work through planned shift-work schedules.

Sleep need is, therefore, the most important limitation during CW, since without its correct management, sleep tends to be totally or partially avoided, resulting in an accumulated sleep debt and consequently in a (sometimes serious) decrease of performance efficiency. The detrimental effects on performance, mood, and efficiency of both partial and total sleep deprivation (TSD) are well established and have been extensively reviewed (25,43).

How can the need for sleep be judiciously satisfied during CW in order to maintain high-quality task performance for prolonged periods? Theoretically, there are four ways to deal with the problem: (a) avoid sleep during CW (i.e., permitting total sleep loss); (b) lessen the requirement for sleep; (c) combine partial sleep loss with occasional recovery napping; and (d) adopt ultrashort sleep/wake schedules or a so-called "prophylactic" napping (12) schedule. A brief analysis of the feasibility of each of these hypothetical modalities will permit a better understanding of the issues under discussion.

Total Sleep Loss

Although numerous studies have reported the possibility of maintaining continuous wakefulness in humans from 24 hr to as long as 264 hr (43), the cost of such stress on the sleep system includes deleterious and sometimes profound effects on performance, vigilance, and mood. In fact, performance on certain tasks has been found to be sensitive to as little as one night of sleep deprivation (13,18). By the fourth day of a study of performance during a sustained military operation with no sleep, all of more than 20 members of a British parachute regiment had withdrawn from the exercise (21). The group showed profound decrements for most tasks, intense sleepiness, negative mood, and, by the third day without sleep, they were judged by observers to be militarily ineffective.

Similar results have been reported by Opstad et al. (47) in a study of Norwegian military academy cadets with sustained field activity of as long as 5 days. Morgan et al. (37) found that task performance began to drop after 18 hr of CW and dropped to 67% of baseline levels toward the end of 48 hr of CW. In summarizing work in this area, Naitoh and colleagues (46) concluded that "the upper limit of human performance for working intensively and continuously . . . [is] 2 to 3 days when tasks are both physical and mental," although the initial detrimental effects of sleep loss may appear even within the first 24 hr of CW.

Lessen the Requirement for Sleep

The strategy of reducing the need for sleep can be approached three ways: (a) by attempting to store sleep in advance of anticipated sleep loss; (b) by enhancing the restorative value of sleep; or (c) by intensifying wakefulness.

Storing Sleep

Accumulating sleep for later use is an intriguing idea. If possible, an individual could merely prolong his or her habitual 8- to 10-hr daily sleep length by several hours before the start of a sustained work scenario. Unfortunately, no study has yet demonstrated that sleep can be voluntarily prolonged much beyond the usual habitual 8 to 10 hr daily sleep length of a nonsleep-deprived person. Even studies of persons deprived of sleep for 8 to 12 days do not report recovery sleep lengths of more than 20 hr. Of course, even a modest prolongation of nocturnal sleep may be of some use for CW scenarios of relatively short (24–36 hr) duration. It has been reported, however, that extending sleep beyond the habitual amount may produce unpleasant physical and mental consequences for both nonsleep-deprived (53) and partially sleep-deprived subjects (22), as well as performance decrements similar to those experienced with total sleep loss. On the other hand, field studies during CW indicate that some individuals may be able to store sleep, although this capacity appears to be extremely rare (51).

Enhancing the Restorative Value of Sleep

Sleep "efficiency" per unit time might be augmented by the use of pharmacologic or natural agents that induce or deepen sleep. Although there are well-known agents that may facilitate sleep onset or produce some benefits for sleep pathologies, no procedures that consistently reduce sleep need have thus far been documented.

Extending Wakefulness

Sleep can be reduced within certain limits; similarly, wakefulness can be extended by the use of nonpharmacologic techniques or pharmacologic stimulants. However, pharmacologic agents to extend wakefulness could only be of relative utility for the short term, as the deleterious and/or rebound effects of various types of stimulants are well-known.

In brief, although the idea of enhancing the efficient utilization of monophasic sleep is conceptually appealing, it is also evident that there are serious limitations to such an approach.

Partial Sleep Loss Plus Recovery Napping

The literature on partial sleep deprivation shows that if sleep is simply reduced in amount, performance can usually be maintained with 60% to 70% of the usual amount of daily sleep, which is approximately 4.5 to 5.5 hr of sleep per day in a habitual 8-hr sleeper [cf. (26)]. When sleep of almost any duration is permitted during CW scenarios, a significant improvement of performance is reported relative to going without sleep. For example, in Haslam's (21) SUSOPs field trial, when only 90 min of sleep was allowed every 24 hr, 50% of the subjects were able to continue for the full 9 days; with 3 hr sleep per day, 91% made it to the final day. In both conditions, performance and mood were somewhat improved, although they were still far from baseline levels.

Various authors (35,47) have reported that naps are disproportionately effective toward maintaining performance when compared with no sleep. In summarizing the results from napping studies during continuous work, Dinges et al. (13) concluded the following:

> Indeed, numerous studies have demonstrated that in terms of behavioral measures of sleepiness, even "a little sleep," in the form of a single nap in a 24-hr period, is better than no sleep at all and, generally, the longer the nap the greater the benefits.

However, Naitoh (44) noted that naps after sleep deprivation might not always be helpful for restoration and maintenance of performance and mood. He suggested that the recovery potential of naps would be influenced by three major factors: duration of prior wakefulness, time of day the nap was taken, and nap duration. Naitoh and co-workers (45) later proposed a model for the prediction of performance during CW periods, based on these three factors (see Chapter 11 for a detailed analysis of recovery effects of naps in various CW scenarios).

Ultrashort Sleep/Wake Patterns and Prophylactic Napping

There appears to be no way to restore performance effectiveness during CW without permitting individuals to sleep, and short periods of sleep, namely naps, are disproportionately effective for recovery of functioning. Despite the benefits of occasional naps, performance and mood under such conditions are still very far from 100% of baseline or optimal levels and therefore the effects of sleep loss are still evident, especially when the nap(s) is taken after a more or less pronounced amount of sleep debt has been accumulated. This is the main limitation of strategies that permit nap sleep only after the person has undergone a sustained period of sleep loss. That is, it is likely that an individual who is already severely sleep deprived by the time a nap is permitted, is more sensitive to further sleep loss than is someone who naps before sleep loss is too severe.

This raises a number of issues that have rarely been considered: Is it more valuable, in terms of maintaining human functioning during CW, to permit people to nap in advance of serious sleep loss in order to prevent sleepiness than it is to permit napping after intense sleepiness in order to recover function? Would such a "preventative" strategy maintain performance closer to 100% efficiency during prolonged work? Could the strategy be accomplished by having subjects systematically sleep for short periods before the accumulation of a sleep debt? In brief, could an individual prophylactically nap to prevent sleepiness by using polyphasic or ultrashort sleep/wake schedules from the very beginning of a CW scenario?

Ultrashort sleep schedules are characterized by multiple alternations of sleep and wakefulness, distributed throughout a 24-hr period; each alternation is generally shorter than 3 hr. Lavie and Scherson (34) were the first to adopt the term "ultrashort sleep," when they studied nine subjects on a 5-min-sleep/15-min-wake (5/15) schedule for a 12-hr period. However, the concept of polyphasic sleep is not new, and the first documented attempt to search for an alternative to the habitual human monophasic (one episode per 24 hr) sleep pattern dates back at least 50 years, to a study by Husband (24). Unfortunately, it was not until the mid-1970s that studies on polyphasic sleep patterns again appeared in the literature, and the primary concern of most of these studies was not an evaluation of fragmented sleep as an alternative to continuous sleep, but rather the relationship among sleep stages, circadian rhythmicity, and the basic rest/activity cycle (28). Nevertheless, these studies demonstrated that adult humans are able to adapt without notable difficulties to relatively prolonged polyphasic napping regimes.

The use of any opportunity for short sleep periods in quasicontinuous operations has been extensively discussed by Dinges, Orne, and colleagues (12). Based on Orne's conceptualization of prophylactic napping as the systematic use of short sleep during CW, they emphasized that

> the important point is to prevent, or at least postpone, the development of sleep debt as much as possible. That is, although the loss of a night's sleep can usually be managed by most individuals, the loss of a night's sleep to an individual who is already sleep deprived is likely to result in profound performance decrements.

Recently, Dinges et al. (13,14) reported the results of a study of the effects of a single 2-hr nap opportunity before (prophylactic) and after (recovery) a night of sleep loss in a 54-hr quasicontinuous work scenario otherwise devoid of sleep opportunities. They found that the prophylactic naps taken 6 hr and 18 hr into the protocol enhanced performance for a prolonged period of time, even though mood was not improved. This occurred despite the fact that the sleep of these prophylactic naps was of a shorter duration and less deep than that of the recovery naps at 30, 42, and 54 hr (9). These results lend considerable support to the view that prophylactic napping is advantageous and suggest that warding off sleepiness or sleep debt accumulation by napping in advance of its build up during prolonged CW is as, if not more, beneficial than attempting to pay it off later by napping more deeply.

If a single short sleep in advance of sustained work is beneficial, then the extent to which humans can repeatedly avail themselves of these benefits will depend on whether they can develop enough flexibility in the sleep process to permit repeated periodic or ultrashort sleep/wake cycles. Thus, the use of prophylactic napping depends heavily on the extent to which humans can engage in polyphasic sleep. It is to this issue that we now turn, in an effort to resolve whether humans have a damped, inherited, and inborn tendency (or capacity) to fragment sleep, as do most animals. Four perspectives will be summarized that suggest that this hypothesis may be true.

ARE HUMANS ABLE TO ADAPT TO FRAGMENTED SLEEP PATTERNS?

Sleep Patterns in Mammals at Risk

Most mammalian species show polycyclic rest/activity patterns, although sleep is preferred at some phases of the 24-hr light/dark cycle more so than others [cf. (4)] (see Chapter 2). Why is typical human sleep an apparent exception to the general rule of the animal kingdom? During evolution, adaptive pressures apparently forced human sleep into a nighttime single episode. Whatever these pressures were, the primary fact is that the human sleep/wake cycle typically follows a circadian or monophasic pattern. Although presumably advantageous, such a monophasic sleep/wake cycle may be neither essential nor the most efficient temporal organization of the human rest/activity pattern. As already discussed, it can pose a serious problem for CW scenarios.

Sleep patterns of mammals living in hazardous environments may permit speculative analogies to the possibility of human ultrashort sleep/wake cycles during CW. Although no systematic study has yet assessed a direct relationship between the level of sleep fragmentation and the habitat demands of a certain species, the general impression is that mammals living under very adverse conditions present higher levels of sleep fragmentation.

For instance, sleep is a particularly vulnerable state in the giraffe, an animal that takes 10 or more sec to stand up. Kristal and Noonan (30) observed that the giraffe lies down for periods of between 3 and 75 min, three to eight times a night. Although most of this time is spent awake, some of the time is spent in what was assumed to be nonrapid eye movement (NREM) sleep (5–30 min) followed by 1 to 10 min of putative rapid eye movement (REM) sleep, for a total of 2 hr of sleep per 24 hr. Horne (23) speculates that, in this animal, sleep is at an "obligatory minimum," with a very limited "facultative" portion.

In a study that did not involve electroencephalographic (EEG) monitoring, Pilleri (49) was able to monitor sound emissions of the blind indus dolphin (*Platanista Indi*), which lives in turbid and muddy waters and has to keep constantly on the move to avoid injury on rocks. The author claimed that sound emissions ceased only during sleep periods, which were very frequent and lasted for only 90 sec at a time, but provided an apparent total sleep of about 7 hr per day. It would appear that sleep in this species is an especially hazardous time. If it were entirely unessential, it should have disappeared, just as has this animal's sight (23). Instead, the scheduling of sleep in this blind dolphin is altered in a remarkable way, namely, an extreme ultrashort sleep pattern is evident. Similarly intriguing cases of fragmented sleep apparently occur in other aquatic mammals, such as the bottle-nosed dolphin (*Tursiops Truncatus*) (41) and the Black Sea porpoise (*Phocoena Phocoena*) (40), where it has been reported that EEG measures show approximately 1-hr periods of sleep involving one hemisphere at a time, separated by 1 hr of wakefulness in both hemispheres; this 3-hr cycle is repeated throughout the night.

Chronobiological Aspects of Sleep

The 24-hr rest/activity cycle and body temperature cycle were the first circadian rhythms in humans to be described and studied. Whereas the temperature cycle (as well as several other physiological rhythms) shows rather precise fluctuations throughout the nycthemeron, the sleep/wake cycle is much more labile or flexible, showing at least two major peaks in sleep propensity. There is a tendency for short sleep episodes (naps) to occur around the time of temperature maximum, approximately halfway between night sleep episodes. It was first proposed by Broughton (3) (see Chapter 5) that afternoon napping reflects an endogenous biological rhythm rather than social or other nonbiological mechanisms. More specifically, cycles may exist for both sleep onset facilitation and for slow wave sleep (SWS) showing a 12-hr (or circasemidian) distribution, with a major peak around the time of temperature minimum, and a minor peak close to the temperature maximum. This minor preferred phase for sleep tendency corresponds to the early to midafternoon period (1400–1600 hr)—a time when most napping occurs in populations in which napping is frequent (see Chapter 9).

This interval also corresponds to a time of peak occurrence of stage 2 sleep in young adults attempting to sleep for 5 min every 20 min in the ultrashort sleep/wake studies of Lavie (e.g., 34). Using such schedules in laboratory studies, Lavie has demonstrated the existence of primary (nocturnal) and secondary (midafternoon) "sleep gates" in which there is a "switching on" of hypnogenic mechanisms with strong facilitation of wake-to-sleep transitions (see Chapter 6). Interestingly, the paradigm also revealed a "forbidden zone" for sleep [i.e., a distinct zone of decreased sleepiness between 2000 and 2200 hr, that was evident even after sleep deprivation (32)].

In addition to the prominent primary and secondary sleep gates, Lavie and Scherson (34) also detected ultradian variations in "sleepability," expressed by fluctuations in the amount of stage 1 sleep, with a frequency of approximately 14.4 cycles/day (period = 100 min). This finding is in agreement with many studies that report oscillations in alertness as well as in physiological functions that show similar periodicities of 90 to 120 min [cf. (29)]. Such findings give further support to Kleitman's (28) hypothesis that the occurrence of REM in sleep is only part of a broader biological rhythm in alertness that operates continuously throughout both sleep and wakefulness, and which he termed the basic rest/activity cycle (BRAC) (29). Kleitman maintained that the adult circadian sleep/wake cycle is a consequence of the coalescence of briefer rest/activity cycles, which have periods of 40 to 60 min in the infant and 90 to 120 min in the adult.

Consistent with Kleitman's idea of ultradian periodicities throughout the nycthemeron, the sleepability data of Lavie (34) suggest that sleep is more easily initiated during certain periodic gates, spaced approximately 1.5 hr from each other. Figure 1 is Lavie's (31) schematic of the variations in the probability of wake-sleep transition $[P_r (w \rightarrow s)]$ throughout 24 hr. As is suggested in the figure, the ultradian cycle of sleepiness has an extremely low amplitude and is not easily detectable, except at the two privileged phases when transition from wake to sleep is greatly facilitated. Thus, it would appear that polyphasic sleep tendency, if not sleep/wake cyclicity, not only is demonstrable in humans, but also coexists with and potentiates circasemidian and/or circadian sleep tendencies.

Social and Cultural Limitations on the Sleep Process

Experimental approaches other than the ultrashort paradigm may also expose putative noncircadian components of the sleep/wake system. One recent interesting example is the "disentrainment" design of Campbell and Zulley (5), in which they eliminated external and self-imposed behavioral

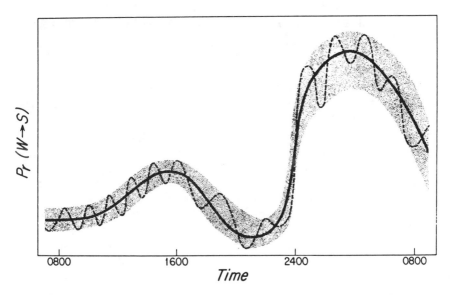

FIG. 1. A schematic representation of the 24-hr circadian, circasemidian, and ultradian variations in the probability of a wake-to-sleep transition. (From ref. 31.)

controls on sleep in a time-free environment, to allow the sleep system to follow a course of least resistance (see Chapter 7). In this situation, subjects exhibited multiple sleep episodes dispersed throughout the nycthemeron, although the highest probability occurred at the temperature minima and maxima. The overall median sleep episode duration was 2.4 hr. It appears, therefore, that sleep recorded from human subjects under these conditions is comparable with polyphasic sleep recorded from animals under standard laboratory conditions.

Anthropological studies on the highly fragmented sleep patterns of nightly active tribes (the Temiars in Indonesia and the Ibans in Sarawak) give further support to the polyphasic hypothesis (48). Although they have different cultures and ways of life, the tribes show similar sleep/wake behavior. Their average nocturnal sleep episode duration ranges between 4 and 6 hr, and nighttime activities (fishing, cooking, watching over the fire, rituals) at any one time involve approximately 25% of the adult members. Daytime napping is very common in both tribes; at almost any time of day, approximately 10% of the adult members can be found asleep. Whatever the cause of these polyphasic sleep patterns, whether the expression of an inborn ultradian rest/activity tendency or other factors, such populations exhibit extremely flexible and fragmentary sleep/wake cycles. The minimal contact with modern civilization could be one of the reasons for the preservation of this possibly ancestral sleep pattern.

STUDIES OF EXTENDED ULTRASHORT SLEEP/WAKE PATTERNS

Laboratory Studies

The first study to evaluate a single monophasic sleep pattern relative to several short sleep episodes was conducted by Husband (24). The hypothesis was that "one might sleep two to three hours, stay awake for a number of hours, and then sleep two or three hours more. This procedure would, in theory, take advantage of two stretches of deep slumber." After 1 month of uninterrupted 8-hr nightly sleep, Husband's subject kept a very regular routine of 3-hr sleep (2300–0200 hr), 3-hr waking (spent in various activities), and a second 3-hr sleep (0500–0800 hr). Thus, the schedule was a compromise between slight sleep deprivation and fragmented sleep. A series of performance, physical, and psychological tests showed no signs of deterioration during or after the 1 month experimental period. The results led Husband (24) to conclude that "there are no consistent differences between interrupted and continuous sleep. . . . It is recognized that the use of only one subject makes the results only suggestive."

After a hiatus of 40 years, the discovery of REM sleep and intrasleep cyclicity stimulated several researchers to study ultrashort sleep/wake schedules. The studies that appeared in the literature generally adopted schedules with a sleep/wake ratio of 1:2 (i.e., one-third of time in a sleep/wake cycle was allowed for sleep and two-thirds of time for wakefulness, resulting in approximately 8 hr of every 24 hr being devoted to total bedtime [TBT] for sleep). Table 1 summarizes the protocols used in these studies.

The sleep episode durations most adopted were of 30 and 60 min. The Lavie and Scherson (34) study of 5-min sleep and 15-min wake and the Lavie (32) and Lavie et al. (33) studies of 7-min sleep and 13-min wake had the shortest sleep/wake cycle and the shortest duration (12 and 36 hr, respectively); as mentioned earlier, the findings relate more to ultradian and circadian cycles of sleepiness. The experiment by Webb and Agnew (55) had the longest sleep/wake cycles (9-, 12-, 18-, 30-, and 36-hr lengths), making it less relevant to ultrashort sleep/wake schedules. All studies adopted regular sleep/wake schedules, with the exception of the Curtis and Fogel (8) study, in which schedules were such that each 24-hr day contained four 2-hr periods in bed, and eight 2-hr periods out of bed; all periods occurred in random sequence.

An important preliminary issue concerns the number of days necessary for a human to adapt to an ultrashort sleep/wake pattern, if indeed this is possible. Since the imposed sleep regimes are quite different from the habitual single nightly sleep episode every 24 hr, it may be expected that subjective adaptation to such unusual schedules would take a rather long time. As a comparable example, it is known that following sleep-displacement during shift-work studies, entrainment to the new schedule requires a min-

TABLE 1. *Laboratory studies of ultrashort sleep/wake patterns*

Study	No. of subjects	Mean SED[a]	Mean WID[b]	Mean TBT[c]	Duration of experiment[d]
Husband (24)	1	180	180–900	6.00	
Curtis and Fogel (8)	6	120	Random	8.00	14
Kelley et al. (27)	1	30	60	8.00	6
Hartley (20)	12	80	300–480	4.00	4
Weitzman et al. (56)	7	60	120	8.00	10
Carskadon and Dement (6, 7)	10	30	60	8.00	6
Moses et al. (38); Lubin et al. (35)	8	60	160	6.50[e]	1.7
Webb and Agnew (55)	2	180	360	8.00	6
Lavie and Scherson (34)	9	5	15	6.00[e]	0.5
Mullaney et al. (42)	10	60	360	3.40[e]	1.7

[a] Sleep episode duration (min).
[b] Wake interval duration (min).
[c] Duration (hr) of total bedtime per 24 hr.
[d] Days.
[e] Plus previous night's sleep.

imum of 3 and an average of 7 days [cf. (17)]. The major limitation of most ultrashort sleep experiments has been the short duration of the experimental periods. Only two experiments lasted for 10 days or more (8,56); four experiments lasted 6 days. All others were for very short periods.

It is obviously very likely that during the first experimental week on such unusual schedules, sleep-related parameters [e.g., total sleep time (TST)], sleep stages, as well as biological rhythms, performance, and behavioral parameters, may all be disrupted or abnormal. Any true comparison with baseline levels will have meaning only after the period necessary for full adaptation (of yet unknown duration, if possible at all). These considerations should be kept in mind in the interpretation of the main findings of these studies.

Sleep Structure

A clear-cut 24-hr distribution of sleep was apparent in the studies of Weitzman et al. (56) and Carskadon and Dement (6,7), with maximum sleep occurring in the late morning and early afternoon hours (0600–1300 hr); neither study explained this apparent shift from baseline (i.e., 2300 to 0700 hr), in the circadian propensity for sleep. Sleep efficiency [calculated as the sum of stages 2, 3, 4, and REM sleep divided by TBT (TST/TBT)] was generally maximal when the temperature was lowest, but overall TST was reduced compared with baseline (6,7,35,38,56). Sleep efficiency was approximately 50%: 50.9% (56), 47.6% (6,7), 47.5% (35,38), and 51.4% (32,33). Consistent with what is generally found in studies of sleep reduction, SWS quantities were not significantly reduced from baseline levels. As could be anticipated from the higher number of sleep onset opportunities, overall stage 1 sleep percentage was increased; conversely, REM sleep per 24 hr was significantly reduced. However, it is interesting to note that in the studies of longer duration, these effects were more severe on the first days and decreased toward the last days of the experiments (6,7,56). It could be hypothesized that if longer periods were allowed on fragmented sleep schedules (e.g., 2 to 3 weeks), the sleep stage percentage measures and sleep efficiency might gradually become more similar to those of baseline nocturnal sleep.

Sleep stages generally progressed in the usual manner, except for the occasions when REM sleep was present. If results are combined across studies, REM sleep occurred in 30% of all naps, but there were differences among studies: 42% of 80 naps (35,38), 26% of 910 naps (6,7), and 35% of 560 naps (56). REM sleep amount revealed a clear circadian rhythm, such that most of it [e.g., 74% (6,7)], occurred in the rising phase of the body temperature curve (i.e., 0200–1000 hr). However, REM sleep onset had no consistent relationship to time of day. Rather, it appeared to be related to the amount of sleep time since the last REM onset. When REM sleep was present, it frequently occurred in close proximity to nap onset (6,7,27,35,38,56). For example, Carskadon and Dement (6,7) found mean REM latencies of 8 min, against an average of 45 min found in baseline recordings.

REM sleep and SWS rarely occurred together. For example, in the study by Carskadon and Dement (6,7), REM and SWS coexisted in only 27 (3%) of 910 naps, and of these, the normal relationship of SWS giving way to REM was evident only eight times. Weitzman et al. (56) reported a sequence reversal of SWS and REM in 50% of 89 naps (total naps = 560) in which the two sleep stages occurred together. In general, REM sleep tended to occur (128 times) during sleep periods that alternated with those containing SWS and rarely appeared (19 times) in consecutive sleep episodes (6,7). SWS, on the other hand, recurred in both consecutive (153 times) and alternate (185 times) sleep periods (6,7). Combining results across studies, subjects were awakened from REM in 18% of the naps and from SWS in 44% of the naps (6,7,35,38,56).

In an interesting study concerning the length and rhythmicity of the REM cycle, Moses et al. (39) reanalyzed data from the studies by Weitzman et al. (56), Carskadon and Dement (6,7) and Moses et al. (35,38). Autocorrelation and r^2 analysis (a measure of the strength of the REM cycle periodicity) were applied to "compressed sleep" (i.e., all sleep minus the wake time between and within sleep periods) of the baseline sleep, the nap sleep during the experimental protocols, and the recovery sleep. Compared with baseline, there were no significant differences in nap REM sleep cycle length (approximately 100 min) in two experiments (35,38,56). The third study (6,7) had significantly shorter cycles (60 min); it appeared that this shortened REM cycle length was owing to the significantly shorter REM episodes in this study. Despite this difference, all three studies obtained the same amount of REM sleep relative to TST.

Relative to baseline sleep, however, nap r^2 values were significantly lower in reports of Weitzman et al. (56) and Carskadon and Dement (6,7), indicating an increased variability in the timing of REM sleep during naps. In other words, the strength of the rhythm was considerably reduced; however, nap r^2 values were significantly higher than those obtained from a random distribution of sleep stages. An interesting finding was that both nap sleep efficiency (TST/TBT) and nap r^2 were significantly correlated across the subjects, demonstrating that those who adapted well to the schedule (i.e., fell asleep easily and remained asleep during the scheduled sleep periods), also had more rhythmic REM sleep. Moses et al. (39) concluded, therefore, that these data offer further support for the view that the REM cycle is a sleep-dependent rhythm (i.e., it appears only when the organism is sleeping) and not an expression of an ongoing BRAC.

In general, sleep stage percentages were remarkably similar in all three studies. Johnson and colleagues (26) compared the sleep data from subjects who adapted well to a gradual reduction of their nocturnal TST from 8 hr to 6 hr with no impairment in performance, with those from their previous nap study (35,38). They found that the time spent per 24-hr period in the various sleep stages was similar between studies, no matter whether TST was reduced by shortening the single period or by fragmenting sleep into naps. Thus, concerning sleep stage percentages and cyclicity measures, the sleep system reacted to these experimentally produced ultrashort sleep/wake (nap) schedules in a relatively predictable manner, no matter what type of nap schedule was adopted.

Recovery SWS showed a nonsignificant increase in all but one (6,7) of the laboratory studies of ultrashort sleep (35,38,56). There is less consensus concerning recovery sleep time, however. Although in the short duration experiments (35,38,42) recovery TST did not show the usual lengthening associated with sleep loss, in the longer duration studies (6,7,56) a striking increase in recovery TST was found—higher than that usually found in TSD studies of similar duration. The disruption of the 24-hr sleep/wake cycle produced by the fragmented sleep schedules may explain the phenomenon: During the recovery nights, the subjects probably delayed awakening time because of the phase shift in TST peak (from early morning to mid-late morning) that occurred during the experiments. These experiments, which lasted 6 to 10 days, were long enough to disrupt the sleep system but still too short to permit full entrainment of subjects to the schedules. In a re-analysis of the data from the study of Carskadon and Dement (6,7), Strogatz (52) showed that, in fact, the circadian rhythms of both TST and body temperature were free running with a period of 25.25 hr, as a result of disruptive schedule, even though no effort was made to shield subjects from time cues.

Behavioral Functioning

In the study by Carskadon and Dement (6,7) sleepiness ratings paralleled the phase shift found in TST. Subjects were sleepiest in the late morning hours and lower ratings were observed in the evening hours. Moreover, pre- and postnap sleepiness ratings appeared to demonstrate regular patterns. Naps with SWS periods were generally followed by more self-reported sleep-iness, whereas greater prenap sleepiness predicted greater amounts of REM sleep in the subsequent nap. Although on the first day, napping did not alleviate subjective drowsiness, throughout the next 4 days the sleepiness measures decreased almost to baseline levels.

In the study by Curtis and Fogel (8), which is the only study that assessed individual ability to fall asleep for short periods at irregular times, there was no evidence that regularity was essential for normal mental functioning. No impairment of cognitive or personality functions was detected. The results also suggested that individual differences in subjects' ability to adjust their sleep to a random schedule may be determined in part by personality factors. Increased ability to nap at irregular times correlated with higher scores on the California Psychological Inventory (19) scales of Intellectual Efficiency, Sense of Well-Being, Dominance, Self-Acceptance, Sociability, Achieve-ment via Conformance, and Psychological Mindedness, and with low scores on the Femininity scale.

Performance was unfortunately not assessed in all studies mentioned in Table 1. According to Lubin et al. (35), ten 60-min naps allotted every 220 min during the 40-hr period (TBT = 10 hr) effectively neutralized the det-rimental effects on performance observed in the two TSD control groups (exercise and forced bed rest). This is particularly interesting since the sub-jects averaged only 3.7 hr of sleep per 24 hr. It was also concluded that forced bed rest was no substitute for sleep.

Significant ameliorative effects on performance during a 42-hr continuous computer-based task session were also found by Mullaney et al. (42) in subjects permitted 1 hr of sleep every 7 hr (TBT = 6 hr), as compared with a totally sleep-deprived group. However, during the second half of the 42-hr period, the most effective performance was obtained by a third group permitted to sleep 6 hr continuously. This is not surprising since in this latter group the temporal placement of sleep was coincident with the subjects' habitual sleep time and subjects were able to accumulate more sleep in a single 6-hr period than in six 1-hr periods.

Very suggestive results come from the study of Hartley (20), who showed that subjects were able to function reasonably effectively for 4 consecutive days with only three 80-min naps per day (TBT = 4 hr per 24 hr). Naps were allowed at 2310, 0530, and 1225 hr. The study revealed that although performance of the nap group was slightly poorer than that of control subjects permitted 8 hr of normal nocturnal sleep, it was much better than the performance of a third group allowed 4 hr of continuous sleep (0100–0500 hr). That is, the same amount of sleep fragmented in three brief episodes throughout the 24 hr produced higher performance relative to monophasic sleep of a comparable duration.

To summarize, only some of the findings are in agreement with the conclusions of Weitzman et al. (56) that "naps are not miniatures of the 8-hr sleep pattern," and also not replicas of the first part of a normal uninterrupted night of sleep. In their disentrainment experiment, Campbell and Zulley (5) found that when naps are taken *ad libitum*, they indeed often appear to be miniatures of the normal 8-hr sleep pattern (i.e., showing typical sleep-stage sequencing and percentages). Reanalysis of "compressed sleep" from ultrashort sleep/wake studies seems to confirm this in part, showing no significant differences in REM sleep cycle and length between nap schedules and baseline (39). The reanalysis by Moses et al. (39) also showed that the reaction of the sleep system to three different nap schedules was apparently similar and predictable. That is, sleep-stage percentage changes seem to be related more to the reduction in TST than to polyphasic sleep per se. It appears that multiple naps penalize REM sleep, but SWS will not be denied regardless of the schedule.

From the perspective of sleep management in the face of demands for prolonged functioning, the most interesting findings are that the sleep system seems to have a high plasticity in terms of scheduling and duration and that humans appear to adapt surprisingly well to schedules of multiple naps. In addition, in some situations short naps have shown higher restorative effectiveness on performance than the same amount of continuous sleep. This suggests again the possibility that human sleep is naturally polyphasic or at least that it can be shaped to a polyphasic schedule. The ultrashort sleep paradigms also provide important information about our theoretical knowledge of the sleep system. In conclusion, sleep during prolonged ultrashort sleep/wake schedules needs further research: "it cannot be accounted for exclusively within the framework of what we know about monophasic nocturnal sleep" (38).

Field Studies

In the beginning of this chapter, it was proposed that ultrashort sleep/wake schedules could represent a feasible and preferable strategy in adapting to CW schedules. This hypothesis was supported by several issues concerning human adaptability to polyphasic sleep and a review of the laboratory studies on napping and ultrashort sleep/wake schedules. We now turn attention to an analysis of the spontaneous reactions of individuals to CW during which no sleep/wake schedule is imposed. During a field trial, would a subject tend to sleep less, totally avoid sleep, or switch to a napping regime? If a polyphasic sleep pattern is chosen, will it benefit the subject's performance? For how long can an individual function on this type of schedule?

Unfortunately, because of the scant literature in this area, these questions cannot be definitively answered at this time. Nevertheless, some indication of what happens to the sleep and functioning of persons undergoing unstructured field trials for many days of prolonged CW has recently appeared (50,51). Because it represents the first systematic study of civilians voluntarily undergoing CW, it is worth examining.

A Study of Racing Across the Ocean

Appropriate field trials designed to assess spontaneous sleep/wake behavior during CW should meet the following requirements: (a) no sleep/wake schedules can be given or suggested, (b) performance must be measured without interfering with the subjects' activity and only minimal collaboration can be requested, and (c) the trial must be as long as possible and involve a substantial number of subjects to permit interindividual comparisons.

After surveying a wide variety of possible CW scenarios, what appeared to be an almost ideal field model of SUSOPs was identified: solo sailing races across the oceans. During a solo race, the yachtsman is exposed for days and weeks to recurrent, unpredictable, and sometimes extenuating demands to be awake, steer and adjust the yacht to varying and occasionally remarkable conditions of sea and wind, and survey the tactics of competitors in order to optimize performance and sustain maximum speed. While he is asleep, the yacht may go off course, either yacht speed or control (by way of the self-steering devices) may not be optimal, and risks of collision with ships increase. In normal conditions of visibility, a yachtsman must check the horizon every 15 min to safely prevent collisions. Motivation to win is very high and incited by valuable prize awards and sponsors' rewards. The latest achievements in technology are continually applied to yacht design to improve speed and performance.

In the first (1960) Observer Single-handed Transatlantic Race (OSTAR), Sir Francis Chichester took 38 days to cross the Atlantic Ocean, whereas in 1988, the fastest yachtsman took only 10 days. Today, the limiting factor is no longer the boat but the sailor and his capacity to wisely administer both mental and physical energy expenditure. Race performance depends, therefore, on an optimal balance between cognitive and motor skills. Interestingly enough, the 1980 OSTAR winner was a 67-year-old yachtsman. A detailed technical, medical, and psychological analysis of a solo race has been reported by Bennet (1).

The general scope of these studies (50,51) was to observe how a group of selected individuals who are extremely motivated to find an efficient sleep strategy under prolonged CW would solve the problem. More specific goals were to analyze (a) the main characteristics of the spontaneous sleep patterns adopted by the sailors, (b) their adaptability to such schedules, (c) possible correlations between the sleep pattern adopted and any particular individual characteristic, and (d) the effects of the adopted sleep patterns on race performance. The hypotheses were that (1) this extreme situation would induce at least some of the yachtsmen to adopt ultrashort sleep/wake schedules, (2) they would adapt to such schedules with relative ease, and (3) the beneficial effects of ultrashort sleep, compared with other sleep strategies, would be revealed by better race performance results.

The study (51) was initiated during the 1980 OSTAR (3,000 nautical miles, $n = 54$ yachtsmen studied) and was later complemented by a study of the 1982 Round Britain Race (RBR) (1,900 miles, $n = 29$) and the 1983 Mini-Transat (MT) (4,300 miles, $n = 16$). The three sailing events (charted in Fig. 2) were of quite different length and type, involving different yacht sizes as well as different sea and wind conditions, inducing different amounts of fatigue, and requiring different degrees of experience. They were purposely chosen to verify whether the findings obtained in the OSTAR would be replicated in the other two races.

Whereas the OSTAR and MT were single handed, RBR was raced by crews of two. OSTAR was a nonstop race, MT had one stopover, and RBR had four intermediate stopovers of 48 hr each. The OSTAR was won in 17 days, the MT in 31 days, and the RBR in 8 days (excluding stopovers); the last competitors to finish each race took 49, 40, and 20 days, respectively. Competitors' ages ranged between 20 and 67 years (means 38.7, 29.7, and 35.6, respectively), and only 4 among the 99 subjects were women.

Procedures

Each competitor was interviewed extensively before and after the races to assess normal sleep habits, the presence of any sleep disturbance, the sleep regimes planned for the race and the patterns effectively adopted, previous sailing experience, and other parameters. The interviewers avoided giving any sort of counseling concerning sleep strategies. A sleep log was given to each sailor and each was asked to report whenever possible the exact moment of beginning and awakening from each sleep episode or, at least, a daily brief report estimating the mean, the minimal, and the maximal duration of sleep episodes and TST each 24 hr. Polygraphic recordings were not performed because of obvious technical difficulties under such situations.

Daily race positions and overall mean speed of each yacht were calculated from satellite data provided by the race organizers, which were compared

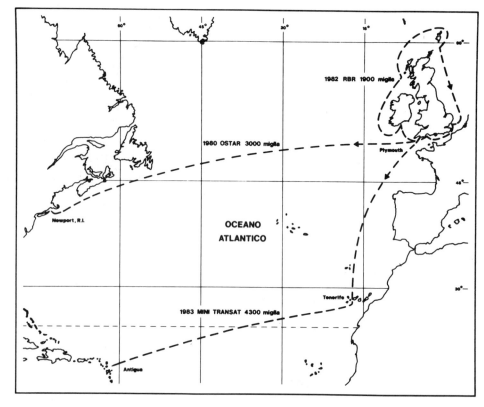

FIG. 2. The routes followed during the races by the single-handed (OSTAR and Mini Transat) and double-crewed (RBR) yachts.

with the subjects' reports. Overall indices of performance for each competitor were essentially based on race results: the mean speed of each yacht and the calculated mean handicap speed [i.e., (mean speed)/(square root of the waterline length)], since the theoretical maximum speed of a vessel is a function of the square root of her waterline length. Yachts' lengths ranged from 7.62 to 17.07 m for OSTAR, and as long as 25.91 m for RBR, but maximum length for MT was limited to 6.5 m.

Results and discussion

For all three races, mean daily TST was 6.33 hr (SD = 1.7), significantly but not markedly shorter than the sailor's average TST at home (7.5 hr). Figure 3 shows that the most frequently adopted daily TSTs were approximately 5 hr (20% of OSTAR sailors) and 8 hr (28%). Mean sleep episode duration (SED) was 2.01 hr for OSTAR, 0.61 hr for MT, and 2.7 hr for RBR.

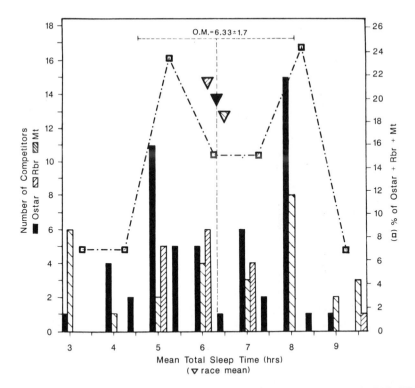

FIG. 3. Distribution of mean total sleep time (TST) for competitors in the OSTAR, RBR, and MT yacht race. Histograms show the number of competitors in each race with each TST. Symbols show the percentage of total competitors (*n* = 99) with each TST. The overall mean (OM) for all competitors is given at top; it was not markedly reduced compared to baseline TST mean (7.5 hr).

Overall mean SED was 2.00 hr (SD = 1.7); individual means ranged from 0.33 hr to 7 hr (Fig. 4).

Therefore, although sailors only slightly reduced daily TST, many divided it up into several short episodes. Sleep fragmentation was far more evident in the MT race (mean SEDs of all competitors ranged from 20 to 60 min), probably because of the reduced dimensions of the yachts involved, which permit faster and hence more frequent maneuvers.

The yachtsmen appeared concerned more with the duration than with the timing of sleep episodes. With the exception of some individuals who were able to adapt to very irregular SEDs, most tried to sleep for the number of minutes (or hours) that, by experience or simply casual impression, were found to be most restorative or to produce the least impairment upon arousal. Most yachtsmen were able to wake up spontaneously at scheduled times without the use of alarms. It was also reported that, even when stable meteorological conditions would permit them to sleep longer, most tried to

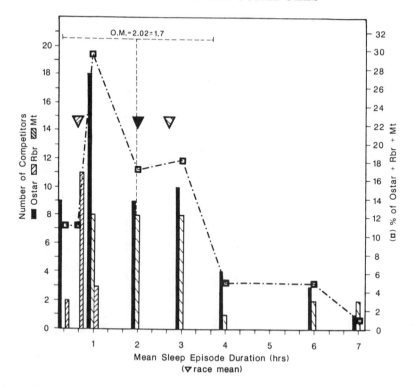

FIG. 4. Distribution of mean sleep episode durations (SEDs) for competitors in the OSTAR, RBR, and MT yacht race. Histograms show the number of competitors in each race with each mean SED length. Symbols show the percentage of total competitors ($n = 99$) with each SED length. The overall mean (OM) for all competitors is given at top.

avoid recuperative oversleeping, since this would produce unpleasant feelings of drowsiness and decreased efficiency upon arousal (i.e., it would increase the sleep-inertia effect). In other words, SED seems to be a stable individual characteristic, and the entrainment of the sleep system to one's own ideal SED appears to play an important role in the restorative value of each sleep episode.

Competitors with a mean SED of 20 min (16.6%) reported that they oscillated between SEDs of 10- to approximately 30-min duration. Interestingly, based on studies of experimentally induced sleep disruption, Bonnet (2) has concluded that 10 min is the minimal duration for sleep to be restorative. Overall, the most frequently adopted SED was approximately 1 hr (33% of OSTAR and 27.6% of RBR subjects); 66.5% of OSTAR competitors had SEDs between 20 min and 2 hr.

Forty-five percent of yachtsmen thought that regularity in timing sleep was important and schedules should be followed as closely as possible, but only 25% thought that this was possible, because of the unpredictability of demands during a race. The recuperative value of sleep per unit time was reported by 30% of the yachtsmen to be higher while they were at sea. Interestingly, all individuals within this group showed ultrashort sleep patterns; only 26% reported that they recuperated better at home.

Although most yachtsmen reported that "training" their sleep patterns some time before the race start is a useful strategy, only 10% were capable of doing so. This was primarily owing to the fact that the sailors were very busy for several weeks immediately prior to the race with preparation of their yachts. Entrainment to the sleep rhythm adopted in the OSTAR took an average of 3.5 days (SD = 1.5), and most sailors reported greater fatigue at the beginning of the race than on the last days before arrival at the finish— after 3,000 to 4,000 miles of sailing. At the conclusion of that race, sleep patterns took an average of 3.6 days (SD = 5.5) to return to the normal monophasic pattern, although some subjects reported taking as long as 3 weeks to reset their sleep. Younger subjects and "evening-types" re-entrained faster (significant correlations). Recovery TST after the race was not significantly increased in comparison with baseline levels.

Analysis of the yachtsmen's sleep patterns revealed that there were four preferred sleep strategies for this type of CW; these are illustrated for OSTAR.

1. *Ultrashort sleep.* This group was very concerned with competitive aspects of the race and represented the most adopted strategy (66.5% of OSTAR competitors); mean SEDs ranged between 20 min and 2 hr.

2. *Navy sleep.* Under SUSOPs, the traditional navy shift-work system is generally 4 hr off/8 hr on. Similar patterns were adopted by 18.5% of OSTAR sailors, with SEDs from 3.5 to 4.5 hr, but generally without rigid timing. They tried to reach a compromise between subjective need for longer sleep episodes and the race requirements for frequent watches.

3. *Sleep accumulation.* A few yachtsmen (7.4%) reported they were able to sleep whenever possible, with no regular SED or timing; they could also, if necessary, sleep in excess in order to store sleep and could then stay awake uninterruptedly for as long as 3 or 4 days.

4. *Monophasic sleep.* Some sailors (7.6%) simply tried to reproduce their habitual sleep habits of a nocturnal 6- to 8-hr-long monophasic sleep, except when in imminent danger. Members of this group were either not interested in race success or were incapable of polyphasic sleep.

Comparison between baseline sleep parameters and the adopted race/sleep patterns revealed no correlation; that is, it was not possible to predict the type of race/sleep schedule based on home sleep. Only sailing experience significantly predicted race/sleep patterns, such that experienced yachtsmen tended to adopt shorter SEDs and TSTs. Seven of the 10 competitors classified as most experienced (by a blind experimenter) adopted mean SEDs of 20 min to 1 hr.

TABLE 2. *Mean sleep episode duration (SED) and total sleep time (TST) of competitors grouped by race result in each of the three races*

Race result[a]	OSTAR		RBR		MT	
	SED	TST	SED	TST	SED	TST
1st	1.3 (1.2)[b]	4.8 (0.7)	1.7 (1.1)	4.8 (2.0)	0.61 (0.2)	5.4 (0.5)
2nd	2.1 (1.7)	6.7 (1.2)	2.5 (1.7)	6.7 (2.3)	0.75 (0.2)	5.6 (0.5)
3rd	2.7 (1.7)	7.5 (1.0)	4.0 (1.8)	8.2 (1.4)	0.54 (0.1)	7.3 (1.3)
Total[c]	2.1 (1.6)	6.3 (1.5)	2.7 (1.8)	6.5 (2.4)	0.61 (0.2)	6.1 (1.3)

[a] Competitors grouped according to yacht speed (1st had the fastest times).
[b] Standard deviations.
[c] All competitors within a race: OSTAR ($n = 54$); RBR ($n = 29$); MT ($n = 16$).

A preliminary analysis of possible relationships between sleep patterns and race performance (mean yacht speed) was done by dividing the competitors into three groups for each race, according to their race result (Table 2). A clear difference in SED and TST was observed among the groups. The fastest group of competitors tended to sleep less and fragment sleep into shorter episodes (OSTAR means: SED = 1.3 hr; TST = 4.8 hr) than did the second fastest group (SED = 2.1 hr; TST = 6.7 hr). Similar differences were observed between the middle and slowest group of competitors (SED = 2.7 hr; TST = 7.5 hr). This finding has been confirmed in RBR, and only for TST in MT, since similar SED values of highly fragmented sleep are found in the three groups in the MT race.

If competitors are grouped together according to their mean SED, and then the mean handicap speed obtained by each group is plotted against SED (Fig. 5), it can be observed that sailors who attained the highest speeds had SEDs between 20 min (best group) and 1 hr, whereas SEDs of 2 hr or more were associated with an abrupt decay in performance. Linear regression analysis performed between mean SED and race performance (mean speed of yacht) for each subject showed a significant negative correlation both for OSTAR ($p < 0.05$) and for RBR ($p < 0.001$) (Table 3). In brief, the shorter a sailor's mean SED, the better his race result. Similar significant results were found by substituting handicap speed for true speed.

This finding is in apparent contrast with what might be expected if the sleep-accumulation strategy, with its high degree of flexibility, were assumed to result in the best race performance. Actually, some subjects in this group did show good results, as can be seen in Fig. 5. Three of the seven sailors sleeping for an average of 4 hr ($n = 4$) or 6 hr ($n = 3$) were classified as sleep accumulators. These three performed better than sailors sleeping for 2 to 3 hr but worse than those with SEDs of 1 hr or less. It is possible that the supposed restorative advantages of having a longer sleep episode compensated in part, but not completely, for the disadvantages of being on watch

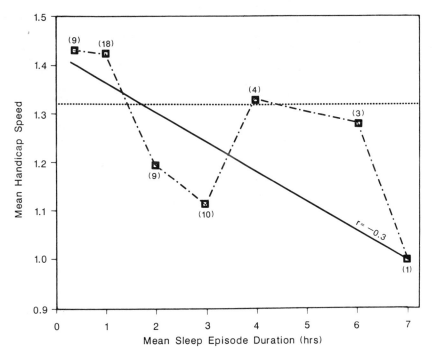

FIG. 5. Mean race performance (handicap speed for each yacht) of competitors as a function of the mean sleep episode durations (SEDs) during OSTAR (*n* = 54). The *horizontal line* represents the overall mean handicap speed. The *solid line* was fitted by linear regression (*p* < 0.05). Note that the best-performing competitors had SEDs between 20 min and 1 hr. The number of sailors within each mean SED group is in brackets.

TABLE 3. *Regression coefficients between race performance (MRS, MHS) and sleep parameters (SED, TST) during each of the races*[a]

Race performance	OSTAR		RBR		MT
	SED	TST	SED	TST	TST
MRS	−0.332[b]	−0.735[a]	−0.579[c]	−0.614[c]	−0.641[d]
MHS	−0.305[b]	−0.606[c]	−0.563[d]	−0.451[b]	—[e]

[a] MRS, mean real-time speed; MHS, mean handicap speed; SED, sleep episode duration; TST, total sleep time.
[b] *p* < 0.05.
[c] *p* < 0.001.
[d] *p* < 0.01.
[e] MHS not calculated because yacht length was limited to 6.5 meters.

less often. However, despite the hypothetical (although not clear and not demonstrated) advantage of the sleep-accumulation strategy, the capacity to store sleep appears to be far more difficult to achieve or at least to be much more rare (7.4%) compared with the ability to fragment sleep (66.5% had SEDs between 20 min and 2 hr).

The analysis of handicap speed as a function of TST (Fig. 6) shows that daily TSTs between 3 and 5.5 hr provided best performance results with highest speeds obtained by those sleeping 4.5 hr per day. It must be added that very few people slept 4.5 hr per day or less (11%) and that within the reduced-sleep group (TST of 3 to 5.5 hr), most (49%) preferred a TST of 5 hr. Subjects sleeping 6 hr or more showed a huge decrease in performance, and the quite high number (28%) of sailors who slept 8 hr per day did not obtain good results.

Linear regression analysis between TST and yacht speed (real time and handicap) showed a highly significant negative correlation ($p < 0.01$) in all three races: the shortest TSTs predicted the best race performance results (Table 3). Provided that TST reductions were not extreme, except for a

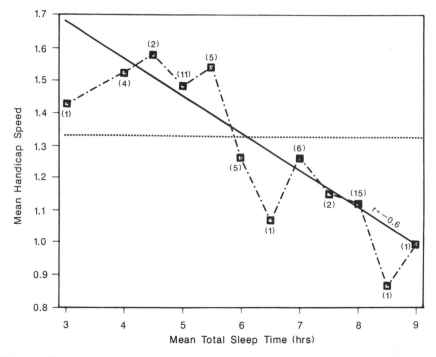

FIG. 6. Mean race performance (handicap speed for each yacht) of competitors as a function of the mean total sleep time (TST) during OSTAR ($n = 54$). The *horizontal line* represents the overall mean handicap speed. The *solid line* was fitted by linear regression ($p < 0.001$). Note that the best-performing competitors had TSTs between 4.5 and 5.5 hr. The number of sailors within each mean sleep episode duration (SED) group is in brackets.

minority of competitors, these findings are in agreement with what is known from studies of partial sleep reduction in normal environments (26); namely, reductions by 1 to 2 hr from habitual TSTs for relatively long periods do not produce impairment in performance, and subjects appear to adapt with relative ease to such reductions. The lab finding of ease of adaptability has, therefore, now been confirmed in the field during extreme sustained work.

Finally, no correlation was found between the level of sleep fragmentation (i.e., the mean number of daily sleep episodes: TST/SED) and race performance. Race performance, therefore, seems more related to a combination of ability to sleep for short periods and capacity to reduce TST to some extent. Competitors tried to meet the need for staying awake and on watch as much and as frequently as they could, not by greatly reducing TST but mostly by fragmenting sleep. More specifically, under this type of CW demand, people who choose to have (and who are able to adapt to) sleep episodes ranging from 10 to 20 min to 1 hr will have higher probabilities of better performance, while obtaining a total of approximately 4.5 to 5.5 hr of sleep per day.

In conclusion, it appears that most highly motivated individuals (and/or people under particularly extreme task demands) adapt to ultrashort sleep/wake patterns without great difficulty, for relatively long periods, and can maintain (at least for the task at hand) high levels of behavioral efficiency when doing so.

PRACTICAL IMPLICATIONS AND SOME METHODOLOGICAL ISSUES RELEVANT TO THE DESIGN OF ULTRASHORT SLEEP SCHEDULES

We have seen that sleep has a polyphasic pattern in most animal species; some primitive and isolated tribes show such sleep patterns; under disentrainment conditions human sleep is polyphasic and similar to animal sleep; the adult human sleep/wake system has at least two major, and probably several minor, pulses of sleep propensity throughout the 24-hr span; humans adapt reasonably well to artificially (laboratory) imposed ultrashort sleep/wake schedules; and, finally, when motivated individuals are free to schedule sleep under extreme CW demands, most tend to adopt ultrashort sleep patterns and this procedure can improve performance compared with longer sleep episodes.

It appears, therefore, that humans are not an exception to the general polyphasic sleep rule observed across phylogeny. When obstructions are removed (social, occupational, or other pressures) or when specific stimuli appear, the endogenous polyphasic sleep tendency is revealed. The evidence strongly suggests that humans are able to adapt to fragmented sleep patterns. However, no systematic study has yet been conducted on the rational design and evaluation of polyphasic sleep/wake patterns (PSWPs) as potential solutions to sleep management under prolonged work requirements. The conceptual, methodological, and practical issues related to such a goal exceed the scope of this chapter. Instead, we conclude by presenting a brief discussion of the major aspects that should be considered when designing PSWPs, paying particular attention to the many issues that still await an answer.

SED, Frequency, and TST

What is the optimal sleep/wake schedule that a subject should adopt to maintain optimal performance during CW or quasicontinuous work scenarios? Longer sleep periods may have greater recuperative value, but the frequency of work episodes is greatly reduced. Shorter sleep periods may improve surveillance but fragment sleep: will this lead to greater cumulative fatigue or sleep deprivation? To what extent will the adoption of so-called anchor sleep, a term introduced by Minors and Waterhouse (36) (i.e., 2 to 4 hr of uninterrupted sleep taken during the primary or secondary sleep gates), combined with a number of shorter sleep episodes, be a reasonable compromise between the putative need for a longer sleep period for optimum recuperation and the CW requirement for quasicontinuous wakefulness to maintain optimal environmental surveillance?

The optimal sleep episode duration for a PSWP may vary among subjects, but it is probably a stable individual characteristic, as discussed previously. It appears that the subjectively derived ideal SED depends on a balance between the positive and negative effects of a nap. These effects may have their basis in the duration of sleep that provides maximal recuperation (positive) and minimal sleep inertia (negative). Which parameters correlate with an individual's ideal SED; which might permit its prediction? Concerning the minimum amount of uninterrupted sleep required for sleep to have beneficial effects on behavior, preliminary evidence suggests that this is in the order of 10 min, as reported by Bonnet (2) and confirmed by the solo sailor field study.

To what extent do PSWPs alter sleep need per 24 hr? Does the total amount of sleep required to function optimally increase, decrease, or remain unchanged by fragmenting sleep into ultrashort periods? It appears that, similar to what was observed in studies of gradual sleep restriction, TST during PSWPs can be reduced to 60% to 70% of baseline levels without interfering with performance effectiveness.

The recuperative value of sleep is generally considered to be a function of the duration of a sleep episode relative to prior wakefulness, its circadian placement, and/or the total amount of sleep obtained per 24 hr. The surprising adaptability to PSWPs and reduced TSTs observed both in the laboratory and field experiments suggests that a fourth parameter might be added: the number of times the sleep system is activated. It appears that every time the organism falls asleep beyond certain minimal durations (i.e., 10 min), some process related to recuperation may be activated, which may not have a linear relation with sleep episode duration. This would explain apparently contradictory (although very interesting) findings, such as those of Hartley (20), who reported that three 80-min naps were more recuperative than a 4-hr continuous sleep.

Sleep Onset and Awakening

One of the main limitations of a schedule of multiple naps involves the increased number of sleep onsets and awakenings, compared with normal monophasic nocturnal sleep. It has been suggested that there is a significant number of individuals who can control sleep onset and that a far greater number than is generally recognized have the potential for learning this skill (15,16). It seems evident that healthy subjects who are able to fall asleep quickly will also be more likely to adapt to PSWPs.

Concerning sleep termination, it is well-known that upon abrupt awakening from naps as well as nocturnal sleep, performance may be severely compromised for periods as long as 15 to 30 min. This ubiquitous sleep-inertia effect appears to depend on the depth of sleep; disorientation and cognitive decrements upon awakening are more severe if the sleep is comprised of NREM stages, especially SWS, and such sleep occurs in relatively close temporal proximity to sleep termination (10,11) (see Chapter 9). Would a PSWP increase the difficulty in functioning upon abrupt awakening? Would postarousal performance decrements simply be accumulated, or can a person learn to reduce sleep inertia and improve performance effectiveness shortly after arousal from naps? This is an important issue, since during CW scenarios critical performance may be required immediately upon awakening.

Temporal Factors

Several chronobiological issues presented in this chapter suggest that in designing a PSWP the timing of sleep periods should respect the underlying dynamics of biological rhythms. At least three major rhythms should be considered: (a) the sleep-propensity cycle (i.e., when transition from wakefulness to sleep is facilitated); (b) the sleep-efficiency cycle (i.e., when the same amount of sleep provides higher recuperative value)—note that the sleep-propensity and sleep-efficiency cycles are not necessarily synchronous; and (c) the sleep-inertia cycle.

It is also known that the human sleep system can be chronobiologically entrained, within certain limits, to artificial zeitgebers [cf. (57)]. Suppose that it were theoretically possible to program an ideal schedule for a given person, which would include several preferred times to nap. The effectiveness of such a chronobiologically appropriate schedule during CW would depend on the extent to which the CW regime permitted the schedule to be closely followed. Is regularity important to entrain the sleep system to a PSWP and maximize its efficiency? Should workers try to keep a proposed schedule as rigidly as possible to synchronize to the new pattern, or conversely, should they be trained to develop sleep flexibility? Finally, how would such an adoption of a multiphasic sleep pattern interfere with (or mutually interact with) the circadian (and other) biological rhythms used in determining the schedule?

Individual Differences

To what extent do individuals vary in their constitutional ability to adopt PSWPs? It is known that interindividual differences exist in a number of sleep-related characteristics. There are natural long and short sleepers, habitual nappers and nonnappers, morning types and evening types, and there are persons who adapt more quickly than others to shift work and jet lag. Are there biological, behavioral, or psychological markers that distinguish among people who will more easily adapt to a PSWP? What physiological parameters might correlate most with such an ability? Clearly, the development of such predictors would have potential utility in the selection of personnel for CW operations or for special sustained tasks.

Training, Education, and Environment

How might workers be trained for PSWPs? How many days are necessary for an individual to be considered fully entrained to a PSWP? After termination of the CW scenario, how quickly should workers return to their normal sleep patterns to avoid undesirable side effects?

Logistics for an optimum CW schedule involving PSWPs should also consider the following: provision for accessible, quiet, and comfortable facilities that permit frequent and efficient napping without loss of time; education of personnel to take advantage of available slack times to nap prophylactically; and training of individuals to learn to fall asleep more rapidly by, for example, the use of relaxation techniques. In brief, the utilization of PSWPs requires a framework in which sleep discipline becomes a legitimized and desirable activity.

This chapter has raised many questions and issues. A final essential one remains. How long can an ultrashort sleep/wake pattern be sustained by an individual without producing deleterious effects (if any) on health?

ACKNOWLEDGMENTS

Some of the field studies summarized in this chapter were partially supported by La Barca Laboratorio Research Society, Bologna, under Project FS 80/82/83. The author wishes to thank Drs. P. Marrino and M. Morosini for their assistance in data collection.

REFERENCES

1. Bennet, G. (1973): Medical and psychological problems in the 1972 singlehanded transatlantic yacht race. *Lancet*, 2:747–754.
2. Bonnet, M. H. (1986): Performance and sleepiness as a function of frequency and placement of sleep disruption. *Psychophysiology*, 23:263–271.
3. Broughton, R. (1975): Biorhythmic variations in consciousness and psychological functions. *Can. Psychol. Rev.*, 16:217–239.
4. Campbell, S. S., and Tobler, I. (1984): Animal sleep: A review of sleep duration across phylogeny. *Neurosci. Biobehav. Rev.*, 8:269–300.
5. Campbell, S. S., and Zulley, J. (1985): Ultradian components of human sleep/wake patterns during disentrainment. *Exp. Brain Res.*, 12(suppl.):234–255.
6. Carskadon, M. A., and Dement, W. C. (1975): Sleep studies on a 90-minute day. *Electroencephalogr. Clin. Neurophysiol.*, 39:145–155.
7. Carskadon, M. A., and Dement, W. C. (1977): Sleepiness and sleep state on a 90-min schedule. *Psychophysiology*, 14:127–133.
8. Curtis, G. C., and Fogel, M. L. (1972): Random living schedule: Psychological effects in man. *J. Psychiatr. Res.*, 9:315–323.
9. Dinges, D. F. (1986): Differential effects of prior wakefulness and circadian phase on nap sleep. *Electroencephalogr. Clin. Neurophysiol.*, 64:224–227.
10. Dinges, D. F., Orne, E. C., Evans, F. J., and Orne, M. T. (1981): Performance after naps in sleep-conducive and alerting environments. In: *Biological Rhythms, Sleep and Shift Work. Advances in Sleep Research, Vol. 7*, edited by L. Johnson, D. Tepas, W. P. Colquhoun, and M. Colligan, pp. 539–552. Spectrum, New York.
11. Dinges, D. F., Orne, M. T., and Orne, E. C. (1975): Assessing performance upon abrupt awakening from naps during quasi-continuous operations. *Behav. Res. Meth. Instr. Comp.*, 17:37–45.
12. Dinges, D. F., Orne, M. T., Orne, E. C., and Evans, F. J. (1980): Voluntary self-control of sleep to facilitate quasi-continuous performance. U.S. Army Medical Research and Development Command, Maryland (NTIS AD-A102264).
13. Dinges, D. F., Orne, M. T., Whitehouse, W. G., and Orne, E. C. (1987): Temporal placement of a nap for alertness: Contributions of circadian phase and prior wakefulness. *Sleep*, 10:313–329.
14. Dinges, D. F., Whitehouse, W. G., Orne, E. C., and Orne, M. T. (1988): The benefits of a nap during prolonged work and wakefulness. *Work and Stress*, 2:139–153.
15. Evans, F. J. (1977): Subjective characteristics of sleep efficiency. *J. Abnorm. Psychol.*, 86:561–564.
16. Evans, F. J., and Orne, M. T. (1976): Recovery from fatigue. U.S. Army Medical Research and Development Command, Maryland (NTIS AD-A100310).
17. Folkard, S., Minors, D. S., and Waterhouse, J. M. (1985): Chronobiology and shift-work: Current issues and trends. *Chronobiology*, 12:31–54.
18. Glenville, M., Broughton, R., Wing, A. M., and Wilkinson, R. T. (1978): Effects of sleep deprivation on short duration performance measures compared to the Wilkinson Auditory Vigilance Task. *Sleep*, 1:169–176.
19. Gough, H. G. (1957): *Manual for the California Psychological Inventory*. Consulting Psychologists Press, Palo Alto.
20. Hartley, L. R. (1974): A comparison of continuous and distributed reduced sleep schedules. *Q. J. Exp. Psychol.*, 26:8–14.

21. Haslam, D. R. (1982): Sleep loss, recovery sleep and military performance. *Ergonomics*, 2:163–178.
22. Herscovitch, J., Stuss, D., and Broughton, R. (1980): Changes in cognitive processing following short-term partial sleep deprivation and subsequent recovery oversleeping. *J. Clin. Neuropsychol.*, 2:301–319.
23. Horne, J. A. (1983): Mammalian sleep function with particular reference to man. In: *Sleep Mechanisms and Functions*, edited by A. Mayes, pp. 262–312. Van Nostrand Reinhold UK, London.
24. Husband, R. W. (1934): The comparative value of continuous versus interrupted sleep. *J. Exp. Psychol.*, 17:792–796.
25. Johnson, L. C., and Naitoh, P. (1974): *The Operational Consequences of Sleep Deprivation and Sleep Deficit* (AGARDograph no. 193). Technical Editing and Reproduction Ltd., London.
26. Johnson, L. C., Naitoh, P., and Moses, J. M. (1977): Variation in sleep schedules. *Waking Sleeping*, 1:133–137.
27. Kelley, J., Laughlin, E., Carpenter, S., Simmons, J., Sidoric, K., and Lentz, R. (1975): A study of ninety minute sleep cycles. *Stanford Rev.*, 3:1–5.
28. Kleitman, N. (1961): The nature of dreaming. In: *The Nature of Sleep*, edited by G. E. W. Wolstenholme and M. O'Connor, pp. 349–364. Churchill, London.
29. Kleitman, N. (1982): Basic rest-activity cycle: 22 years later. *Sleep*, 5:311–317.
30. Kristal, M. B., and Noonan, M. (1979): Note on sleep in captive giraffes (*giraffa camelopardalis reticulata*). *S. Afr. J. Zool.*, 14:108.
31. Lavie, P. (1985): Ultradian rhythms: Gates of sleep and wakefulness. In: *Ultradian Rhythms in Physiology and Behavior*, edited by H. Schulz and P. Lavie, pp. 148–164. Springer-Verlag, Berlin.
32. Lavie, P. (1986): Ultrashort sleep-waking schedule. III. Gates and "forbidden zones" for sleep. *Electroencephalogr. Clin. Neurophysiol.*, 63:414–425.
33. Lavie, P., Gopher, D., and Wollman, M. (1987): Thirty-six hour correspondence between performance and sleepiness cycles. *Psychophysiology*, 24:430–438.
34. Lavie, P., and Scherson, A. (1981): Ultrashort sleep-waking schedule. I: Evidence of ultradian rhythmicity in "sleepability." *Electroencephalogr. Clin. Neurophysiol.*, 52:163–174.
35. Lubin, A., Hord, D., Tracy, M. L., and Johnson, L. C. (1976): Effects of exercise, bedrest and napping on performance decrement during 40 hours. *Psychophysiology*, 13:334–339.
36. Minors, D. S., and Waterhouse, J. M. (1980): Does 'anchor sleep' entrain the internal oscillator that controls circadian rhythms? *J. Physiol.*, 308:92P–93P.
37. Morgan, B. B., Brown, B. R., and Alluisi, E. A. (1974): Effects of sustained performance of 48 hours of continuous work and sleep loss. *Hum. Factors*, 16:406–414.
38. Moses, J. M., Hord, D. J., and Lubin, A. (1975): Dynamics of nap sleep during a 40 hour period. *Electroencephalogr. Clin. Neurophysiol.*, 39:627–633.
39. Moses, J. M., Naitoh, P., and Johnson, L. C. (1978): The REM cycle in altered sleep-wake schedules. *Psychophysiology*, 15:569–575.
40. Mukhametov, L. M., and Polyakova, I. G. (1981): EEG investigation of the sleep in porpoises (*phocoena phocoena*). *J. High. Nerv. Act. USSR*, 31:333–339.
41. Mukhametov, L. M., Supin, A. Y., and Polyakova, I. G. (1977): Interhemispheric asymmetry of the electroencephalographic sleep patterns in dolphins. *Brain Res.*, 134:581–584.
42. Mullaney, D. J., Kripke, D. F., and Fleck, P. A. (1983): Sleep loss and nap effects on sustained continuous performance. *Psychophysiology*, 20:643–651.
43. Naitoh, P. (1969): Sleep loss and its effects on performance. Report 68-3. U.S. Naval Medical Neuropsychiatric Research Unit, San Diego, CA.
44. Naitoh, P. (1981): Circadian cycles and restorative power of naps. In: *Biological Rhythms, Sleep and Shift-Work. Advances in Sleep Research, Vol. 7*, edited by L. C. Johnson, D. Tepas, W. P. Colquhoun, and M. Colligan, pp. 553–580. Spectrum, New York.
45. Naitoh, P., Englund, C., and Ryman, D. H. (1982): *Restorative Power of Naps in Designing Continuous Work Schedules*. Report 82-25. Naval Health Research Center, San Diego, CA.
46. Naitoh, P., Englund, C., and Ryman, D. H. (1983): Extending human effectiveness during sustained operations through sleep management. Report 83-13. Naval Health Research Center, San Diego, CA.

47. Opstad, P. K., Ekanger, R., Nummestad, M., and Raabe, N. (1978): Performance, mood, and clinical symptoms in men exposed to prolonged, severe physical work and sleep deprivation. *Aviat. Space Environ. Med.,* 49:1065–1073.
48. Petre-Quadens, O. (1983): The anthropology of naps. In: *Proceedings of 4th International Congress of Sleep Research, Bologna,* p. 89.
49. Pilleri, G. (1979): The blind Indus dolphin (Platanista Indi). *Endeavor,* 3:48–56.
50. Stampi, C. (1985): Ultrashort sleep-wake cycles improve performance during one-man transatlantic races. In: *Sleep 1984,* edited by W. P. Koella, E. Ruther, and H. Schulz, pp. 271–272. Gustav Fischer Verlag, Stuttgart.
51. Stampi, C. Polyphasic sleep strategies improve prolonged sustained performance: A field study on 99 sailors. *Work and Stress (in press).*
52. Strogatz, S. H. (1986): *The Mathematical Structure of Human Sleep-Wake Cycle. Lecture Notes in Mathematics No. 69.* Springer-Verlag, Berlin/New York.
53. Taub, J. M., and Berger, R. J. (1973): Performance and mood following variations in length and timing of sleep. *Psychophysiology,* 10:559–570.
54. Deleted at proofs.
55. Webb, W. B., and Agnew, H. W., Jr. (1975): Sleep efficiency for sleep-wake cycles of varied length. *Psychophysiology,* 12:637–641.
56. Weitzman, E., Nogeire, C., Perlow, M., et al. (1974): Effects of a prolonged 3-hour sleep-wakefulness cycle on sleep stages, plasma cortisol, growth hormone and body temperature in man. *J. Clin. Endocrinol.,* 38:1018–1030.
57. Wever, R. A. (1979): *The Circadian System in Man.* Springer-Verlag, Berlin.

Sleep and Alertness: Chronobiological,
Behavioral, and Medical Aspects of
Napping, edited by
D. F. Dinges and R. J. Broughton.
Raven Press, Ltd., New York © 1989.

9

Napping Patterns and Effects in Human Adults

David F. Dinges

Unit for Experimental Psychiatry, The Institute of Pennsylvania Hospital and University of Pennsylvania, Philadelphia, Pennsylvania 19139-2798, USA

Napping is a common practice among many, but not all, healthy human adults. What influences the decision to nap? Is it increased sleep need, chrono-biologic pressure, postprandial sleepiness, ambient temperature, disturbed nocturnal sleep, age, gender, physical condition, personality features, or some combination of these factors? When and where are naps typically taken? How long do they last and what kind of sleep is obtained? What effect does a nap have on mood, behavior, and subsequent sleep? In short, what role does napping play in the overall sleep/wake economy of human functioning and what factors mediate that role? This chapter addresses such issues.

Because the focus is on understanding characteristics, timing, causes, and effects of naps in human adults, the review will be concerned with studies of napping by adults who have their major sleep period located during the nocturnal phase of the nycthemeron. Studies of napping by adults who are deprived of a major sleep period because of prolonged work or have their major sleep period displaced to other circadian phases owing to shift work are reviewed in subsequent chapters, along with cultural differences in nap-ping prevalence and napping among patients with sleep disorders. This first section reviews studies that have attempted to assess napping behavior among adults who otherwise sleep at night and remain awake during the day.

INCIDENCE AND PREVALENCE OF NAPPING

Many of the early widely cited studies of sleep behavior made no mention of napping (18,72–74). The reasons for this are unclear but probably have more to do with benign neglect on the part of the investigators than with the actual absence of naps in the populations studied. Fortunately, there were two comprehensive early studies of sleep and wakefulness that included nap behavior. Kleitman and colleagues (58) had 36 adults (ages 19–55 years) keep daily sleep logs for an average of 179 days each. Tune (109), in the most extensive study using daily sleep logs ever conducted (approximately 26,000 days recorded), had 509 adults (ages 20–80 years) complete sleep logs for an average of 52 days each [see also (107,108) for subsets of these data]. Although neither study found napping to be a widespread daily activity in the samples investigated, it also was not an uncommon occurrence. Kleitman et al. (58) observed that although 39% of their U.S. subjects never napped, 33% napped at least once a week, and 11% napped four or more times per week. Tune (109) found that among adults in the United Kingdom the mean number of naps per week was just less than one.

Since these benchmark investigations, there has been a total of 23 studies of sleep/wake behavior that included assessments of napping; most took place after 1970. Table 1 summarizes the studies with regard to data on napping. The majority ($n = 15$) were surveys. Overall, subjects were studied in nine countries and five areas of the continental United States. A grand total of 10,440 adults participated; 9,492 (91%) were surveyed, 928 (9%) completed daily logs, and 20 had ambulatory monitoring. The age range of participants was 15 to 94 years, but the majority of studies ($n = 14$) were concerned with young adults (ages 17–35 years). Although not all reports contain similar information on napping parameters, enough of them provide data in basic categories to permit a composite picture. Within these categories there was surprisingly good overall agreement between the survey results and findings from the daily log studies, despite the fact that logs are more valid and reliable measures of sleep/wake activity.

Ten studies contained information on the mean number of naps per week. Means for the four survey studies ranged from 1.6 to 1.9 naps per week (Grand $M = 1.75$), and those for the six daily log studies ranged from 0.9 to 3.0 naps per week ($M = 1.81$). Thus, one to two naps per week per subject appears to be the average across all studies. This result was not inflated by some subjects napping more than once a day, as multiple daily napping was not observed in any study. There were wide differences among studies, however, in the proportion of subjects who actually napped at least once a week. The figure ranged from 36% to 80% for the 12 surveys and from 33% to 84% for the six daily log studies. Interestingly, the overall mean percentage was 61% for both types of studies. Thus, the majority of subjects nap weekly or more frequently. The remainder nap less frequently or not at all.

Of the 11 surveys that reported information on the proportion of subjects not napping, the range was 7% to 65%, relative to a range of 10% to 39% for the five daily log studies reporting this parameter. Again, however, the

TABLE 1. Results from studies inquiring about napping when evaluating sleep patterns in adult populations

Study	Type	Location	Number of subjects	Age range (years)	Mean naps/week	% napping weekly	Mean nap length[a]	% not napping	% napping freq.[b]	Mean noct. sleep[a]	Mean 24-hr sleep[a]
Kleitman et al. (58)	Daily log	U.S.-Illinois	36	19–55	—	33%[c]	—	39%	11%	7.47	—
O'Connor (80)	Survey	U.S.-Florida	157	17–29	—	—	1.00[d]	65%	7–28%	7.30	—
	Survey	U.S.-Florida	79	30–49	—	—	1.00[d]	—	—	7.23	—
Tune (109)	Daily log	U.K.-England	509	20–80	0.9[c]	—	0.58[c]	—	—	7.43	7.50[c]
Johns et al. (54)	Survey	Australia	249	20–24	—	53%	0.67[c]	—	—	7.90[c]	—
Taub (95)	Survey	Mexico	257	20–89	—	79%	1.32	21%	79%	8.16	9.12
Lawrence (62)	Survey	U.S.-Oklahoma	505	19–23[c]	1.8[c]	42%–79%	1.19[c]	—	4%	7.46	7.86[c]
Lawrence and Shurley	Survey	U.S.-Oklahoma	196	20[e]	—	76%	1.25[c]	—	7%	7.85[c]	—
(64)	Survey	Guatemala	196	27[e]	—	52%	1.25[c]	—	27%	7.47[c]	—
White (118)	Survey	U.S.-Florida	102	18–27	1.7	—	1.37[c]	—	—	7.85	7.90
	Daily log	U.S.-Florida	102	18–27	1.6	—	1.62[c]	—	—	7.50	8.08
Webb (110)	Daily log	U.S.-Florida	89	18–24[c]	1.5[c]	58%–84%[c]	—	16%	42%	7.67	—
Evans and Orne (39)	Survey	U.S.-Pennsyl.	430	18–37	—	61%	1.21	7%	14%[c]	7.43	—
Evans and Orne (40)	Survey	U.S.-Pennsyl.	469	18–35[c]	—	55%	1.22	11%	12%[c]	7.33	—
Kunken (60)	Survey	U.S.-New York	471	18–22[c]	1.9	72%	1.10	28%[c]	17%	—	6.90
Ogunremi (81)	Survey	Nigeria	194	21[e]	—	80%	1.50[c]	—	80%[c]	6.78[c]	—
Dinges et al. (31)	Survey	U.S.-Pennsyl.	956	18–31	1.6	55%	1.50	12%	13%[c]	7.29	—
	Daily log	U.S.-Pennsyl.	72	18–31	2.3	80%	1.22	10%	15%[c]	7.09	7.52
Webb (113)	Daily log	U.S.-Florida	80	50–60	1.6[c]	55%[c]	1.14[c]	26%	19%	6.98[c]	—
Okudaira et al. (82)	Ambulant	Ecuador	20	80–94	—	60%	—	—	60%[c]	—	—
Soldatos et al. (88)	Survey	Greece	1,068	19–65	—	47%	1.32	—	24%	—	—
Lugaresi et al. (69)	Survey	Italy	4,170	15–94	—	36%[f]	—	—	—	7–8	—
Webb and Aber (114)	Daily log	U.S.-Florida	40	63[e]	3.0[c]	65%[c]	1.0–1.9	25%	50%[c]	8.16[c]	—

[a] Hours.
[b] Defined as four or more naps/week.
[c] Calculated from information in report.
[d] Mode.
[e] Mean.
[f] Occasionally.

overall mean percentage of subjects not napping across studies was virtually identical for surveys (24%) and daily logs (23%). At the other end of the spectrum were figures for subjects who napped frequently (i.e., at least 4 per week). Across studies, the range was 4% to 80% for 11 surveys (M = 27%) and 11% to 60% for five daily log studies (M = 33%). The greater variability among studies in this dimension was undoubtedly attributable to the fact that both siesta (e.g., Mexico) and nonsiesta cultures (e.g., United States) were included. Overall, the studies in Table 1 suggest that napping appears to be a common and at times prevalent feature of adult sleep/wake activity.

Duration

The duration of naps observed across studies was remarkably consistent. The range of mean nap length was 0.67 to 1.50 hr for 14 surveys and 0.58 to 1.60 hr for five daily log studies. The overall mean was 1.21 hr for each class of study. Virtually no study found nap durations of less than 0.25 hr (15 min) or greater than 2.0 hr to be common. For example, O'Connor (80) observed that 68% of naps reported in surveys of younger adults (17–49 years) were between 30 min and 1.5 hr in duration, but only 11% were 15 min or less [see also (31,62)]. Similarly, Webb (113) found that 65% of naps recorded by older adults (50–60 years) completing 2-week daily sleep diaries were between 15 and 119 min in duration, but only 5% were less than 15 min [see also (114)].

It would appear that the apocryphal short catnap (less than 15 min)—long attributed to some successful persons [e.g., Stegman (92), the late Salvador Dali]—is either not reported in surveys and sleep logs or not commonly practiced by most persons who do nap. There have been few studies of the effects of such very short naps on performance and mood (see below), probably because sleep of a few minutes' duration comprises mostly stage 1, which has not been shown to enhance functioning in sleep-deprived subjects. On the other hand, many a parent of a 2- or 3-year-old child has observed that a 10-min sleep by the child during an automobile ride home can prevent a longer nap from being taken when the child is home and in bed. There is a need for studies of short-duration naps (and nappers) to provide information on their effects.

Age and Sex

In general, the incidence of napping increases in adulthood with age (see Chapter 3) for both siesta and nonsiesta cultures. In the comprehensive study of British adults, Tune (107,109) found increases in both the frequency and duration of naps from the third to eighth decade of life. Taub (95) also observed that the frequency and duration of naps taken by Mexican adults decreased slightly between 20 and 50 years of age but increased thereafter, up to 90 years. Similarly, Soldatos et al. (88) reported that among Greeks the mean age of nappers was significantly greater than that of nonnappers. Although they did not describe the relationship between napping and age in their study of Italians, Lugaresi et al. (69) did note that reports of post-prandial sleepiness increased steadily across adulthood, up to 64 years of age.

Gender differences in nap behavior are less certain to exist than are age-related changes. Although studies of adults, especially young adults, have observed that women report having somewhat more regular sleep/wake schedules than men and having sleep onset and offset phase-advanced by 0.5 to 1.0 hr relative to men (31,42,62), differences have rarely been found in studies of daytime napping. Two survey studies (United States and Greece) suggested that there are slightly more men than women among habitual nappers relative to their proportions in the samples surveyed (31,88). Soldatos et al. (88) also noted that women reported averaging slightly shorter nap durations than men. The most extensive gender-related differences observed to date were reported by Webb (113) in a daily log study of 50- to 60-year-old employed adults. Men napped more frequently and for a shorter duration than women, resulting in comparable amounts of nap sleep between sexes. Most studies, however, have not observed significant differences between men and women at any age in either the incidence or duration of naps (89,95,108,113,114,118). It would appear, therefore, that sex has little to do with the tendency to nap and gender differences present at any given time in a population probably have more to do with the regularity of sleep/wake cycles and other behavioral parameters than with gender.

Culture and Opportunity

There are obvious cultural differences in the extent to which daytime napping is encouraged (79) (see Chapter 12). This is evidenced among the studies in Table 1 when considering the range of percentages of persons napping frequently (4%–80%). Although geographical location and socio-cultural factors play major roles in the extent to which persons will nap during the daytime, it is also possible to examine the influence of work itself on nap behavior within a culture.

Unlike the major circadian sleep period, voluntary napping appears to be easily suppressed, if no opportunity to nap is available or there are sanctions against napping. Daytime work demands probably account for the reduced levels of napping in nonsiesta cultures more than does a presumed inability to nap, a need to nap, or an inclination to nap. Evidence for this, however, is limited to studies of college students. In surveys, Dinges et al. (31) and

Daiss et al. (22) found that the highest percentage (46% and 37%, respectively) of college student nonnappers cited "no time to nap" and "napping interferes with work or study" as the primary reasons naps were not taken. Lawrence (62) observed among the college students she surveyed that twice as many nonnappers (32%) as nappers (16%) held jobs while in school. She also found that more U.S. college students (76%) reported napping than did Guatemalan students (52%), despite the latter's siesta culture and proximity to the equator, but that many more Guatemalan students (79%) were employed than were U.S. students (30%) (64). Although limited to the college population, such data suggest that daytime work can mask the desire and ability to nap.

Ironically, daily work can also contribute to the tendency to nap. In the only study of work and sleep/wake behavior in older adults, Webb and Aber (114) compared sleep diaries of 20 employed and 17 retired subjects (mean age 63 years). Retirees recorded obtaining significantly more nocturnal sleep on weekdays than the employed ($p < 0.01$); this accounts for the high average nocturnal sleep duration for this study (Table 1). There was also a trend for retirees to report fewer naps than the employed, which may have resulted from the increased nocturnal sleep associated with retirement. Thus, to the extent that work requirements reduce the amount of nocturnal sleep obtained, they may increase the need for compensatory daytime napping; but whether naps are taken will depend on the opportunity to do so.

Climate and Season

Climate, particularly relationship to tropical zones, has long been anecdotally associated with siesta. The data in Table 1 provide some evidence for this. The percentage of persons napping frequently (i.e., four naps or more per week) was highest (60%–80%) in studies of cultures near the equator that permit a daily siesta period: Mexico (95), Nigeria (81), Ecuador (82). Conversely, the lowest incidence of napping reported among the 23 studies in Table 1 was observed by Tune (109) for persons living in England, which is located farthest from the equator (50°–55° N) among the sites listed.

The heat of the midday sun has traditionally been assumed to be the reason that napping is so prevalent in siesta cultures. Webb (111) suggested that it is the reason for the midday sleep tendency evident in many aspects of human functioning (see Chapters 4 and 12). But not every equatorial culture allows siesta time, and among those that do, napping is not universally the most prevalent activity during siesta (64). Moreover, both the incidence and duration of daytime napping in siesta cultures appears to be independent of wide seasonal variations in ambient temperature for a majority of persons napping (95). The association between midday heat and napping may exist simply because the ambient temperature makes wake activity (work) difficult, especially in agrarian cultures. It is doubtful that heat actively promotes sleep or sleepiness, however, because temperature extremes are more likely to disturb sleep than to enhance it (67). On the other hand, by forcing cessation of work at midday, tropical heat may permit daytime sleepiness to express itself in the form of naps. Chronobiological modulation and shifts in basal sleep need may underlie the daytime sleepiness.

To the extent that napping is more prevalent when nocturnal sleep is reduced, variations in nocturnal sleep associated with seasons should result in changes in daytime napping. Only the study by White (118) made such comparisons in a nonsiesta culture (i.e., college students in Florida). Both sleep log and survey data revealed that significantly more nocturnal sleep was obtained during winter relative to summer months, but significantly fewer daytime naps were taken in the winter. Sleep per 24 hr was therefore comparable in summer and winter seasons, and naps appeared to compensate for nocturnal sleep loss. It is far more likely that these seasonal differences came about as a result of changes in work/rest (school versus no school) schedules associated with seasons than as a function of climatic changes.

Nap Context

There are virtually no systematic data on the contexts in which napping is most common, including nap environment (e.g., light and noise level), nap contingencies (e.g., mode of awakening), and napper disposition (e.g., degree of volition for naps). To what extent do the majority of naps resemble nocturnal sleep, which typically involves a voluntary desire to sleep, presleep rituals, a change of clothing, and sleeping on a bed in a dark quiet room?

Perhaps the most important issue is the extent to which daytime napping may be involuntary—a person falling asleep when intending to remain awake. Involuntary napping has been observed among night-shift workers (106) (see Chapter 10) and is of course common in sleep disorders (Chapter 13). It is not known what proportion of napping reported in Table 1 was involuntary, or even whether subjects actually report involuntary daytime sleep episodes in surveys and logs. Clearly, more data are needed on the proportion of involuntary daytime naps among adults, their dissimilarities from naps that are voluntarily taken, and their prevalence as a function of age, opportunity, and other factors relevant to sleep/wake dynamics. This issue, more than any other, highlights the need for studies involving ambulatory monitoring of sleep/wake activity in conjunction with self-report measures.

The environment in which a nap is taken is important because it affects both the nap sleep infrastructure and the effects of the nap on subsequent wakefulness. Two laboratory studies have addressed this issue. One involved 2-hr naps on chairs and on beds in the morning, afternoon, and evening (52), and the other involved 1-hr afternoon naps in sleep-conducive (on a bed in a dark quiet room) and alerting (on a chair in a lighted room with moderate noise) environments (26). In both studies, naps taken on comfortable chairs compared with beds were associated with significant reductions in time spent in slow wave sleep (SWS) and reciprocal increases in stage 1 sleep. In the 2-hr nap study, the chair also resulted in a significant reduction in the percentage of rapid eye movement (REM) sleep and an increase in wakefulness (52), whereas the 1-hr nap study found no differences in either the ability to fall asleep or total sleep time (TST) between sleep-conducive and alerting environments (26). The shifts from deep to light sleep were especially prominent for afternoon and evening (versus morning) naps (52) and appeared to result from increased fragmentation of sleep continuity (26).

These effects were evident in the nap sleep of both habitual nappers and nonnappers, and each group rated the nap in the alerting environment as less satisfying and less refreshing than the one taken on a bed in a dark quiet room (31). For young adults, therefore, naps on beds in sleep-conducive environments result in deeper sleep and greater postnap subjective benefits. From what little is known about contextual effects on nap sleep, it appears that nap sleep infrastructure is affected by environmental factors in the same manner that nocturnal sleep can be disturbed by such factors.

NAP SLEEP INFRASTRUCTURE

Historically, there have been four issues addressed in studies of daytime nap sleep infrastructure: (a) whether daytime nap sleep stage sequencing is a temporal miniaturization or a recycling of nocturnal sleep; (b) whether time-of-day (circadian phase) affects nap sleep stages and sleep efficiency; (c) whether sleep infrastructure changes as a function of variations in nap duration; and (d) whether sleep stages are related to the behavioral effects of naps or to their effects on nocturnal sleep infrastructure. This fourth question will be dealt with in the final sections of the chapter, where the full range of nap effects is discussed.

Laboratory studies of napping conducted from 1963 to date have included polysomnographic monitoring of sleep stages, resulting in a wealth of data on sleep infrastructure of naps. Results have been remarkably consistent and straightforward. Daytime naps are not miniatures of nocturnal sleep compressed in time and do not involve a recycling of nocturnal sleep. Rather, nap sleep infrastructure depends on the time of day the nap is taken and the duration of the nap, assuming that the person has slept the night before.

The most robust finding is that REM sleep shows a pronounced circadian pattern (see Chapter 5 for more extensive discussions of the chronobiology of sleep stages). Latency to REM sleep is shorter and REM sleep amount is greater when naps are taken in the morning between 0800 and 1200 hr relative to the afternoon or evening hours (34,35,55,56,97). Thus, morning naps tend to have sleep infrastructures similar to the final hours of nocturnal sleep (116). Assuming comparable nap durations, afternoon and evening naps have more SWS than do morning naps, with evening naps having the most SWS and the least REM sleep (35,71,76). Afternoon naps have sleep infrastructures unlike both the beginning and end portions of nocturnal sleep (55,115).

Whether afternoon and evening naps will contain REM sleep in persons who have normal nocturnal sleep depends on the length of the nap (62,96). If the nap is approximately 30 min in duration, REM sleep does not occur in otherwise healthy persons. If it is 1 hr in duration, REM sleep may occur, but not in all subjects. If the nap is 2 hr or more in duration, REM sleep is very likely [and there obviously will be greater amounts of nonrapid eye movement (NREM) sleep stages as well]. However, even when REM sleep does appear in naps taken between noon and midnight, it rarely occurs before 50 min after sleep onset (55,56,62,66,71,97). To the extent that most adult daytime napping occurs between noon and midnight (see below) and averages approximately 1 hr in duration (Table 1), REM sleep is not likely to compose a large portion of naps. Whether, as in noctural sleep, it is the stage of sleep from which naps are most often spontaneously terminated remains to be determined.

Although naps have been anecdotally reported to be more efficient than nocturnal sleep in terms of their benefits for wakefulness relative to the time invested, they are actually less efficient than nocturnal sleep in terms of the TST obtained as a proportion of the total time in bed (TIB) attempting to nap (i.e., TST/TIB). Nocturnal sleep efficiency is almost always more than 80% and often more than 90%. Nap sleep efficiency at any time of day (0700–2300 hr) is rarely more than 80%. In general, the shorter the nap is, the lower its efficiency. Sleep efficiency for 1-hr afternoon or evening nap periods is typically between 55% and 70% (4,26,38,45,62). It is somewhat higher (65%–80%) for 2-hr naps (24,41,63). Daytime nap sleep efficiency decreases further with age (66). The lowered sleep efficiency of naps relative to nocturnal sleep appears to be a result of differences in sleep duration rather than in sleep onset or sleep consolidation. That is, for most persons studied, there is no apparent difficulty in going to sleep or sustaining sleep during daytime naps. There is, however, a preferred time of day for naps.

TEMPORAL PLACEMENT OF NAPS

As discussed elsewhere, daytime napping challenges the notion of a monophasic sleep/wake cycle (see Chapter 5) and casts doubt on a common assumption, based largely on results by Taub and Berger (99,100), that any change in the timing of the typical monophasic sleep/wake cycle may be detrimental. Aside from the widespread occurrence of naps during the daytime supplemental to nocturnal sleep, there are serious questions as to the magnitude of effects resulting from modest shifts in the temporal placement of nocturnal sleep (8,49,59,112). In fact, as will become evident, modest shifts in the timing of nocturnal sleep appear to be associated with the occurrence of naps, and, like nocturnal sleep, daytime naps reveal phase-preferred positions in the circadian cycle.

Napping has been assumed to most likely occur in the afternoon, after the midday meal. Very few studies have evaluated the timing of naps, however. Tune (107) reported the first systematic data on the timing of naps in adults aged 20 to 80 years, showing that most napping in England occurred between 1200 and 1800 hr (especially 1400–1600 hr). Surveys of college students by Lawrence (62) in the United States, Taub (95) in Mexico, Lawrence and Shurley (64) in Guatemala, and Ogunremi (81) in Nigeria also revealed the afternoon as the time when the majority of young adults (60%–80%) napped. Napping among college students has received a great deal of attention not only because this is an adult population that is easily studied, but also because the typical student lifestyle allows considerable flexibility in work/rest schedules.

The high incidence of napping among college students in part reflects the opportunity this group has, relative to many noncollege populations, for daytime sleep. The most extensive data on the timing of naps and nocturnal sleep have come from intensive study of college students by Dinges et al. (31,32). Figure 1 displays the distribution of almost 3,000 sleep onsets during the 24-hr period by 58 regular nappers and 14 regular nonnappers who completed detailed daily sleep logs for an average of 35 days each. No distinction was made for length of sleep (i.e., major sleep period versus nap sleep). It is obvious that there are two prominent 4 to 5-hr zones in which 85% of sleep onsets occur. The first is the nocturnal zone (2300–0400 hr) and the second can be considered the nap zone (1400–1800 hr), since nappers had 15% of their sleep onsets at this time. The fact that another 15% of napper's sleep onsets occurred diffusely at times other than these two zones is consistent with other data showing that college students have highly variable sleep/wake cycles (117).

Figure 2 displays the temporal distribution of all logged sleep onsets and offsets on nap days for 22 habitual nappers during a 35-day period (32). Naps are evident by the slight (1–2 hr) phase shift between the sleep onset and offset distributions in the center of the figure. They appear to be located midway in the wakefulness period. The median time for nap sleep onset (1630 hr) was approximately 8 hr after the median time for nocturnal sleep offset (0830 hr), and the median time for nap sleep offset (1745 hr) was 8 hr before the median time for nocturnal sleep onset (0145 hr). It is noteworthy that the two sleep onset zones correspond closely to the "primary and sec-

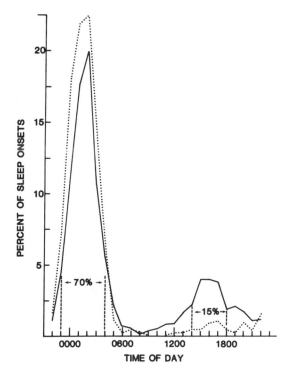

FIG. 1. Frequency distribution of all sleep onset times logged during a 35-day period by 72 healthy young adults, 58 regular nappers and 14 nonnappers. (···) 434 onsets (nonnappers); (—) 2,562 onsets (nappers).

ondary sleep gates" observed by Lavie (61) in ultradian studies of sleep/wake structure (see Chapter 6). Moreover, Fig. 2 reveals two time zones during which sleep rarely occurred, at approximately 1200 and 2300 hr, respectively. These are similar to the "forbidden zones" for sleep of Lavie (61) and the "wake-maintenance zones" of Strogatz (94).

The timing of naps in the college population also coincides with reports of sleepiness in this group. In a recent survey of inhibitory motor experiences among 707 students (78), no mention was made of napping, but subjects were asked if they had difficulty staying awake during the day and if so, the hours of the day when this occurred. The question was answered affirmatively by 39% of the subjects. Figure 3 illustrates the pattern reported. Of nearly 600 hr mentioned by subjects, the 5-hr afternoon period from 1300 to 1800 hr composed more than 62% of all responses of the times of greatest sleepiness. The modal response was 1600 hr, which is also the modal time for nap sleep onsets observed in other studies in this age group (31), the time when sleep tendency as assessed by sleep latency tests is most likely to be increased (see Chapter 4), and the time that human accidents and errors display a secondary peak (25,75).

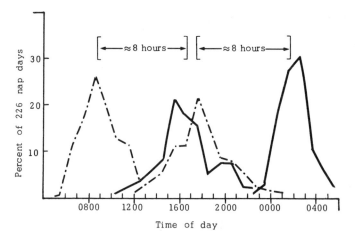

FIG. 2. Frequency distribution of all sleep onset (—) times and sleep offset (·—·) times logged on nap days only, during a 35-day period by 22 healthy young adult nappers. Brackets identify the median times of each distribution. Nap onsets during the afternoon occur approximately 8 hr after nocturnal sleep offsets, and nap offsets occur approximately 8 hr before nocturnal sleep onsets.

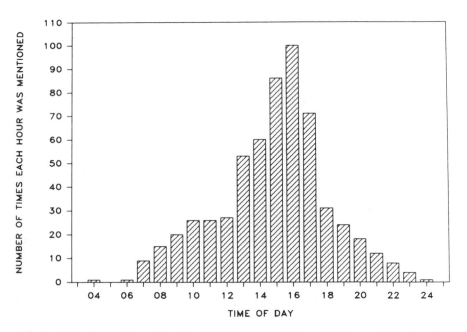

FIG. 3. Histogram of times identified by 276 young adults as the hours of the day they have difficulty staying awake.

Although such data suggest that napping is caused by increased sleepiness in the mid- to late afternoon, it is possible that the occurrence of naps during these hours represents time available for sleep rather than chronobiological tendency to sleep at this time (circadian phase). This issue was evaluated by Dinges et al. (28) using the 35-day daily sleep log data of 58 regular nappers (Fig. 1). Data for weekdays were separated from data for weekends. In college students, as in most other adult groups, weekends are considerably more free of environmental restrictions on time usage than are weekdays. Students obtained more nocturnal sleep on weekends and phase-delayed their nocturnal sleep by 30 min to 2 hr relative to weekdays (28). Similar nocturnal sleep increases and phase delays on weekends have been reported in other studies of adults (42,53,54,62,64,81,93,113,118).

Although some trends toward differences between weekday and weekend naps have been reported (97,113), the comparisons made by Dinges et al. (28) have revealed a difference that is particularly important. In their study, 58 regular nappers logged 471 weekday naps and 201 weekend naps. When calculated as a proportion of the number of weekdays and weekend days in the 35-day diary period, however, napping was as likely to occur on weekends (32% of the time) as on weekdays (33%). This was unexpected since the students also averaged more nocturnal sleep and phase-delayed nocturnal sleep on weekends compared with weekdays. One might therefore expect that they would be napping less on weekends. Table 2 displays weekday and weekend sleep times for night and daytime nap sleep reported by these subjects. Although an average of 44 min more of nocturnal sleep was obtained on weekends, weekend naps were not significantly longer, unlike Webb's (113) observation in older adults.

TABLE 2. *Mean (SD) times for nocturnal sleep (prior to nap days) and nap sleep on weekdays and weekends in 58 nappers during a 35-day period*

	Weekdays (471 naps)	Weekends (201 naps)	Mean diff. (min)	Paired t56
Nocturnal TST	6:40[a] (1:45)	7:24[a] (1:47)	44	4.78[b]
Nap TST	1:10 (0:52)	1:15 (0:53)	5	1.10
Midnocturnal sleep time	5:03 (1:46)	5:50 (1:30)	47	5.67[b]
Midnap sleep time	16:37 (3:00)	17:13 (2:44)	36	2.57[c]
Time from midnocturnal to midnap sleep	11:43 (2:52)	11:25 (2:44)	18	1.22

TST, total sleep time.
[a] Hr:min.
[b] $p < 0.0001$.
[c] $p < 0.02$ (two-tailed).

It is important to note that the nocturnal sleep times presented in Table 2 were derived exclusively from sleep on nights immediately prior to nap days. The statistically significant phase shift in nocturnal sleep time from weekday to weekend is evident in the mean difference of 47 min in mid-nocturnal sleep times (time of sleep onset subtracted from time of awakening, divided by 2). Surprisingly, midnap sleep time also revealed a significant phase delay, averaging 36 min from weekday to weekend ($p < 0.02$). The shift in both night and day sleep resulted in no significant difference in the time from midnocturnal to midnap sleep. On both weekdays and weekends, this value averaged approximately 11.5 hr.

Thus, it appears that daytime napping by college students in the midafternoon is not simply the result of environmental restriction, but rather is more or less temporally bound to a phase of the circadian cycle and the placement of nocturnal sleep. Using midsleep as a marker, naps occurred nearly 180° (24 hr = 360°) after nocturnal sleep and, together with night sleep, formed a biphasic sleep/wake cycle. Thus, the temporal placement of a nap was less bound by sidereal time than by biological time. These results are consistent with studies of SWS reappearance in extended sleep (44), and with chronobiological models of sleep/wake regulation that posit an endogenously generated second sleep phase midway in the circadian cycle (12) (see Chapter 5).

Finally, the data for weekday versus weekend naps suggest that the precise hours of the afternoon when naps will occur probably depend on the timing of the prior nocturnal sleep period. For example, college students in general have a relatively more variable and delayed nocturnal sleep phase, often going to sleep after 0200 hr (31), and their nap phase occurs in mid- to late afternoon (1400–1800 hr). Early afternoon naps (1300–1500 hr) are more commonly reported in working populations and the elderly, who have less phase-delayed nocturnal sleep relative to students (95).

FACTORS PRESUMED TO PRECIPITATE NAPPING

Chronobiological Regulation

The temporal placement of nap sleep in the afternoon appears to be the result of a biphasic biological rhythm in sleep regulation. In addition to the sleep parameters reviewed elsewhere in this volume, many aspects of human performance and mood also show transient afternoon shifts consistent with increased sleepiness (7,20,25,29,31,75,87,98). But is this increased sleep propensity in the midday solely a function of chronobiological regulation? If so, why is napping not more prevalent than the average (across studies in Table 1) of less than two naps per week? Why does a quarter or more of the population in some cultures never nap? In short, why is nap sleep tendency less pervasive than nocturnal sleep tendency? The answer rests in part in the chronobiology of activation, specifically, the fact that naps and afternoon performance dips tend to occur as transient events in an otherwise rising circadian function of behavioral and physiological activation (87) (see Chapter 5). Simply put, it is much easier to skip this relatively brief window of afternoon sleep tendency and still feel and function well later in the day than it is to skip the longer period of nocturnal sleep.

If napping is a reflection of a chronobiologically regulated sleepiness tendency, then it should also be affected by factors that influence basal level of sleepiness. As we shall see, there is considerable evidence that variations in sleep need are associated with napping.

Sleep Reduction and Sleep Need

If daytime napping is influenced by nocturnal sleep reduction and the consequent increase in diurnal sleep pressure, then naps should be more common when nocturnal sleep is reduced or disturbed for any reason. Napping in response to lost nocturnal sleep has been referred to as replacement or compensatory napping (39). The most common technique used to assess it has been to correlate nocturnal sleep length or indices of disturbance with nap sleep incidence and length. Among British adults, Tune (109) observed a low positive correlation ($r = 0.18$, $p < 0.001$) between the frequency of nocturnal awakenings and the number of midday naps, which he interpreted as a compensatory response of napping. Consistent with this, Johns et al. (54) reported from a survey of Australian college students that the duration of day sleep was positively correlated with total time awake during night sleep ($r = 0.29$, $p < 0.01$) and negatively correlated with the duration of night sleep ($r = -0.31$, $p < 0.01$). Similarly, using sleep logs in U.S. college students, White (118) found that for data obtained in the summer, hours of naps per week was negatively related to sleep length per night ($r = -0.46$, $p < 0.001$), and positively related to variability in nocturnal sleep length ($r = 0.45$, $p < 0.002$). However, only the latter relationship was found to be significant for data from the winter ($r = 0.38$, $p < 0.01$). In contrast, among adults in a siesta culture (i.e., Mexico), Taub (95) reported a positive correlation ($r = 0.27$, $p < 0.01$), with age partialed out, between the duration of afternoon sleep and the duration of nocturnal sleep, suggesting that longer naps were associated with longer nocturnal sleep lengths and therefore were not compensatory.

Thus, in studies of nonsiesta cultures, the highest coefficients suggesting that naps occur to replace or compensate nocturnal sleep loss were for correlations conducted within subjects by White (118). The question of how nap incidence is affected by nocturnal sleep is perhaps best addressed through within-subject comparisons, since these are not influenced by (un-

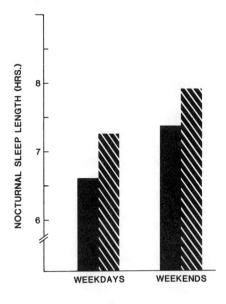

FIG. 4. Median nocturnal sleep amounts reported in a 35-day sleep log by 47 young adult nappers. (■) nights before nap days; (◩) nights before nonnap days.

wanted) between-subject variance in sleep need per 24 hr. In the only non-correlational approach to date, Dinges et al. (28) used a within-subject approach to determine the degree to which nocturnal sleep reduction influences daytime nap tendency. Using daily sleep log data from 47 nappers, they compared the average amount of nocturnal sleep on nights prior to nap and prior to nonnap days. This was done separately for weekdays and weekends, since nocturnal sleep increased on weekends (Table 2). Figure 4 displays the resulting median nocturnal sleep lengths. Analysis of variance confirmed what the figure shows. There was a main effect for weekday versus weekend nocturnal sleep length ($F_{1,138} = 22.7, p < 0.001$) and a main effect for nap days versus nonnap days ($F_{1,138} = 15.4, p < 0.001$), but no significant interaction ($F_{1,138} = 0.1$).

The results of Dinges et al. (28) suggest that nap days among college students are preceded by nights of reduced sleep, relative to nonnap days, regardless of whether the naps occur on weekdays or weekends. In this sense, the naps were compensatory, serving to replace lost nocturnal sleep. Interestingly, detailed examination of the daily sleep diary maintained by these subjects revealed that nap days were also characterized by earlier (30–60 min) morning awakenings ($p < 0.01$) relative to nonnap days (32). The awakenings tended to be by alarm clock rather than spontaneous, suggesting that they were planned. Consistent with this observation is the fact that subjects rated their sleepiness after awakening as greater on mornings of nap days relative to nonnap days ($p < 0.001$). No other behavioral variable, including dietary, exercise, health, academic, and extracurricular factors, was different between nap and nonnap days in these subjects.

These data seem to indicate that naps in the young adult population are frequently preceded by nights of reduced nocturnal sleep because of staying up late and/or awakening early (50). To the extent that subjects are not obtaining as much nocturnal sleep as is needed to sustain daytime alertness—and the weekday median in Fig. 4 of just more than 6.5 hr is disturbingly low—sleep need or sleep debt can be argued to precipitate the tendency toward midafternoon napping.

What is problematic in this formulation is that there was no interaction between nocturnal sleep amount and day of the week. That is, naps were not only as prevalent on weekends as on weekdays, but sleep the night before nap days in both instances was reduced relative to sleep before nonnap days, despite the fact that substantially more nocturnal sleep was obtained overall on weekends. For example, in Fig. 4, nocturnal sleep preceding nonnap weekdays is not significantly different from nocturnal sleep preceding weekend nap days. Why would the same nocturnal sleep amount promote napping on weekends but not on weekdays? Apparently, the amount of sleep obtained depended on the opportunity to sleep and on activity demands. Moreover, it is possible that in all cases except weekend nights before nonnap days, when the median sleep length logged is almost 8 hr, the amount of nocturnal sleep generally obtained by the students is inadequate to ward off daytime sleepiness.

If napping is a reflection of sleep need as well as a secondary phase for sleep tendency, then there should be a positive relationship between nocturnal sleep length and daytime nap sleep; that is, subjects who need high amounts of sleep should have longer nocturnal sleep and nap more during the day. Conversely, those who need less sleep to feel alert should sleep less at night and nap less during the day. There is considerable support for this.

As noted above, Taub (95) found a low but significant correlation between the duration of afternoon sleep and of nocturnal sleep among Mexicans. Evans and Orne (39) also identified a group of college students who frequently napped but without reduction in nocturnal sleep relative to nonnap days, and without daytime fatigue. They were 12% of all subjects and 26% of nappers (31). These subjects were dubbed appetitive nappers [see also (38)] because they appeared to nap for psychological reasons unrelated to fatigue (see section on individual differences). But, as will be argued, it is equally likely that they napped because they have increased sleep need. It is noteworthy that appetitive nappers can take replacement naps but rarely do so because they are careful not to reduce their nocturnal sleep (31,38,39).

Other recent studies in adolescents and young adults have provided support for napping being more prevalent in subjects who have increased nocturnal sleep need. Levine et al. (65) compared nocturnal sleep efficiency and daytime sleepiness, as assessed by sleep latency tests, of 47 older (30–

80 years) adults with 129 healthy young adults (18–29 years), 76 of whom were college students. They reported the following:

> The mean latency of young subjects (particularly college students) was shorter than that of the older subjects, with the differences occurring between the sleepiest 80% of each distribution. Among the college students, those with higher nocturnal sleep efficiencies (the previous night) were sleepier the following day than those with lower sleep efficiencies. The relation between nocturnal sleep efficiency and daytime sleepiness suggests that the increased sleepiness of average young adults is due to mild sleep restriction.

It is especially noteworthy that their subjects who had elevated nocturnal sleep efficiencies (TST/TIB) (i.e., minimal, disturbed, or fragmented sleep), showed the more dramatic afternoon increase in sleepiness, especially at 1600 hr. Similarly, Broughton et al. (14) reported in extended sleep studies in young adults that the delayed SWS peak that occurs some 12.5 to 13.0 hr after the first peak (i.e., in the nap zone) shows a strong trend to correlate positively with the magnitude of initial SWS peak and has a significant negative correlation with wakefulness after sleep onset, with wakefulness plus REM sleep, and with stage 1 sleep. Thus, in extended sleep, the greater the sleep depth and efficiency during the usual nocturnal sleep period, the deeper the sleep in the subsequent nap zone.

Strauch and Meier (93) recently reported results from 93 high school students (10–14 years) who were surveyed six times, at 2-year intervals, until ages 20 to 24 years. At each age, they compared the subjects who indicated a desire for more sleep with those who indicated they were obtaining sufficient sleep. The frequency of naps increased in both groups across the years, but there were significantly more naps reported taken among subjects who indicated they needed more sleep at survey years 18 to 22 and 20 to 24, suggesting that daytime napping in young adults is in response to increased sleep pressure from insufficient nocturnal sleep, relative to sleep need per 24 hr.

Dinges et al. (31) compared how much nocturnal sleep 31 young adults estimated they regularly obtained, how much they needed, and how much they would like to get with their actual sleep amounts on nap and nonnap days during a 35-day daily sleep log study. Table 3 displays the results of these comparisons. The only daily log sleep mean that is not significantly different from the amount of sleep subjects estimated they needed was TST/24 hr on nap days. This value differed by less than 10 min from the amount of sleep subjects indicated they would like to get. Nocturnal sleep amounts on nights prior to nap and prior to nonnap days were consistently less than (by 0.5–1.75 hr) subjective estimates of sleep needed and sleep they would like to obtain. Although it is not surprising that subjects would indicate that they desire more sleep ($M = 8:21$) than they typically obtain [see also (39,118)], it is remarkable that only nap days yield the sleep they estimate

TABLE 3. *Mean differences between subjective sleep length estimates and sleep lengths recorded in daily logs for 31 healthy young adults*

	Mean (SD)	35-Day sleep log		
		Nocturnal TST before nonnap days 7:14[a] (0:35)	Nocturnal TST before nap days 6:34 (0:56)	24-hr TST on nap days 7:52 (0:59)
Estimated nocturnal sleep averaged in past year	7:26 (1:04)	−0:12	−0:52[b]	+0:26[c]
Estimated nocturnal sleep needed	7:44 (1:07)	−0:30[c]	−1:10[b]	+0:08
Estimated nocturnal sleep you would like to get	8:21 (0:56)	−1:07[b]	−1:47[b]	−0:29[c]

TST, total sleep time.
[a] Hr:min.
[b] $p < 0.0001$.
[c] $p < 0.05$ (two-tailed).

they need to function optimally. Moreover, the effect was evident in the data from replacement nappers, appetitive nappers, and nonnappers, suggesting that it is not the result of individual differences in the tendency or opportunity to nap.

It is possible, of course, that these young adults do not actually need more sleep, but rather have an enhanced ability to sleep and as a result they obtain more sleep by napping, but they misattribute this extra sleep to their sleep need [cf. (86)]. Although this explanation may hold for the subgroup that Evans and Orne (39) identified as appetitive nappers, it seems unlikely that the intense subjective feeling that more sleep is needed in otherwise healthy persons has no validity relative to ability to feel and function alertly. As we have seen, napping appears to be a major way in which young adults meet their sleep requirement. In this sense it serves a homeostatic function. It appears that the function may be to fulfill the daily quota of SWS (41,56,105), to increase daily TST (31), or both. The fact that it is chronobiologically regulated to maximum propensity midway in the wakefulness cycle and coincident with many other aspects of increased diurnal sleepiness at this time, suggests that it is a second phase of the circadian cycle at which sleep need is expressed.

Despite evidence that it is often associated with increased sleep need and is biorhythmically regulated, napping continues to be attributed to postprandial sleepiness, poor sleep hygiene, health, and personality differences. The data relevant to these factors are now examined.

Postprandial Sleepiness

Because the experience of increased lethargy and sleepiness can follow the consumption of meals, napping has often been attributed to postprandial sleepiness, as have the results of studies (6) showing increased errors in the afternoon at the time napping is common. It appears that some foods can increase subjective sleepiness in some individuals, particularly within 20 to 30 min after ingestion (2,3,17,91). Such effects also have been reported 2 hr after consumption of high-carbohydrate meals (90) and consumption of tryptophan (48,85). The effects are thought to occur through the brain's uptake of neurotransmitter precursors, such as tryptophan (which accelerates the synthesis and release of serotonin) following insulin-produced reductions in competing amino acids (43,119).

The question at issue, however, is whether postprandial effects account for a large portion or even a significant minority of daytime naps. Postprandial sleepiness is not the inevitable consequence of every meal, even those that contain carbohydrates (91). Moreover, the effect appears to be mediated by a variety of factors, including the length of time an individual has fasted, the meal size and composition (specific nutrients), the time of day it is consumed, and the gender and age of the consumer (2,3,17,90,91). For example, older adults have been reported to be more susceptible than younger adults to postprandial sleepiness following high-carbohydrate lunches (90). In addition, the effect of some psychoactive food substances, such as caffeine and alcohol, interact with the basal level of daytime sleepiness as determined by the amount of nocturnal sleep (70). It is conceivable that, for reasons not yet known, the more dramatic effects of certain foods for producing postprandial sleepiness depend on increased sleep need, which also increases the tendency for daytime napping.

Given the temporal patterning of naps in midafternoon, lunch should be disproportionately more likely to induce sleepiness than breakfast or supper, if postprandial sleepiness accounts for napping. Some siesta cultures, such as Mexico, do indeed eat their largest meal at lunch (95), and the effects of food may contribute to the tendency to nap. Lugaresi et al. (69) observed that 43.5% of Italian "good sleepers" reported postprandial sleepiness. It is noteworthy, however, that in a study of Greeks, Soldatos et al. (88) found that despite much napping, less than 1% of Greek nappers reported being drowsy after a heavy lunch. Among siesta cultures, therefore, postprandial sleepiness is not as consistent a phenomenon as is napping. Some siesta cultures (e.g., Spain) eat their main meal at night, whereas some nonsiesta cultures (e.g., West Germany) eat their main meal at lunch. Finally, similar increases in postingestion napping are not found after morning and evening meals.

The few laboratory data that exist suggest that the midday sleep tendency is present independent of food intake, making it difficult to attribute most napping to food consumption. Blake, reported by Colquhoun (19), conducted two studies on the effects of varying lunch time on letter cancellation performance, which typically shows a postlunch dip. The performance dip was present regardless of whether food was consumed, and the timing of the meal and interval between the meal and test had no effect on the dip. Using a constant routine protocol and the multiple sleep latency test (MSLT) procedure as an index of daytime sleep tendency, Carskadon and Dement (15) studied daytime sleepiness in five 16- to 17- and six 62- to 75-year-old subjects. The constant routine involved being confined to bed at a 45° angle and being fed equal portions of food and water at 1-hr intervals throughout the day. The MSLT continued to show an afternoon dip (1530 hr) during the constant-routine day similar to that seen on the baseline day (1330 hr), albeit one test later; and the pattern was present in both age groups. The authors (15) concluded that "the midafternoon tendency to fall asleep is not a postprandial consequence." Thus, to the extent that the afternoon dip in aspects of performance and the MSLT reflect the same chronobiological tendency that underlies napping, postprandial sleepiness does not appear to be the cause of napping but may exacerbate the tendency toward afternoon sleepiness in some individuals.

Personality and Behavioral Differences

There are well-documented individual differences in napping behavior, but it is often assumed that underlying the distinction between nappers and nonnappers is a difference in personality, mental health, sleep hygiene/sleep disorders, or some other facet of functioning. More often than not, the bias is in the direction of a deficit or defect producing the nap behavior. Although, as was discussed previously, sleep need may promote much napping in young adults and young adult nappers may have more variable sleep/wake schedules, there are virtually no reliable data that nappers are more likely to be pathological in any way than are nonnappers.

To the extent that persons with hypersomnia (e.g., with sleep apnea or narcolepsy) are more likely to nap, there will be more nappers among patients with sleep disorders (see Chapter 13). There do not, however, appear to be more disturbed sleepers in the normal population among habitual nappers than among nonnappers. In a study of college student nappers and nonnappers, for example, there was an identical proportion of subjects (11% in each group) who reported sleep complaints (27). In a survey study of Italians, Lugaresi et al. (69) did find that more poor sleepers (17.5%) habitually napped in the afternoon than did good sleepers (8.8%), but it is not certain whether these differences were confounded by age-related variations in sleep quality and napping. Both increased napping and increased disturbed sleep occur with advancing age (108).

In a survey study of Mexicans, Taub (95) found no differences in reports of disturbed sleep between nappers and nonnappers. Similarly, among Greeks, Soldatos et al. (88) found no differences in health status among nappers and nonnappers. In short, among otherwise healthy adults, napping is neither necessarily a reflection of sleep problems nor a palliative for them. Rather, it appears to be affected adversely in the same way as nocturnal sleep by age, sleep disturbance, and other factors disruptive of sleep (31,95).

In terms of personality, habitual nappers and nonnappers have not been found to differ on objective testing (4,31,62). Although there have been observations of increased anxiety scores (5) and elevated scores on depression-dejection subscales from adjective checklists (22) in nappers relative to non-nappers, these results have not been reliably replicated between studies and have not been found elsewhere (4,31). The lack of reliable differences between nappers and nonnappers in sleep disturbance, incidence of psychopathological patterns, and personality dimensions is also replicated within subjects. The nap and nonnap days of habitual nappers are distinguishable only in terms of voluntarily reduced nocturnal sleep amounts on nights before nap days and increased daytime sleepiness on nap days (31), although even in this area there are some striking interindividual differences.

Evans and Orne (38,39) distinguished between two types of nappers among the approximately 55% of college students who nap "sometimes too often." The larger group comprised 78% of regular nappers and were called replacement nappers because they reported napping only when tired. As already noted, this was directly associated with reduced nocturnal sleep the night before nap days. The smaller group (22%) were considered appetitive nappers because they reported napping without regard to daytime fatigue, napped significantly more often than replacement nappers, and did so without evident differences in nocturnal sleep amounts between nap and nonnap days (31,39). Evans et al. (38) concluded that replacement nappers napped to "compensate for accumulated sleep deficit," whereas appetitive nappers napped primarily for "psychological restorative functions seemingly unrelated to sleep needs." Thus, napping seemed to serve different functions for these two groups. This conclusion was punctuated by the fact that there were differences in nap sleep infrastructure between the two groups. Appetitive nappers appeared to have more stage 1 sleep during their naps relative to replacement nappers and a subgroup (4%) of nonnappers who reported that naps produced unpleasant physical and mental consequences (38).

Although the nap sleep infrastructure differences between appetitive nappers and other subjects have not been consistently replicated (31; A. D. Bertelson, 1980, personal communication), the behavioral aspects of their nap behaviors (e.g., frequency of napping, have been reliable). It is clear, however, that appetitive nappers can have replacement naps, and most nonnappers can nap if time were available or they felt the need (31). Consequently, these napper-type distinctions may not represent stable and permanent personality or behavioral traits. It is probably more appropriate to speak of replacement and appetitive nap patterns rather than nappers. On the other hand, there are self-reported differences among nappers and nonnappers in the extent to which they can fall asleep in various contexts, and these have been related to other personality dimensions, such as hypnotizability. For example, low hypnotizable subjects who are nonnappers report difficulty falling asleep in different contexts relative to most other subjects (36,37).

Behavioral differences are also evident in certain nonnappers. The subset of nonnappers who reported that naps produced unpleasant physical and/or mental aftereffects (4% of all college adults) were more likely to be bothered by background sounds, have difficulty falling asleep in an alerting environment, drink coffee, and have a blood relative with insomnia (31). In laboratory nap studies, these subjects also had significantly higher body temperatures prior to nap onsets than did nappers. Although these nonnappers did not differ in sublingual temperature between nap and nonnap days, nappers had significantly lower prenap temperatures on lab nap days relative to lab nonnap days. These prenap temperature drops were accompanied by increases in self-reported sleepiness and calmness (31). It was as though nappers were psychologically and physiologically preparing themselves to nap. More research is needed on the significance of these preparatory responses and their relationship to the infrastructure and effects of naps.

EFFECTS OF NAPS

The appropriateness and desirability of daytime napping is often debated in terms of the effects it has on mood, performance, and subsequent nocturnal sleep. Data on the effects of naps reveal that they are neither consistently beneficial nor detrimental. Rather, both positive and negative effects can result, depending on such factors as the extent to which the subject habitually naps, the nap duration, time of day of napping, and abruptness of awakening. These factors are discussed in terms of the manner in which naps affect four classes of outcome variable: mood, behavior, wake psychophysiology, and nocturnal sleep. We begin with the last of these because it is often the basis for discouraging persons from napping.

Nocturnal Sleep

Do daytime naps affect nocturnal sleep, and if so, are the effects undesirable? Karacan et al. (56) conducted the first systematic study showing that afternoon naps reduce the amount of SWS during the night that follows. Although the authors did not conclude that napping was inappropriate, their results have often been cited as the basis for discouraging persons from napping. In fact, the study was focused on the extent to which there is a psychobiological need (each 24-hr day) for certain sleep stages, not on whether the sleep-stage shifts following naps at different times of day were deleterious.

Consistent with other studies, Karacan et al. (56) found that REM sleep predominated in a 2-hr morning nap (start 0800 hr) and had no effect on subsequent nocturnal sleep. Afternoon naps (starting at 1600 hr) contained much more SWS and were associated with reduced nocturnal SWS. They did not appear to affect either nocturnal sleep onset or nocturnal TST. The authors speculated that "stage 4 fulfillment is more effective during afternoon naps than during night sleep." Although recent evidence casts doubt on whether naps are more effective than nocturnal sleep in fulfilling daily SWS amounts, there is increasing evidence that the sleep of an afternoon nap may indeed be involved in the control of SWS (24,41,105). These findings are consistent with data described previously on the use of naps to fulfill TST need and suggest that afternoon napping is a response to the homeostatic components of sleep (i.e., SWS/24 hr and TST/24 hr). That napping is also temporally bound to the afternoon (see previously) and chronobiologically controlled (see Chapter 5) does not diminish its potential role in sleep homeostasis (see Chapter 14). Rather, it is consistent with the view that sleep regulation involves both biorhythmic and homeostatic components (9,21).

An issue related to the effects of afternoon naps on nocturnal sleep concerns the extent to which they can be used to phase-adjust the major sleep period or other circadian markers, following an acute phase-shift in the sleep/wake cycle. Although napping is relatively common among flight crews (47) and shift workers (see Chapter 10), both of whom experience phase shifts in circadian rhythms, there are surprisingly few data on the effects of naps in facilitating phase adjustment. In the only study conducted to date, Monk et al. (77) found that naps neither helped nor hindered a 6-hr phase adjustment. They may, however, help relieve cumulative sleep pressure resulting from such circadian shifts. Clearly, more data are needed.

Another way in which napping has been thought to produce ill effects is through its impact on the ability to fall asleep and remain asleep the following night. There is no doubt that repeated daytime napping can make it more difficult for some persons to fall asleep or stay asleep at night and, to this extent, napping can reflect poor sleep hygiene. The relevant issue, however, is the extent to which most napping necessarily results in poor sleep hygiene, not whether it is a symptom of poor sleep hygiene in some persons. Neither survey studies (27,80) nor extensive daily sleep log data (1,31,58) have provided any evidence that daytime napping is more often associated with nocturnal sleep difficulties and complaints than is nonnapping. The conclusion from the first systematic study of napping by Kleitman et al. (58) has yet to be refuted: "We found that naps taken during the waking period, in adults at least, have no effect on the ease of going to sleep, nor on motility during, or length of, (nocturnal) sleep.'' Consequently, prohibitions against daytime napping, from the perspective of its effects on nocturnal sleep, are not supported by data from otherwise healthy sleepers.

Mood and Performance

Upon Awakening from a Nap

Kleitman (57) observed that ''immediately after getting up, irrespective of the hour, one is not at one's best.'' It is a paradoxical phenomenon— being more impaired upon awakening from sleep than upon going to sleep— that has been documented for a wide array of behavioral tasks. Originally described for awakenings from nocturnal sleep as ''sleep drunkenness'' (10,11), it has come to be known by other names, including ''postdormital disorientation'' (1a) and ''sleep inertia'' (68). The former terms are now used primarily for sleep pathology (see Chapter 13), whereas the latter is widely applied to describe transient awakening impairment in healthy persons (see Chapters 8, 10, and 11).

Sleep inertia is ubiquitous. It not only occurs after awakening from sleep at anytime of the 24-hr day, but also after awakening from afternoon naps in nonsleep-deprived healthy adults (26). Characterized by transient performance decrement and/or confusion, it is especially evident when awakening is abrupt. The phenomenon is of relatively brief duration (5–15 min) in nonsleep-deprived subjects. Postnap impairment and disorientation can last longer and be more severe, however, if the nap sleep follows a prolonged period of wakefulness (30). It appears that the greater the depth of nap sleep, especially the amount and intensity of NREM sleep stages, the more severe the sleep inertia (30), similar to the relationship observed for nocturnal sleep (10,11). Since sleep depth is a function of sleep pressure resulting from prior wakefulness and circadian phase, nap sleep inertia will vary in relation to these factors, as well as the duration of the nap (26,30).

To the extent that awakenings from daytime naps, even naps in alerting environments, result in a transient sleep inertia (26), napping may be considered to yield negative effects on behavior and mood. On the other hand, the duration of these effects is usually quite brief, giving way to performance and mood comparable to the prenap period. They are usually so brief that they are ignored in many investigations of the effects of naps in adults maintaining night sleep.

Nevertheless, sleep inertia may account for some of the dysphoria reported after daytime naps, especially afternoon naps. That this may not be very widespread or significant a factor in the decision to nap is evidenced by the finding that unpleasant aftereffects are not the reason most nonnappers indicate that they do not nap (31). In fact, there is little evidence that sleep inertia is even the basis for the complaints of the small minority of nonnappers who report unpleasant mental and physical aftereffects from naps, since they report these consequences long after sleep inertia effects on performance have disappeared (31). Sleep inertia may become a significant concern, however, in the decision to nap after prolonged periods of sleep loss and when the need to perform effectively after awakening is paramount.

Following a Nap

As sleep inertia rapidly dissipates, mood and performance following a nap return to or exceed prenap levels. It is often assumed that a major reason many persons nap is to enhance mood and/or performance, either by increasing psychophysiological activation shortly after a nap or by extending activation into the later evening to attenuate or delay the circadian decline in alertness. The latter effect, and its use in sustained operations, has been referred to by Orne as "prophylactic napping" (33).

There have been more than a dozen studies of the effects of daytime naps on mood, performance, and psychophysiological activation. Most have been conducted on adults under age 40 and involved 1-hr (or less) afternoon naps (1200–1800 hr). All involved polysomnographic monitoring of nap sleep and some quantitative assessment (e.g., sleep log) of nocturnal sleep on nights before and after the nap. Although between-subject designs are more common, both within-subject and mixed-model designs have been used. The effects of naps have been compared with nonnap bed rest conditions and wake periods of control activity. In general, the findings of these studies have been quite consistent.

Mood variables, especially self-reports of sleepiness, fatigue, and activation, have been found quite consistently to improve after naps. Although the positive effects of naps on alertness reports have been found in some studies to be only slightly greater than those observed for bed rest without sleep (4,22,84), other investigations have documented significantly greater mood enhancement after naps compared with both bed rest (98) and control wake periods that do not involve bed rest (31,38,101,103). When increased subjective alertness has not been observed, the subjects under study were either those who typically had adverse reactions to naps (31,39) or subjects who were undergoing (or about to undergo) nocturnal sleep restriction (33,45).

The evidence that naps enhance subjective reports of alertness must be tempered by the fact that such self-report measures are easily influenced by demand characteristics (83), and they often correlate poorly with objective measures of sleepiness (13,51). Nevertheless, the fact that the laboratory results are generally consistent with what subjects report about naps in their daily sleep logs (31) suggests that naps do improve feelings of alertness in persons who are also sleeping at night. But to what extent is the improvement a function of the nap per se or a product of time? Subjects may indeed feel somewhat better after a nap to the extent that they slept during a circadian phase of increased sleep pressure, have reduced their sleep need, or both. Whether or not one naps, however, feelings of alertness will generally continue to increase into the early evening hours.

The effects of naps on performance are less evident than those observed for self-reports of activation/alertness. In three studies of nap effects, Taub consistently found statistically significant performance improvements during the hour after naps. In an initial study, Taub et al. (103) reported that both 30-min and 2-hr afternoon naps enhanced reaction time (RT) performance [see also (96,102,104) for reports of these data]. There was no difference in the RT enhancement between the two nap durations. In a study of 2-hr morning (0935–1135 hr) versus 2-hr evening (2135–2335 hr) naps, Taub et al. (101) found that both nap conditions equally improved RT speed and memory retention relative to the no-nap condition [see also (97)]. In a third study, Taub (98) observed that 90-min afternoon naps improved choice RT, short-term memory, and vigilance performance relative to bed rest.

Other studies of performance shortly after naps have not observed improved performance relative to prenap performance levels and no-nap control conditions (4,22,26,31,84), although there have been suggestions of a trend in this direction (38). The same inconsistencies appear for psychophysiological measures of activation, such as EEG frequency, heart rate, and electrodermal activity, after naps (31,102). It is unclear why performance and psychophysiological enhancements shortly after naps have only been observed in Taub's studies. Procedural factors might account for the discrepancies, but regardless of what the reason may be, it is important to keep in mind that performance before naps in all of these studies was quite high, so it is not clear that significant improvement was possible beyond the prenap level.

Although performance enhancement is not always evident shortly after a daytime nap, it may be more conspicuous hours later, especially if prolonged wakefulness follows the nap. Godbout and Montplaisir (46) reported increased RT performance in the second half of a 10-min RT task 2 hr after 20-min daytime naps, but not 3 min after the naps. Gillberg (45) observed significantly better RT performance at 0700 hr, after a 24-hr period in which a 1-hr nap was allowed either in the evening (2100 hr) or at night (0430 hr), relative to no nap. Thus, the beneficial effects of the 2100-hr and 0430-hr naps were evident in RT performance 10 hr and 1.5 hr later, respectively. Similar delayed effects on RT and other aspects of performance were observed by Dinges et al. (33,33a) following a 2-hr afternoon nap opportunity prior to 2.5 days without sleep. Again, the positive effects of the afternoon nap were not evident in performance until 10 hr after the nap, suggesting that the nap prophylactically prevented some of the performance deterioration that typically results from sustained wakefulness.

Psychophysiological measures of alertness have also shown this delayed beneficial effect from naps. In Gillberg's (45) study, sleep latency tests showed trends similar to the performance data. Carskadon and Dement (16) conducted a study of 45-min nap periods at 1300 hr each day during a 7-day period involving nocturnal sleep restricted to 5 hr. MSLTs were conducted postnap at 0930, 1130, 1300, 1530, 1730, and 1930 hr each day; the latter three assessments thus occurred between 1.75 and 5.75 hr after each daily nap. Prenap MSLTs did not show reduced sleep tendency relative to basal scores, regardless of whether naps were allowed. Postnap MSLTs, however, showed that the group permitted a nap maintained or exceeded basal sleep latency scores for these times and did not show the cumulative decrement evident in the group not permitted to nap. Thus, the naps helped maintain alertness as assessed by MSLT for a period approximately 2 to 6 hr later. When considered with the performance data, it appears that afternoon naps have beneficial effects on performance and physiological sleep tendency for as long as 12 hr after they are taken.

Finally, there is no consistent evidence that the effects of naps on mood and performance beyond the sleep inertia period are directly associated with a specific stage of sleep. Taub et al. (103,104) reported that 30-min and 2-hr naps produced the same positive effects, despite great differences in nap sleep infrastructure. Bertelson (4) found no relationship between nap stages and postnap mood or performance. On the other hand, Taub (97) reported greater postnap RT performance improvement was associated with more stage 4 sleep during naps ($r = 0.37$). Davidson et al. (23) observed a similar relationship between SWS and RT performance for naps after sleep deprivation ($r = 0.58$). In terms of mood after naps, Evans et al. (36) and Dinges et al. (31) found that appetitive nappers, who napped most often and had significantly more stage 1 sleep during naps than a subset of nonnappers, who had significantly more SWS during their naps, also had the greatest nap benefits.

Such results imply that both individual differences and factors that influence nap sleep infrastructure will determine postnap benefits. If the nap sleep is not consolidated owing, for example, to environmental disturbances (27), then SWS will be reduced and postnap benefits may be reduced. On the other hand, it remains uncertain to what extent particular sleep stages are necessary in order for a nap to be perceived as increasing subjective alertness and performance.

CONCLUSIONS

Naps, especially afternoon naps of approximately 1-hr duration, are a common feature of the sleep of healthy adults in many countries. They appear most often to occur as a function of an endogenously timed period of sleep facilitation to fulfill sleep need per 24 hr. If nocturnal sleep is inadequate, daytime napping is more prevalent. Napping typically occurs approximately 12 hr after sleep the night before nap days, and there is evidence to suggest that it is phase-dependent on nocturnal sleep placement. In addition to sleep need and chronobiological regulation, the opportunity to nap and cultural biases concerning naps play significant roles in the prevalence of napping behavior. On the other hand, there is little evidence to suggest that food ingestion, personality factors, and climate are significant determinants of napping. Naps generally do not adversely affect nocturnal sleep or indicate disordered sleep at night. Although sleep inertia can occur after awakening from naps, as it can after nocturnal sleep, naps generally have later overall beneficial effects on mood and performance. Thus, the available data on nap patterns and nap effects in otherwise healthy adults suggest that napping is a normal, appropriate, and beneficial feature of adult sleep/wake patterns.

ACKNOWLEDGMENTS

The review and substantive evaluation on which this chapter is based were supported in part by grants MH-19156 and MH-44193 from the National Institute of Mental Health, U.S. Public Health Service and, in part, by a grant from the Institute for Experimental Psychiatry Research Foundation.

REFERENCES

1. Aber, R., and Webb, W. B. (1986): The effects of a limited nap on night sleep in older subjects. *Psychology and Aging*, 1:300–302.
1a. Association of Sleep Disorders Centers (1979): Diagnostic classification of sleep and arousal disorders, 1st ed., prepared by the Sleep Disorders Classification Committee. *Sleep*, 2:1–137.
2. Bell, I. R., Blair, S. R., Owens, M., Guilleminault, C., and Dement, W. C. (1976): Postprandial sleepiness after specific foods in normals. *Sleep Res.*, 5:41.
3. Bell, I. R., Rosekind, M., Hargrave, V., Guilleminault, C., and Dement, W. C. (1977): Lunchtime sleepiness and nap sleep after specific foods in normals. *Sleep Res.*, 6:45.
4. Bertelson, A. D. (1979): Effects of napping and bedrest on performance and mood. Doctoral dissertation, Ohio State University.
5. Bertelson, A. D., Fisher, D. S., and Herrmann, S. C. (1982): Personality differences between nappers and non-nappers. *Sleep Res.*, 11:118.
6. Bjerner, B., Holm, A., and Swensson, A. (1955): Diurnal variation in mental performance: A study of three-shift workers. *Br. J. Ind. Med.*, 12:103–110.
7. Blake, M. J. F. (1967): Time of day effects on performance in a range of tasks. *Psychosom. Sci.*, 9:349–350.
8. Bonnet, M. H., and Alter, J. (1982): Effects of irregular versus regular sleep schedules on performance, mood and body temperature. *Biol. Psychol.*, 14:287–296.
9. Borbély, A. A. (1982): A two process model of sleep regulation. *Hum. Neurobiol.*, 1:195–204.
10. Broughton, R. J. (1968): Sleep disorders: Disorders of arousal? *Science*, 159:1070–1078.

11. Broughton, R. J. (1973): Confusional sleep disorders: Interrelationship with memory consolidation and retrieval in sleep. In: *A Triune Concept of the Brain and Behavior*, edited by T. J. Boag and D. Campbell, pp. 115–127. University of Toronto Press, Toronto.
12. Broughton, R. (1975): Biorhythmic variations in consciousness and psychological functions. *Can. Psychol. Rev.*, 16:217–239.
13. Broughton, R. (1982): Performance and evoked potential measures of various states of daytime sleepiness. *Sleep*, 2:S135–S146.
14. Broughton, R. J., De Koninck, J., Gagnon, P., and Dunham, W. Circadian, circasemidian and ultradian human sleep-wake rhythms: A brief review and further data from extended sleep studies. In: *Sleep and Biological Rhythms: Basic Mechanisms and Application to Psychiatry*, edited by J. Montplaisir, R. Godbout, and H. H. Jasper. Oxford Press, Oxford and New York (*in press*).
15. Carskadon, M. A., and Dement, W. C. (1985): Midafternoon decline in MSLT scores on a constant routine. *Sleep Res.*, 14:292.
16. Carskadon, M. A., and Dement, W. C. (1986): Effects of a daytime nap on sleepiness during sleep restriction. *Sleep Res.*, 15:69.
17. Christie, M. J., and McBrearty, E. (1978): Diurnal variation, and the effects of lunch on capillary blood glucose, heart period and deep body temperature, in male and female subjects. *Soc. Psychophysiol. Res. Abstracts*, 15:275.
18. Cohen, E. L. (1944): Length and depth of sleep. *Lancet*, 830–831.
19. Colquhoun, W. P. (1971): Circadian variations in mental efficiency. In: *Biological Rhythms and Human Performance*, edited by W. P. Colquhoun, pp. 39–107. Academic Press, London.
20. Colquhoun, W. P. (1982): Biological rhythms and performance. In: *Biological Rhythms, Sleep, and Performance*, edited by W. B. Webb, pp. 59–86. John Wiley, Chichester.
21. Daan, S., Beersma, D. G. M., and Borbély, A. A. (1984): The timing of human sleep: A recovery process gated by a circadian pacemaker. *Am. J. Physiol.*, 246(Regulatory, Integrative Comp. Physiol. 12):R161–R178.
22. Daiss, S. R., Bertelson, A. D., and Benjamin, L. T., Jr. (1986): Napping versus resting: Effects on performance and mood. *Psychophysiology*, 23:82–88.
23. Davidson, J. R., Radomski, M. W. M., Heselgrave, R. J., Angus, R. G., and Montelpare, W. J. (1986): Sleep structure of a nap following 40 hours of sleep deprivation and its relation to performance upon awakening. *Sleep Res.*, 15:213.
24. Dinges, D. F. (1986): Differential effects of prior wakefulness and circadian phase on nap sleep. *Electroencephalogr. Clin. Neurophysiol.*, 64:224–227.
25. Dinges, D. F. (1989): The nature of sleepiness: Causes, contexts and consequences. In: *Perspectives in Behavioral Medicine: Eating, Sleeping, and Sex*, edited by A. Stunkard and A. Baum. Lawrence Erlbaum, Hillsdale, NJ.
26. Dinges, D. F., Orne, E. C., Evans, F. J., and Orne, M. T. (1981): Performance after naps in sleep-conducive and alerting environments. In: *Biological Rhythms, Sleep and Shift Work. Advances in Sleep Research, Vol. 7*, edited by L. Johnson, D. Tepas, W. P. Colquhoun, and M. Colligan, pp. 539–552. Spectrum, New York.
27. Dinges, D. F., Orne, E. C., and Orne, M. T. (1982): Napping: Symptom or adaptation? *Sleep Res.*, 11:100.
28. Dinges, D. F., Orne, E. C., and Orne, M. T. (1983): Napping in North America: A siesta rhythm. Paper presented at 4th International Congress of Sleep Research, Bologna, Italy.
29. Dinges, D. F., Orne, M. T., and Orne, E. C. (1984): Sleepiness during sleep deprivation: The effects of performance demands and circadian phase. *Sleep Res.*, 13:189.
30. Dinges, D. F., Orne, M. T., and Orne, E. C. (1985): Assessing performance upon abrupt awakening from naps during quasi-continuous operations. *Behav. Res. Meth., Instrum., Comput.*, 17:37–45.
31. Dinges, D. F., Orne, M. T., Orne, E. C., and Evans, F. J. (1980): Voluntary self-control of sleep to facilitate quasi-continuous performance. U.S. Army Medical Research and Development Command, Maryland (NTIS AD-A102264).
32. Dinges, D. F., Orne, M. T., Orne, E. C., and Evans, F. J. (1981): Behavioral patterns in habitual nappers. *Sleep Res.*, 10:136.
33. Dinges, D. F., Orne, M. T., Whitehouse, W. G., and Orne, E. C. (1987): Temporal placement of a nap for alertness: Contributions of circadian phase and prior wakefulness. *Sleep*, 10:313–329.

33a. Dinges, D. F., Whitehouse, W. G., Orne, E. C., and Orne, M. T. (1988): The benefits of a nap during prolonged work and wakefulness. *Work and Stress*, 2:139–153.

34. Endo, S. (1977): Circadian aspects of REM sleep and NREM sleep. *Waking and Sleeping*, 1:218–219.

35. Endo, S., Yamamoto, T., Kobayashi, T., and Fukuda, H. (1979): The effects of REM sleep deprivation on naps. *Sleep Res.*, 8:148.

36. Evans, F. J. (1977): Hypnosis and sleep: The control of altered states of awareness. *Ann. NY Acad. Sci.*, 296:162–174.

37. Evans, F. J. (1977): Subjective characteristics of sleep efficiency. *J. Abnorm. Psychol.*, 86:561–564.

38. Evans, F. J., Cook, M. P., Cohen, H. D., Orne, E. C., and Orne, M. T. (1977): Appetitive and replacement naps: EEG and behavior. *Science*, 197:687–689.

39. Evans, F. J., and Orne, M. T. (1975): Recovery from fatigue. U.S. Army Medical Research and Development Command, Maryland (DTIC A100347).

40. Evans, F. J., and Orne, M. T. (1976): Recovery from fatigue. U.S. Army Medical Research and Development Command, Maryland (NTIS AD-A100310).

41. Feinberg, I., March, J. D., Floyd, T. C., Jimison, R., Bossom-Demitrack, L., and Katz, P. H. (1985): Homeostatic changes during post-nap sleep maintain baseline levels of delta EEG. *Electroencephalogr. Clin. Neurophysiol.*, 61:134–137.

42. Feldman, M. J. (1970): Sleep habits of college students. *Psychophysiology*, 7:294.

43. Fernstrom, J. D., and Wurtman, R. J. (1971): Brain serotonin content: Increase following ingestion of carbohydrate diet. *Science*, 174:1023–1025.

44. Gagnon, P., De Koninck, J., and Broughton, R. (1985): Reappearance of electroencephalographic slow waves in extended sleep with delayed bedtime. *Sleep*, 8:118–128.

45. Gillberg, M. (1984): The effects of two alternative timings of a one-hour nap on early morning performance. *Biol. Psychol.*, 19:45–54.

46. Godbout, R., and Montplaisir, J. (1986): The performance of normal subjects on days with and days without naps. *Sleep Res.*, 15:71.

47. Graeber, R. C. ed. (1986): Crew factors in flight operations: IV. Sleep and wakefulness in international aircrews. NASA Tech. Memorandum 88231.

48. Hartmann, E., and Spinweber, C. L. (1979): Sleep induced by L-tryptophan: Effect of dosages within the normal dietary intake. *J. Nerv. Ment. Dis.*, 167:497–499.

49. Hauri, P. (1979): What can insomniacs teach us about the function of sleep? In: *The Functions of Sleep*, edited by R. Drucker-Colin, M. Shkurovich, and M. B. Sterman, pp. 251–271. Academic Press, New York.

50. Hawkins, J. (1988): Napping and sleep length: A categorical self-report system. *Sleep Res.*, 17:85.

51. Herscovitch, J., and Broughton, R. (1981): Sensitivity of the Stanford Sleepiness Scale to the effects of cumulative partial sleep deprivation and recovery oversleeping. *Sleep*, 4:83–92.

52. Ichihara, S., Miyasita, A., Inugami, M., et al. (1979): Comparison of diurnal naps on a chair with on a bed. *Clin. Electroencephalogr. (Osaka)*, 21:293–302.

53. Johns, M. W., Bruce, D. W., and Masterton, J. P. (1974): Psychological correlates of sleep habits reported by healthy young adults. *Br. J. Med. Psychol.*, 47:181–187.

54. Johns, M. W., Gay, T. J. A., Goodyear, M. D. E., and Masterton, J. P. (1971): Sleep habits of healthy young adults: Use of a sleep questionnaire. *Br. J. Prev. Soc. Med.*, 25:236–241.

55. Karacan, I., Finley, W. W., Williams, R. L., and Hursch, C. J. (1970): Changes in stage I-REM and stage 4 sleep during naps. *Biol. Psychiatr.*, 2:261–265.

56. Karacan, L., Williams, R. L., Finley, W. W., and Hursch, C. J. (1970): The effects of naps on nocturnal sleep: Influence of the need for stage 1 REM and stage 4 sleep. *Biol. Psychiatr.*, 2:391–399.

57. Kleitman, N. (1963): *Sleep and Wakefulness*. University of Chicago Press, Chicago.

58. Kleitman, N., Mullin, F. J., Cooperman, N. R., and Titelbaum, S. (1937): *Sleep Characteristics. How They Vary and React to Changing Conditions in the Group and the Individual*. University of Chicago Press, Chicago.

59. Knowles, J. B., Cairns, J., and MacLean, A. W. (1978): Acute shifts in the sleep-wakefulness cycle: 2. Effects on performance. *Sleep Res.*, 7:159.

60. Kunken, K. J. (1977): College students' sleep patterns: Selected factors affecting the length and quality of sleep. Master's thesis, Cornell University.
61. Lavie, P. (1986): Ultrashort sleep-waking schedule. III. Gates and "forbidden zones" for sleep. *Electroencephalogr. Clin. Neurophysiol.*, 63:414–425.
62. Lawrence, B. E. (1971): A questionnaire and electroencephalographic study of normal napping in a college occupied population. Master's thesis, University of Oklahoma.
63. Lawrence, B. E. (1976): Electroencephalographic sleep patterns during long and short afternoon naps. Paper presented at the American Psychological Association, Washington, D.C.
64. Lawrence, B., and Shurley, J. T. (1972): Comparison of sleep habits of American and Guatemalan students. *Sleep Res.*, 1:95.
65. Levine, B., Roehrs, T., Zorick, F., and Roth, T. (1988): Daytime sleepiness in young adults. *Sleep*, 11:39–46.
66. Lewis, S. A. (1969): Sleep patterns during afternoon naps in the young and elderly. *Br. J. Psychiatry*, 115:107–108.
67. Libert, J. P., Di Nisi, J., Fukuda, H., Muzet, A., Ehrhart, J., and Amoros, C. (1988): Effect of continuous heat exposure on sleep stages in humans. *Sleep*, 11:195–209.
68. Lubin, A., Hord, D., Tracy, M. L., and Johnson, L. C. (1976): Effects of exercise, bedrest and napping on performance decrement during 40 hours. *Psychophysiology*, 13:334–339.
69. Lugaresi, E., Cirignotta, F., Zucconi, M., Mondini, S., Lenzi, P. L., and Coccagna, G. (1983): Good and poor sleepers: An epidemiological survey of the San Marino population. In: *Sleep/Wake Disorders: Natural History, Epidemiology, and Long Term Evolution*, edited by C. Guilleminault and E. Lugaresi, pp. 1–12. Raven Press, New York.
70. Lumley, M., Roehrs, T., Asker, D., Zorick, F., and Roth, T. (1987): Ethanol and caffeine effects on daytime sleepiness/alertness. *Sleep*, 10:306–312.
71. Maron, L., Rechtschaffen, A., and Wolpert, E. A. (1964): Sleep cycle during napping. *Arch. Gen. Psychiatry*, 11:503–508.
72. Masterton, J. P. (1965): Patterns of sleep. In: *The Physiology of Human Survival*, edited by O. G. Edholm and A. L. Bacharach, pp. 387–397. Academic Press, New York.
73. McGhie, A. (1966): The subjective assessment of sleep patterns in psychiatric illness. *Br. J. Med. Psychol.*, 39:221–230.
74. McGhie, A., and Russell, S. M. (1962): The subjective assessment of normal sleep patterns. *J. Ment. Sci.*, 108:642–654.
75. Mitler, M. A., Carskadon, M. A., Czeisler, C. A., Dement, W. C., Dinges, D. F., and Graeber, R. C. (1988): Catastrophes, sleep and public policy: Consensus report of a committee for the Association of Professional Sleep Societies. *Sleep*, 11:100–109.
76. Miyasita, A., Inugami, M., and Niimi, Y. (1979): Spontaneous skin potential responses during diurnal naps. *Sleep Res.*, 8:64.
77. Monk, T. H., Moline, M. L., and Graeber, R. C. (1986): The effect of a brief post-lunch nap on circadian adjustment to a 6h phase advance in routine. *Sleep Res.*, 15:278.
78. Morrison, A. R., and Dinges, D. F. (1988): Reports of inhibitory motor experiences in a normal young adult population. In: *Sleep '86*, edited by W. P. Koella, F. Obal, H. Schulz, and P. Visser, pp. 409–410. Gustav Fischer Verlag, New York.
79. Murray, E. J. (1965): *Sleep, Dreams, and Arousal*. Appleton-Century-Crofts, New York.
80. O'Connor, A. L. (1964): Questionnaire responses about sleep. Master's thesis, University of Florida.
81. Ogunremi, O. (1978): The subjective sleep pattern in Nigeria. *Sleep Res.*, 7:163.
82. Okudaira, N., Fukuda, H., Ohtani, K., Nishihara, K., Endo, S., and Torii, S. (1981): Sleep patterns in healthy elderly persons in Vilcabamba, Ecuador. *Sleep Res.*, 10:116.
83. Orne, M. T. (1962): On the social psychology of the psychological experiment: With particular reference to demand characteristics and their implications. *Am. Psychol.*, 17:776–783.
84. Percival, J. E., and Woodrow, K. (1983): The effects of 1/2 hour and 1/4 hour appetitive, afternoon naps upon vigilance performance and Stanford sleepiness scale ratings. Paper presented at 4th International Congress of Sleep Research, Bologna, Italy.
85. Porter, J. M., and Horne, J. A. (1981): Bedtime food supplements and sleep: Effects of different carbohydrate levels. *Electroencephalogr. Clin. Neurophysiol.* 51:426–433.
86. Rechtschaffen, A. (1979): The function of sleep: Methodological issues. In: *The Functions*

of Sleep, edited by R. Drucker-Colin, M. Shkurovich, and M. B. Sterman, pp. 1–17. Academic Press, New York.

87. Rutenfranz, J., and Colquhoun, W. P. (1979): Circadian rhythms in human performance. *Scand. J. Work Environ. Health*, 5:167–177.

88. Soldatos, C. R., Madianos, M. G., and Vlachonikolis, I. G. (1983): Early afternoon napping: A fading Greek habit. In: *Sleep, 1982*, edited by W. P. Koella, pp. 202–205. Karger, Basel.

89. Spiegel, R. (1981): *Sleep and Sleepiness in Advanced Age*. SP Medical and Scientific Books, New York.

90. Spring, B., Maller, O., Wurtman, J., Digman, L., and Cozolino, L. (1982/83): Effects of protein and carbohydrate meals on mood and performance: Interactions with sex and age. *J. Psychiatr. Res.*, 17:155–167.

91. Stahl, M. L., Orr, W. C., and Bollinger, C. (1983): Postprandial sleepiness: Objective documentation via polysomnography. *Sleep*, 29–35.

92. Stegman, H. M. (1932): The nap as a pick-me-up. *Hygeia*, 10:541–543.

93. Strauch, I., and Meier, B. (1988): Sleep need in adolescents: A longitudinal approach. *Sleep*, 11:378–386.

94. Strogatz, S. H. (1986): *The Mathematical Structure of the Human Sleep-Wake Cycle. Lecture Notes in Mathematics No. 69*. Springer-Verlag, Berlin/New York.

95. Taub, J. (1971): The sleep-wakefulness cycle in Mexican adults. *J. Cross-Cultural Psychol.*, 44:353–362.

96. Taub, J. M. (1977): Napping behavior, activation and sleep function. *Waking and Sleeping*, 1:281–290.

97. Taub, J. M. (1979): Effects of habitual variations in napping on psychomotor performance, memory and subjective states. *Int. J. Neurosci.*, 9:97–112.

98. Taub, J. M. (1982): Effects of scheduled afternoon naps and bedrest on daytime alertness. *Int. J. Neurosci.*, 16:107–127.

99. Taub, J. M., and Berger, R. J. (1973): Performance and mood following variations in length and timing of sleep. *Psychophysiology*, 10:559–570.

100. Taub, J. M., and Berger, R. J. (1974): Acute shifts in the sleep-wakefulness cycle: Effects on performance and mood. *Psychosom. Med.*, 36:164–173.

101. Taub, J. M., Hawkins, D. R., and Van de Castle, R. L. (1978): Temporal relationships of napping behavior to performance, mood states and sleep physiology. *Sleep Res.*, 7:164.

102. Taub, J. M., and Tanguay, P. E. (1977): Effects of naps on physiological variables, performance and self-reported activation. *Sleep Res.*, 6:117.

103. Taub, J. M., Tanguay, P. E., and Clarkson, D. (1976): Effects of daytime naps on performance and mood in a college student population. *J. Abnorm. Psychol.* 85:210–217.

104. Taub, J. M., Tanguay, P. E., and Rosa, R. R. (1977): Effects of afternoon naps on physiological variables, performance and self-reported activation. *Biol. Psychol.* 5:191–210.

105. Tilley, A., Donohoe, F., and Hensby, S. (1987): Homeostatic changes in slow wave sleep during recovery sleep following restricted nocturnal sleep and partial slow wave sleep recovery during an afternoon nap. *Sleep*, 10:600–605.

106. Torsvall, L., Åkerstedt, T., Gillander, K., and Knutsson, A. (1986): 24h recordings of sleep/wakefulness in shift work. In: *Night and Shift Work: Longterm Effects and Their Prevention*, edited by M. Haider, M. Koller, and R. Cervinka, pp. 37–41. Verlag Peter Lang, Frankfurt am Main.

107. Tune, G. S. (1968): Sleep and wakefulness in normal human adults. *Br. Med. J.*, 2:269–271.

108. Tune, G. S. (1969): The influence of age and temperament on the adult human sleep-wakefulness pattern. *Br. J. Psychol.*, 60:431–441.

109. Tune, G. S. (1969): Sleep and wakefulness in 509 normal human adults. *Br. J. Med. Psychol.*, 42:75–80.

110. Webb, W. B. (1975): *Sleep: The Gentle Tyrant*. Prentice Hall, Englewood Cliffs, NJ.

111. Webb, W. B. (1978): Sleep and naps. *Specul. Sci. Tech.*, 1:313–318.

112. Webb, W. B. (1978): Review of sleep displacement studies by J. M. Taub and R. J. Berger. *Sleep Res.*, 7:423–425.

113. Webb, W. B. (1981): Patterns of sleep in healthy 50–60 year old males and females. *Res. Commun. in Psychol. Psychiatr. Behav.*, 6:133–140.

114. Webb, W. B., and Aber, W. R. (1984/85): Relationships between sleep and retirement-nonretirement status. *Int. J. Aging Hum. Dev.*, 20:13–19.
115. Webb, W. B., and Agnew, H. W., Jr. (1967): Sleep cycling within twenty-four hour periods. *J. Exp. Psychol.*, 74:158–160.
116. Webb, W. B., Agnew, H. W., Jr., and Sternthal, H. (1966): Sleep during the early morning. *Psychon. Sci.*, 6:277–278.
117. Webb, W. B., and Dube, M. G. (1981): Temporal characteristics of sleep. In: *Handbook of Behavioral Neurobiology. Biological Rhythms*, Vol. 4, edited by J. Aschoff, pp. 499–522. Plenum Press, New York.
118. White, R. M., Jr. (1975): Sleep length and variability: Measurement and interrelationships. Doctoral dissertation, University of Florida. [Univ. Microfilms Int., Ann Arbor, MI, 1977 (76-4283).]
119. Wurtman, R. J., and Fernstrom, J. D. (1974): Effects of the diet on brain neurotransmitters. *Nutr. Rev.*, 32:193–200.

Sleep and Alertness: Chronobiological,
Behavioral, and Medical Aspects of
Napping, edited by
D. F. Dinges and R. J. Broughton.
Raven Press, Ltd., New York © 1989.

10

Shift Work and Napping

*†Torbjörn Åkerstedt, *Lars Torsvall, and ‡Mats Gillberg

*IPM and Stress Research, Karolinska Institute, and
†National Institute for Psychosocial Factors and Health, and
‡National Defense Research Institute, Stockholm, Sweden

As evidenced in many chapters of this book, napping is a widespread phenomenon and occurs under many conditions of life. This chapter discusses its occurrence in relation to shift work. Attention is focused on the pattern of napping, its determinants, and its related benefits and disadvantages.

SHIFT WORK

The concept of shift work essentially means that an extended production process is covered by two or more teams (shifts) that relieve each other according to some type of schedule. The term applies to permanent as well as rotating shifts and to diurnal schedules as well as nocturnal or combined (nycthemeral) ones. Typical work hours may be 0600 to 1400 hr for the morning shift, 1400 to 2200 hr for the afternoon shift, and 2200 to 0600 hr for the night shift. Roster work, which is more irregular and may contain more overlaps or gaps between shifts, also belongs under this heading. Work hours that are only slightly displaced from the daytime and do not constitute a part of an extended production process (i.e., "displaced" work hours) belong elsewhere.

Maurice (41) estimated that approximately 20% of the work force in the industrialized world is engaged in shift work. In a detailed Swedish survey of 45,000 individuals (46), it was found that 5.1% were engaged in nycthemeral shift or roster work, 0.5% in permanent night work, and 9.3% in diurnal shift or roster work. Another 14.4% had displaced or unscheduled work hours primarily occurring at times other than at night.

NAPPING BEHAVIOR

Survey Studies

Many, but not all, of the major survey studies of shift work and sleep have specified the prevalence of napping. Tune (57), in a classic study, found a higher frequency of napping in shift workers (three shifts) than in day workers, but only 0.7 compared with 0.5 naps per week. The average duration of a nap was 1.1 hr compared with 30 min, which yielded a longer total sleep time (per 24-hr period) for shift workers than for day workers. The pattern during different shifts was not investigated, however. Several later studies quite clearly demonstrated that napping occurs mainly in relation to the night shift (7,9,33), sometimes also in relation to the morning shift. For example, the latter study (33) demonstrated that 33% of a group of three-shift workers took a nap after the night shift. The same proportion did so after the morning shift, whereas virtually no naps occurred in relation to the afternoon shift. Typical nap durations were just less than 2 hr, and the morning shift nap was begun 2.5 hr earlier than the night shift nap (which began at 1949 hr).

A more detailed illustration of the napping pattern is given in Fig. 1, which summarizes part of a time-budget study of three-shift workers (5). It shows that in connection with the morning shift the times of retiring and rising are very clearly defined; all rise and retire at approximately the same time. The afternoon shift gives a more heterogeneous picture. Some individuals apparently stay up a while after work and the time of awakening spans the interval from 0600 to 1200 hr; very little napping occurs. Sleep after the night shift is bimodal and demonstrates a rather short main sleep followed by extensive afternoon napping.

Aside from afternoon leisure time napping, there may in some occupations actually be time for a small nap during work hours. Andersen (9) reported that 50% of the policemen in his study took a small nap during the night shift. Kogi (34) found that nighttime naps were taken by 40% to 50% of Japanese shift workers on nycthemeral schedules. These naps appear to be fully sanctioned and are made possible by on-site bedroom facilities. Despite unavoidable environmental disturbances, the durations of these naps vary between 2 and 3 hr.

Thus, survey studies indicate that napping is very common in shift workers. How stable is this pattern? Very few data are available to answer this question. We attempted to address it by readministering the questionnaire from our study of three-shift workers cited previously (7). During the 18 months that passed between the administrations, 30% of the three-shift workers had been transferred to day or two-shift work (for economic reasons). In the first administration, 49% of the three-shift workers were classified as nonnappers, whereas the remaining 51% reported napping habitually (between once a month and daily) in connection with either night shifts, morning

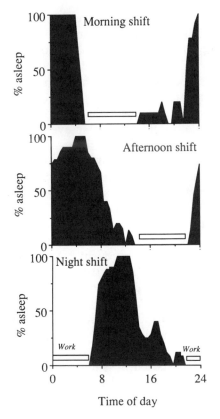

FIG. 1. Percentage of three-shift workers (*n* = 17) who were asleep at different times of day in connection with morning, afternoon, and night shifts. (Data from ref. 5.)

shifts, or leisure time. No napping occurred in connection with afternoon shifts. Of the original nonnappers, 81% of those who stayed in three-shift work continued to be nonnappers. Of those who changed to day or two-shift work, 95% remained nonnappers. Of the original nappers, 56% of those who continued in three-shift work remained nappers. Among those who changed to day work, the corresponding figure was only 31%. The results give a rather clear impression that napping behavior is not a very stable habit if environmental conditions are changed, and it is not therefore a strong trait of most individuals.

Electroencephalographic Studies

Although a number of electroencephalographic (EEG) studies have been carried out on the sleep of shift workers, most have been concerned solely with the main sleep episode. A particular problem is the spontaneous nature of napping in shift workers, which makes it unamenable to conventional laboratory/hospital techniques for EEG recording. This may be solved through continuous 24-hr ambulatory monitoring techniques. We carried out such a study of 25 rotating shift workers in a process industry (55). We selected the 24-hr period that started with the night (2200–0600 hr) and afternoon (1400–2200 hr) shifts, respectively. The results showed that no subject took a nap in connection with the afternoon shift. After the main sleep following night work, eight subjects took a nap with an average duration of 45 min. These naps tended to occur at approximately 1500 hr and were characterized by slow wave sleep (SWS), with little or no rapid eye movement (REM) sleep.

To our surprise we also found that five subjects took a nap during night work hours. This nap was unauthorized and to a great extent involuntary. It occurred during the second half of the night shift when sleepiness was at its peak and was taken in a sitting position at the actual work site. Figure 2 illustrates the sleep pattern in connection with the night shift for one of the workers. During work he falls asleep three times. After work he has a relatively short main sleep episode, followed by a subsequent (voluntary) nap. Incidentally, the diaries of the shift workers confirmed the sleep length data and the frequency of voluntary napping; they did not, however, reveal any of the involuntary napping during work hours.

In one of the first EEG studies that included napping, Foret and Lantin (23) recorded the sleep of train drivers, at home and in dormitories. The daytime sleep after night work varied from full duration (8 hr) to short naps, with a decrease in duration the later sleep was started. It is not clear, however, how much the results were influenced by external factors, such as having to go back to work, poor daytime sleep environment, or by any (uncontrolled) prior sleep. The results were corroborated in a second study (22). A detailed analysis of the EEG showed that the proportion of REM sleep during the night increased with increasing sleep duration. In addition, morning sleep contained disproportionately high amounts of REM sleep. SWS retained its normal proportions in both sleep periods.

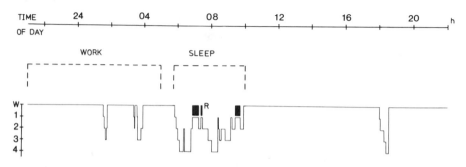

FIG. 2. Hypnogram for the 24-hr ambulatory EEG recording of one three-shift worker in connection with a night shift. The write-out was based on visually scored data. (Data from ref. 55.)

Torii et al. (52) studied a group of four nurses on a three-shift system. The change from day to night shift work took place on the same day and left only 6 hr free, between 1700 and 2300 hr, before work was resumed. On this transition day, their habits were to take a 3 hr nap before and after the night shift, followed by a normal night sleep 2300 to 0600 hr. This habitual schedule was studied in the sleep laboratory. The results showed that total sleep time for normal night sleep was 6.9 hr; both naps reached 2.3 hr and the subsequent night sleep was 6.8 hr. REM sleep was almost completely absent during the naps despite their considerable length. SWS content was reduced during the naps, but when values for the two nap periods were combined, they exceeded that of baseline sleep (100 versus 81 min).

In another study of scheduled sleep, Matsumoto et al. (39) recorded the sleep of guards working a 24-hr shift. The guards took a scheduled 3.5-hr nap during the night. Control night sleep averaged 8.7 hr long compared with an average 3.1-hr nighttime nap and a 4.1-hr postwork (daytime) recovery sleep. REM sleep was associated with increasing sleep length whereas SWS during the nighttime matched that of baseline sleep (67 versus 73 min) and exceeded that at recovery sleep (46 min). In an updated version of the study (40), the authors reported normal night sleep averaging 8.1 hr, the nighttime nap averaging 3.2 hr, and subsequent recovery (day) sleep averaging 4.0 hr; SWS amounted to 96, 63, and 52 min of each sleep period, respectively.

Thus, of the studies that have been conducted on polygraphic monitoring of sleep in shift workers, there is clear evidence that both voluntary and involuntary napping occurs. The results do, however, vary greatly with respect to sleep duration and composition of sleep stages. The probable reason for this is the very different conditions for groups studied.

REASONS FOR NAPPING

As discussed in several other chapters in this volume, the major reasons for napping are to reduce sleepiness or for simple enjoyment. In addition, the first factor is influenced by sleep loss, biological rhythmicity, and exposure to monotony. On the whole, napping in shift work is regulated by the same factors. Shift work does, however, through its disruptive effects on sleep, offer a rather special setting in which to study napping. As will be seen, this setting brings forward some aspects that are not always considered in discussions of nap mechanisms.

Sleep Loss

In the prior section it was made clear that napping mainly occurred after night or, possibly, morning shifts. This must, of course, be seen against the background of the total 24-hr sleep/wakefulness pattern. Most survey studies show that the duration of the main sleep episode is reduced by 2 to 3 hr after the night shift and sometimes by as much in connection with the morning shift; the latter effect depends on how early the shift starts (1,9,11,33,57). As may be expected in view of such shortened sleep, most shift workers also complain of not getting sufficient sleep and feeling fatigued.

EEG studies of sleep of shift workers have mainly involved recording after the night shift (23,38,51,56). These studies, carried out in the workers' normal sleep environments, have confirmed the finding of reduced sleep and shown that the loss mainly affects REM and stage 2 sleep. Stages 3 and 4 sleep (SWS) seldom seem affected. Another frequently observed phenomenon is that REM sleep tends to be rather evenly distributed throughout the sleep period, instead of the gradual increase usually found during night sleep.

When the results from the above studies are compared with those that have assessed napping, it is clear that those shifts that cause a sleep deficit are also the ones that are most likely to be associated with napping. Furthermore, a few studies have demonstrated that, even if average main sleep length is reduced, there is still a large variability among individuals and this variability is directly related to napping behavior. Figure 3, employing data from our own study of three-shift workers, illustrates how the proportion of subjects napping decreases with increasing duration of the main sleep episode (7). The relationship is the same for both the morning and afternoon shifts.

The composition of the main sleep also may be of importance in determining whether naps will occur. Using the same reasoning as above, we find that napping occurs in association with those shifts that cause deficits of stage 2 and REM sleep. The results are somewhat surprising, since it is SWS that has been associated with the homeostatic aspect of sleep (13,21,30). In the only study in which nappers and nonnappers have been compared in respect to postshift main sleep, no significant differences were found in the amounts of different sleep stages (55).

The picture does not get much clearer if the sleep lost in the main sleep

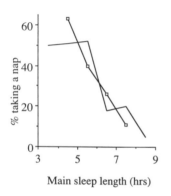

FIG. 3. The relation between the length of the major sleep episode and the proportion of three-shift workers taking a nap (n = 282). Lowest n in any sleep length category was not <15. (—) night shift; (□) morning shift. (Data from ref. 7.)

period is compared with what is recovered during the nap. In our survey of three-shift workers, we found no correlation between the amount of sleep deficit and that obtained during the nap (7). Sakai and Kogi (45) did find a clear negative correlation between the nap during the night shift and the duration of the subsequent day sleep. This observation is supported by the EEG study of security guards by Matsumoto et al. (39). The guards who had a 3-hr nap during the night shift had considerably shorter subsequent day sleep than did a control group of night nurses who did not take a nap during work.

Torii et al. (52) found that the night nurses who napped before the night shift slept much less (2 hr) than what is usually observed after a night shift without preceding naps. SWS content, however, behaved more according to expectation; that is, prework SWS and postwork SWS were reduced. Together they slightly exceeded that of normal night sleep.

These results indicate that lack of sleep increases the probability of napping and that there may be some (negative) relationship between the sleep deficit and subsequent nap duration. To what extent the content of sleep is of importance is unclear (see Chapter 9). Possibly, however, the recent techniques of period-amplitude analysis (21) and spectral analysis (13), which yield measures much closer to the actual EEG quantification than does conventional visual analysis, might provide additional information.

In one experimental study, we allowed subjects to sleep 8, 4, 2, or 0 hr during the night (always being awakened at 0700 hr), followed by a new opportunity for sleep at 1100 hr (with spontaneous ending of sleep) (4). The daytime sleep turned out to mainly consist of naps (<4 hr of sleep) whose durations correlated negatively with prior sleep length. In no way, however, did the nap compensate for all the sleep lost (whether as total duration or sleep stages). When spectral power density was computed, however, it was found that each nap ended at approximately that moment at which the night-time power density loss had been recovered. This seems to indicate that there are measurable properties of the EEG that may represent the homeostatic regulation of napping in shift work (and other areas as well).

Is Sleepiness Reduced?

If recovery from sleep loss is a major reason for napping in shift work, then it should be possible to observe positive effects on wakefulness or performance. In traditional research on napping such effects are well established (see Chapters 9 and 11). In shift work there is no doubt that sleepiness/fatigue is a major problem. The sleepiness is mainly confined to the night shift and may reach levels that overcome the individual during work (8,35,44). Performance is reduced (12,14,51), and the risk for accidents is increased (25,26,29). The reduced alertness during night work is also reflected in EEG parameters, particularly as intrusions of alpha (8–12 Hz) and theta (4–8 Hz) activity in the waking EEG (54).

The availability of studies evaluating napping and wakefulness in shift work is rather limited and based mainly on questionnaire data. Ishibashi et al. (31) compared a group that took a single long continuous sleep after a night shift with a group that combined a shorter main sleep with a later nap (both groups obtained the same total sleep). The authors found that the split group had lower levels of rated fatigue.

Similarly, we found in our survey study of train drivers that the longer the nap they took *before* night work, the less fatigue they felt during the drive. Those who only slept for a short while had more complaints about their ability to fall asleep and maintain sleep whenever they slept during nonnight hours. For example, they also had the shortest sleep after the night shift. Interestingly, the maximum number of complaints was not found among those who took no naps but rather among those who did nap for only a short time. Apparently, the nonnappers contained a considerable proportion of subjects who simply did not need to nap (i.e., possibly habitual short sleepers).

Studies of napping during the night shift also show interesting and positive effects. Sakai and Kogi (45) found that 96% of the shift workers in their study had a very positive attitude toward night-shift napping. The most frequently cited reason for this positive view was the experience of increased performance capacity and (perceived) improvement in safety on the job.

To the best of our knowledge, the positive effects of napping have never been experimentally demonstrated under conditions of shift work. It has been demonstrated in the laboratory, however, that monophasic sleep may be broken up into naps without any negative effects on functioning (27,37). Total nap time per 24 hr does not even have to add up to the normal amount of sleep time obtained during monophasic sleep. Furthermore, Opstad et al. (43) noted that naps interjected during 92 to 120 hr of continuous combat training reduced the profound loss in behavioral efficiency of military cadets. Haslam (28) made similar observations.

In addition, a few laboratory studies tried to evaluate short evening naps as prophylaxis against fatigue during the night. Angiboust and Gouars (10) scheduled a 2-hr nap from 2200 to 2400 hr for an experimental group. Another group (control) who did not get the nap suffered a substantial decrement in vigilance performance at 0300 and 0500 hr when this was tested. The nap group, on the other hand, kept their performance almost at daytime levels. Gillberg (24) obtained similar results for reaction time performance from 1-hr naps started at 1900 hr.

In the most extensive study of prophylactic napping to date, Dinges and colleagues (18,19) observed that a 2-hr nap opportunity at either 1500 or 0300 hr during the first 18 hr of wakefulness (in a 56-hr sustained work scenario) prevented some of the drop in reaction time and cognitive performance, as well as oral temperature, seen in subjects who did not nap until 30 hr or more of wakefulness. There were two other noteworthy outcomes of this study. First, the positive effects on performance and temperature were not present in subjective self-report measures of mood and sleepiness; that is, subjects did not appear to be phenomenally aware of performing better than they would have had they taken no nap in advance of sleep loss. Second, although the earlier prophylactic naps involved less TST and SWS than later naps (15), they still afforded considerable performance benefit. Thus, napping in advance of sleep loss appears to aid performance more than mood, even though the nap may not involve much deep sleep (for further discussion of these issues see Chapters 8, 9, and 11). There seems to be no reason why these findings should not apply also to shift workers.

Apparently, then, napping will alleviate the reduced alertness that often occurs in shift work. This plays a part in the afternoon nap after a night shift, although the impetus then is acute sleepiness. It also plays a part in more calculated strategies to prevent anticipated fatigue during night shifts. This is a reason for napping that clearly differs from the traditional reason (i.e., as a response to acute sleepiness after sleep loss).

Biological Rhythmicity

A very obvious factor in sleep/wake regulation is biological rhythmicity. General aspects of this subject are discussed elsewhere in this volume (Chapters 5, 6, and 7). Suffice it to reiterate here that the ability (or "drive") to sleep is strongest in the early morning and weakest during the late afternoon and early evening. This pattern is directly related to the trough and peak, respectively, of the body temperature rhythm. In addition, there seems to be a 12-hr or even more rapidly oscillating rhythm that provides a gate to sleep also close to the peak (i.e., during the afternoon). This gate may be particularly obvious if the individual carries an extra load of sleep loss.

Biorhythmic considerations have several implications for the sleep/wake pattern in shift work. First and foremost, the circadian rhythm causes severe sleepiness during the night shift (6). Second, sleep after the night shift is curtailed by the rising circadian rhythm (3). Third, the afternoon dip in alertness steers the nap to this time. When naps are not taken at the critical lows in alertness there is, apparently, an increased risk for reduced performance capacity and even dozing off incidents (12,25,26,29,44).

One should also keep in mind that the need for the nap and the opportunity for it must coincide. Thus, in connection with the afternoon shift there is little reason to expect a need for a nap during the relatively short free time between rising from night sleep and going to work. In connection with morning or night shifts, there may be a need for an (afternoon) nap, but social demands may well prevent it from occurring. The time when the afternoon gate to sleep is open is rather limited, and if the opportunity is passed over, it may be difficult to initiate a nap during the evening when the circadian rhythm is promoting wakefulness (59). This difficulty of prophylactic evening sleep is frequently encountered in shift workers (8,9).

Individual Differences

One very important individual aspect of napping is age; both at the beginning and end of our life span sleep is polycyclic (see Chapters 3 and 4). In shift work a similar pattern was observed by Bjerner and colleagues (11). They found that, particularly after the night shift, the frequency of napping increased with increasing age. This was related to a reduced ability of older shift workers to maintain sleep. A similar observation has been made by Sakai and Kogi (45).

In our survey study of three-shift workers and their napping behavior (7), we found three distinct types: (a) those who napped only in connection with the morning shift (18%), (b) those who napped only in connection with the night shift (18%), and (c) those who napped in connection with both the morning and night shifts (15%). The rest (49%) did not nap at all. As may have been expected, the four groups differed with respect to "diurnal type" (53). Those who napped in the afternoon after a night shift were clearly morning types. Those who napped in the afternoon after a morning shift were clearly evening types. The two remaining groups were in-between these diurnal types. The reason for the different napping patterns was, apparently, that morning types were too alert in the morning after the night shift to obtain sufficient amounts of main sleep, whereas the evening types were too alert to sleep well before the morning shift. Therefore both types compensated with an afternoon nap. It also should be noted that age was involved in these phenomena; the post-night shift nappers were 7 to 8 years older on average than the other groups.

Other Reasons for Napping

Very little is known about pleasure-seeking motives for napping in shift work. It seems, however, that the disruptive nature of shift work on sleep and wakefulness pattern would demand that sleep be used most efficiently, thus precluding luxuries such as appetitive or recreational naps. This remains to be empirically demonstrated, however.

Another reason for napping in shift work (and elsewhere) is to avoid boredom or an unpleasant situation. To what extent this occurs in shift work is completely unknown.

A CAVEAT: NEGATIVE EFFECTS OF NAPPING

The previous discussion has strongly emphasized the benefits of napping to the shift worker. The effects need not, however, be undividedly positive. Naps may increase fatigue and cause sleep disturbances. Unpleasant effects of naps were reported by some of the shift-working policemen in the study by Andersen (9). Experimental results to the same effect were reported in a study of a small subset of healthy young adult nonnappers who find naps physically or mentally unpleasant (20). The reason for these effects being especially pronounced in some individuals is unknown. Part of the explanation, however, might be the well-known sleep inertia that follows upon awakening from sleep.

Sleep Inertia

Sleep inertia was first studied by Langdon and Hartman (36). Since then it has been demonstrated for a variety of tasks, including simple reaction time (61), short-term memory (49), and cognitive performance (47,60). This inertia, however, is rather quickly dissipated and followed by the recuperative phase of the nap. Webb and Agnew (58) estimated the inertia span to be 1 to 5 min. Wilkinson and Stretton (60) gave 15 min as their estimate. Dinges et al. (16) found the time varied between 5 and 35 min. Interestingly, reaction time performance was directly related to the sleep stage at awakening, stage 4 yielding maximum reaction times. Cognitive performance (backward subtraction), on the other hand, was related to the total amount of SWS and other factors associated with sleep depth (17). For example, cognitive performance deteriorated most when sleep was dominated by stages 3 and 4 sleep.

Although sleep inertia is normally rapidly dissipated, it appears that it may be extended under certain circumstances. Naitoh (42) studied the effects of 2-hr naps in a sleep deprivation design. He found that when the nap was taken at 1200 hr after 53 hr of waking, the benefits were considerable on a number of measures. When it was taken at 0400 hr (8 hr earlier), however, the effects were detrimental for 6 hr afterward. It was suggested that this was owing to the effects of the circadian trough together with the increased need for sleep after the extended period of waking. The latter may have induced an extremely deep level of sleep, leading to a pronounced inertia. It was suggested that the nap may have been too short in relation to the need for sleep. In a similar study, profound sleep inertia effects were evident but not prolonged (18). We also have seen dramatic inertia effects from 64 hr of waking on recovery sleep (2). These effects dissipated toward the morning when most of the need for sleep had been satisfied.

These studies indicate clearly that sleep inertia exists. Whether it constitutes a significant problem in shift work is unclear. If it does, it would probably be in relation to schedules that allow individuals to sleep while being on call. Conceivably, emergency situations that must be handled a few minutes after awakening might be dealt with in an improper manner. This would have to be demonstrated empirically, however.

Disturbed Postnap Sleep

Since the nap plays an important role in reducing the need for sleep, it stands to reason that a postnap sleep might be difficult to initiate and maintain. Karacan et al. (32) found that a 2-hr afternoon nap clearly affected the ensuing night sleep (i.e., sleep latency was increased and the amount of SWS was reduced). In individuals with a weak sleep mechanism, it is possible that naps could be instrumental in disturbing the subsequent night's sleep. In fact, sleep restriction during the day is now an established behavioral treatment of insomnia (48). Feinberg et al. (21) demonstrated, using period-amplitude analysis of the EEG, that night sleep after a 2-hr afternoon (1700–1900 hr) nap suffers a reduction of EEG delta wave content that closely corresponds to that obtained in the nap. Associated effects were increased sleep latency and decreased duration of total sleep time. Naitoh (42) found that naps taken during sleep deprivation reduced much of the SWS rebound during recovery sleep after 64 hr of sleep deprivation.

In their study of nurses on occasional night shifts, Torii et al. (52) found that the naps taken before and after night work reduced both total sleep time and SWS during subsequent night sleep. In addition, REM sleep and sleep latency were increased. These findings seem to indicate that the shift-work napping reduced the quality of night sleep, at least in the sense that it became more superficial and took longer to initiate.

The findings previously discussed seem to suggest that napping in connection with a night shift may certainly help recovery from sleep loss but also may have negative effects on subsequent night sleep. Such night sleep, however, occurs only after the final night shift in a series and may not necessarily be of major importance to the shift worker. The effects of shift-work naps on subsequent sleep are essentially unknown. However, Tepas (50) found that permanent night-shift workers who napped in the afternoon had a shorter main sleep duration and generally more sleep complaints than did nonnappers. He suggested that the nap may play a part in the development of night-shift sleep problems and that it may be useful to recommend only limited napping or even to advise individuals who need to nap not to take up shift work. This notion will, however, need more experimental support.

CONCLUDING DISCUSSION

The studies reviewed in this chapter are clearly convergent in respect to the description of napping behavior in shift work. Napping is widespread. It occurs in relation to night and morning shifts (i.e., when sleep has been disturbed by the shift system). Reasons for napping among shift workers are response to fatigue and prevention of fatigue, and the timing of naps appears to be determined by circadian and ultradian sleep regulation. The latter is also reflected in different napping behaviors for morning- and evening-type individuals. Occasionally, napping may have negative effects on the shift workers' sleep and/or wakefulness.

How can this knowledge benefit the shift worker? First, if the next major sleep period is going to be taken at night, it is probably a good idea not to nap. This will ensure a good night's sleep of considerable recuperative value. If, however, the next night is going to be one of work, it would seem sensible to prevent some of the night shift fatigue by taking an afternoon (or evening) nap. This is of increasing importance the more that vigilance is required to carry out the tasks. Second, it is not possible to estimate the optimal amount (length) of the nap because the content (sleep structure) of the nap appears to be more important than the duration. Moreover, an individual shift worker would have to try naps of different durations to determine which has the optimal effects for him or her. If there are difficulties initiating a nap, one might consider reducing the sleep duration of the preceding major sleep period (e.g., during the day after the night shift). This suggestion may seem strange, since it would cause a sleep deficit. However, if this procedure helps to initiate and maintain an afternoon nap, the net effect would presumably be to lessen night-shift fatigue.

ACKNOWLEGMENTS

This work was supported by the Swedish Work Environment Fund and the Swedish Medical Research Council. We also wish to thank David Dinges for valuable comments on the manuscript.

REFERENCES

1. Aanonsen, A. (1964): *Shift Work and Health.* Universitetsforlaget, Oslo.
2. Åkerstedt, T., and Gillberg, M. (1979): Effects of sleep deprivation on memory and sleep latencies in connection with repeated awakenings from sleep. *Psychophysiology*, 16:49–52.
3. Åkerstedt, T., and Gillberg, M. (1981): The circadian variation of experimentally displaced sleep. *Sleep*, 4:159–169.
4. Åkerstedt, T., and Gillberg, M. (1986): Sleep duration and the power spectral density of the EEG. *Electroencephalogr. Clin. Neurophysiol.*, 64:119–122.
5. Åkerstedt, T., and Torsvall, L. (1976): Tidsanvandning och skiftarbete. (Use of time in

shift work). Stress Research Reports No. 49. Dept. of Stress Research, Karolinska Institute, Stockholm.

6. Åkerstedt, T., Torsvall, L., and Gillberg, M. (1982): Sleepiness and shift work: Field studies. *Sleep*, 5:S95–S106.

7. Åkerstedt, T., and Torsvall, L. (1985): Napping in shift work. *Sleep*, 8:105–109.

8. Åkerstedt, T., Torsvall, L., and Froberg, J. E. (1983): A questionnaire study of sleep/wake disturbances and irregular work hours. *Sleep Res.*, 12:358.

9. Andersen, J. E. (1970): *Three-Shift Work—A Socio-Medico Investigation*, vol. I, pp. 134–159. Teknisk Forlag, Copenhagen.

10. Angiboust, R., and Gousars, M. (1972): Tentative d'évaluation de l'efficacité operationelle du personnel de l'aeronautique militaire au cours de veilles nocturnes. In: *Aspects of Human Efficiency*, edited by W. P. Colquhoun, pp. 151–170. The English Universities Press, London.

11. Bjerner, B., Holm, A., and Swensson, A. (1948): *On Matt och Skiftarbete* (*Night and Shift Work*). Statens Offentliga utredningar, Stockholm.

12. Bjerner, B., Holm, A., and Swensson, A. (1955): Diurnal variation of mental performance. A study of three-shift workers. *Br. J. Ind. Med.*, 12:103–110.

13. Borbély, A. A., Baumann, F., Brandeis, D., Strauch, H. J., and Lehmann, D. (1981): Sleep deprivation: Effect on sleep stages and EEG power density in man. *Electroencephalogr. Clin. Neurophysiol.*, 51:483–493.

14. Browne, R. C. (1949): The day and night performance of teleprinter switch board operators. *Occup. Psychol.*, 23:121–126.

15. Dinges, D. F. (1986): Differential effects of prior wakefulness and circadian phase on nap sleep. *Electroencephalogr. Clin. Neurophysiol.*, 64:224–227.

16. Dinges, D. F., Orne, E. C., Evans, F. J., and Orne, M. T. (1981): Performance after naps in sleep-conducive and alerting environments. In: *Biological Rhythms, Sleep and Shift Work*, edited by L. C. Johnson, D. I. Tepas, W. P. Colquhoun, and M. J. Colligan, pp. 539–552. Spectrum, New York.

17. Dinges, D. F., Orne, M. T., and Orne, E. C. (1985): Sleep depth and other factors associated with performance upon abrupt awakening. *Sleep Res.*, 14:92.

18. Dinges, D. F., Orne, M. T., Whitehouse, W. G., and Orne, E. C. (1987): Temporal placement of a nap for alertness: Contributions of circadian phase and prior wakefulness. *Sleep*, 10:313–329.

19. Dinges, D. F., Whitehouse, W. G., Orne, E. C., and Orne, M. T. (1988): The benefits of a nap during prolonged work and wakefulness. *Work and Stress*, 2:139–153.

20. Evans, F. J., Cook, M. R., Cohen, H. D., Orne, E. C., and Orne, M. T. (1977): Appetitive and replacement naps: EEG and behavior. *Science*, 197:687–689.

21. Feinberg, I., March, J. D., Floyd, T. C., Jimison, R., Bossom-Demitrack, L., and Katz, P. H. (1985): Homeostatic changes during post-nap sleep maintain baseline levels of delta EEG. *Electroencephalogr. Clin. Neurophysiol.*, 61:134–137.

22. Foret, J., and Benoit, O. (1974): Structure du sommeil chez des travailleurs à horaires alternants. *Electroencephalogr. Clin. Neurophysiol.*, 37:337–344.

23. Foret, J., and Lantin, G. (1972): The sleep of train drivers: An example of the effects of irregular work schedules on sleep. In: *Aspects of Human Efficiency*, edited by W. P. Colquhoun, pp. 273–281. The English Universities Press, London.

24. Gillberg, M. (1984): The effects of two alternative timings of a one-hour nap on early morning performance. *Biol. Psychol.*, 19:45–54.

25. Hamelin, P. (1981): Les conditions temporelles de travail des conducteurs routiers et la sécurité routière. *Trav. Hum.*, 44:5–21.

26. Harris, W. (1977): Fatigue, circadian rhythm and truck accidents. In: *Vigilance*, edited by R. R. Mackie, pp. 133–147. Plenum Press, New York.

27. Hartley, L. R. (1974): A comparison of continuous and distributed reduced sleep schedules. *Q. J. Exp. Psychol.*, 26:8–14.

28. Haslam, D. R. (1982): Sleep loss, recovery sleep, and military performance. *Ergonomics*, 25:163–178.

29. Hildebrandt, G., Rohmert, W., and Rutenfranz, J. (1974): 12 and 24 hour rhythms in error frequency of locomotive drivers and the influence of tiredness. *Int. J. Chronobiol.*, 2:175–180.

30. Horne, J. A. (1985): The functional significance of human slow-wave sleep. In: *Sleep 84*, edited by W. P. Koella, E. Ruther, and H. Schulz, pp. 171–173. Gustav Fischer Verlag, Stuttgart.
31. Ishibashi, T., Uiura, T., Kitagawa, M., and Tainaka, H. (1982): Sleep patterns and fatigue symptoms in permanent night workers and shift workers. *J. Hum. Ergol.*, 11(suppl.):317–323.
32. Karacan, L., Williams, R. L., Finley, W. W., and Hursch, C. J. (1970): The effects of naps on nocturnal sleep: Influence of the need for stage 1 REM and stage 4 sleep. *Biol. Psychiatr.*, 2:391–399.
33. Knauth, P., Landau, K., Droge, C., Schwitteck, M., Widynski, M., and Rutenfranz, J. (1980): Duration of sleep depending on the type of shift work. *Int. Arch. Occup. Environ. Health*, 46:167–177.
34. Kogi, K. (1981): Comparison of resting conditions between various shift rotation systems for industrial workers. In: *Night and Shift Work—Biological and Social Aspects*, edited by A. Reinberg, N. Vieux, and P. Andlauer, pp. 155–160. Pergamon Press, Oxford.
35. Kogi, K., and Ohta, T. (1975): Incidence of near accidental drowsing in locomotive driving during a period of rotation. *J. Hum. Ergol. (Tokyo)*, 4:65–76.
36. Langdon, D. E., and Hartman, B. (1961): Performance upon sudden awakening. U.S. Air Force School of Aerospace Medicine (SAM report 62-17). Brooks Air Force Base, Texas.
37. Lubin, A., Hard, D. J., Tracy, M. L., and Johnson, L. C. (1976): Effects of exercise, bedrest, and napping on performance decrement during 40 hours. *Psychophysiology*, 13:334–339.
38. Matsumoto, K. (1978): Sleep patterns in hospital nurses due to shift work: An EEG study. *Waking Sleeping*, 2:169–173.
39. Matsumoto, K., Matsui, T., Kawamori, M., and Kogi, K. (1982): Effects of nighttime naps on sleep patterns of shift workers. *J. Hum. Ergol.*, 11(suppl.):279–289.
40. Matsumoto, K., and Morita, Y. (1987): Effects of nighttime naps and age on sleep patterns of shift workers. *Sleep*, 10:580–589.
41. Maurice, M. (1975): *Shift Work*. ILO, Geneva.
42. Naitoh, P. (1981): Circadian cycles and restorative power of naps. In: *Biological Rhythms, Sleep and Shift Work*, edited by L. C. Johnson, D. I. Tepas, W. P. Colquhoun, and M. J. Colligan, pp. 553–580. Spectrum, New York.
43. Opstad, P. K., Ekanger, R., Nummestad, M., and Raabe, N. (1978): Performance, mood, and clinical symptoms in men exposed to prolonged, severe physical work and sleep deprivation. *Aviat. Space Environ. Med.*, 49:1065–1073.
44. Prokop, O., and Prokop, L. (1955): Ermudung und Einschlafen am Steuer. *Zentralbl. Verkehrs-Medizin, Verkehrs-Psychologie und Angrenzende Gebiete*, 1:19–30.
45. Sakai, K., and Kogi, K. (1986): Conditions for three-shift workers to take night time naps effectively. In: *Night and Shift Work: Longterm Effects and Their Prevention*, edited by M. Haider, M. Koller, and R. Cervinka, pp. 173–180. Verlag Peter Lang, Frankfurt am Main.
46. SCB. (1982): *Oregelbundna och Obekvama Arbetstider (Irregular and Uncomfortable Work Hours)*. National Bureau of Statistics, Liber, Stockholm.
47. Scott, J., and Snyder, F. (1968): "Critical reactivity" (Pieron) after abrupt awakenings in relation to EEG stages of sleep. *Psychophysiology*, 4:370.
48. Spielman, A. J., Saskind, P., and Thorpy, M. J. (1983): Sleep restriction treatment of insomnia. *Sleep Res.*, 12:286.
49. Stones, M. J. (1977): Memory performance after arousal from different sleep stages. *Br. J. Psychol.*, 68:177–181.
50. Tepas, D. I. (1982): Shift worker sleep strategies. *J. Hum. Ergol. (Tokyo)*, 11(suppl.):325–336.
51. Tilley, A. J., Wilkinson, R. T., Warren, P. S. G., Watson, W. G., and Drud, M. (1982): The sleep and performance of shift workers. *Hum. Factors*, 24:624–641.
52. Torii, S., Okudaira, N., Fukuda, H., et al. (1982): Effects of night shift on sleep patterns of nurses. *J. Hum. Ergol.*, 11(suppl.):233–244.
53. Torsvall, L., and Åkerstedt, T. (1980): A diurnal type scale: Construction, consistency, and validation in shift work. *Scand. J. Work Environ. Health*, 6:283–290.
54. Torsvall, L., and Åkerstedt, T. (1987): Sleepiness on the job: Continuously measured EEG changes in train drivers. *Electroencephalogr. Clin. Neurophysiol.*, 66:502–511.

55. Torsvall, L., Åkerstedt, T., Gillander, K., and Knutsson, A. (1986): 24h recordings of sleep/wakefulness in shift work. In: *Night and Shift Work: Longterm Effects and Their Prevention*, edited by M. Haider, M. Koller, and R. Cervinka, pp. 37–41. Verlag Peter Lang, Frankfurt am Main.
56. Torsvall, L., Åkerstedt, T., and Gillberg, M. (1981): Age, sleep and irregular work hours: A field study with EEG recording, catecholamine excretion, and self-ratings. *Scand. J. Work Environ. Health*, 7:196–283.
57. Tune, G. S. (1969): Sleep and wakefulness in a group of shift workers. *Br. J. Ind. Med.*, 26:54–58.
58. Webb, W. B., and Agnew, Jr., H. (1964): Reaction time and serial response efficiency on arousal from sleep. *Percept. Mot. Skills*, 18:783–784.
59. Webb, W. B., and Agnew, Jr., H. W. (1975): Sleep efficiency for sleep/wake cycles of varied length. *Psychophysiology*, 12:636–641.
60. Wilkinson, R. T., and Stretton, M. (1971): Performance after awakening at different times of night. *Psychonom. Sci.*, 23:283–285.
61. Williams, H. L., Morlock, Jr., H. C., and Morlock, J. V. (1966): Instrumental behavior during sleep. *Psychophysiology*, 2:208–216.

Sleep and Alertness: Chronobiological,
Behavioral, and Medical Aspects of
Napping, edited by
D. F. Dinges and R. J. Broughton.
Raven Press, Ltd., New York © 1989.

11

Napping and Human Functioning During Prolonged Work

*Paul Naitoh and †Robert G. Angus

*Naval Health Research Center, San Diego, California 92318, USA, and †Defence and Civil Institute of Environmental Medicine, Downsview, Ontario, Canada

This chapter focuses on sleep management and on the usefulness of napping for maintaining and recovering function during and after prolonged work. Sleep management is the determination of how to satisfy sleep needs of people who work under demanding conditions. The concept has been in existence for more than 50 years (46). Sleep logistics is a military application of sleep management, which includes balancing the cost of losing manpower during sleep against the gains of increased alertness and morale (27,63). Napping is defined here as short sleep periods (less than 50% of an individual's average nocturnal sleep) during the day or night. In the context of sleep management, naps are used to replace rather than supplement normal nocturnal sleep in an effort to counteract the effects of sleep loss on performance and mood. This chapter begins with a brief review of the effects of partial sleep loss on mood and performance. Readers interested in the effects of total sleep loss are referred to reviews on the topic (36,41,59,85,91).

PARTIAL SLEEP DEPRIVATION STUDIES

The first partial sleep deprivation study was conducted by Smith (70). She reduced her sleep from 8 hr per night to 5.5, 3.5, and 1.5 hr for 3 consecutive nights. The immediate effect was improved task performance. Decreased task performance occurred only after a recovery sleep. Smith suggested that usually untapped reserves were being employed and exhausted during sleep deprivation.

Other partial sleep deprivation studies also show little decrement in performance (21,69,79,80). However, these findings may relate to the use of insensitive tests. Wilkinson (85) suggested that undemanding tasks of short duration may not be sensitive to partial sleep loss. For example, the loss of one complete night of sleep usually has no effect on performance during the first 5 min on many tasks, but produces clear impairment after 15 to 45 min (83,84).

More sensitive tests have shown impairment after partial sleep deprivation (25,74,75,85,87). Wilkinson et al. (87) allowed subjects 0, 1, 2, 3, 5, or 7.5 hr of sleep per night on 2 consecutive nights. After the first night, subjects who were allowed 2 hr or less of sleep showed impaired performance on a 1-hr test of vigilance and a test of addition. Those with 3 hr of sleep maintained baseline performance through the following day. However, after the second night, even subjects who had slept for 5 hr each night showed impaired performance.

In a subsequent study subjects were permitted 4, 6, or 7.5 hr of sleep per night for 4 consecutive nights (25). Those allowed 4 hr of sleep showed impaired performance (days 3 and 4) on both vigilance and addition tasks but not on a digit span memory task. A recent study showed that performance on a 20-min unprepared simple reaction time task deteriorated after a night with 4 hr sleep (75). Thus, more than 4 hr of nocturnal sleep is required to maintain performance on some tasks but not on others, at least for a period of 1 to 4 days.

Friedmann et al. (21) had subjects gradually reduce their amount of sleep during a 6- to 8-month period. Increases in self-reported daytime sleepiness and impaired task performance occurred when subjects reduced sleep to less than 4 to 5 hr per day. This result indicates that "obligatory sleep" length (37) was the same for long sleepers (who usually slept 9 hr per night) as for short sleepers (who habitually slept 6 hr). Thus, long sleepers could reduce their sleep by 4 hr (56%), whereas short sleepers could only reduce their sleep by 1 hr (17%).

It has also been reported that long sleepers do not need additional sleep beyond their habitual nightly sleep after 36 hr of continuous wakefulness, whereas normal and short sleepers increase their usual sleep lengths by 25% and 33%, respectively (38). This suggests that long sleepers are getting sleep beyond obligatory sleep, and short sleepers are getting obligatory sleep only.

Partial sleep deprivation studies have also been conducted outside laboratory environments. For example, Haslam (30,31) reported the results of a 9-day exercise in which soldiers were allowed 0, 1.5, or 3 hr of sleep each night. Sleep periods commenced at 0230 hr. Each morning, performance was evaluated with a battery of many short cognitive tests, and numerous other tests were conducted throughout the day. All 22 subjects in the no-sleep condition withdrew from the exercise after the fourth night. In the 1.5-hr sleep group, 12 of 23 (52%) completed the 9-day exercise, as did 20 of 22 (91%) in the 3-hr sleep group. Survivors in the 1.5-hr sleep group were judged by military observers to be effective with respect to physical tasks for 6 days, whereas survivors in the 3-hr sleep group were considered effective for physical tasks for all 9 days.

Studies of partial sleep deprivation suggest, therefore, that although mood may suffer before performance, some aspects of human functioning can be maintained for a few days with 50% of the usual amount of nocturnal sleep length. However, if a nap is defined as a period of sleep that is less than 50% of a person's average nocturnal sleep length, then a single nap in a 24-hr period otherwise devoid of sleep opportunities would not be considered enough sleep to fully maintain performance or alertness.

NAPPING STUDIES OF SLEEP MANAGEMENT

Research into partial sleep deprivation and performance has looked primarily at the effects of limiting nocturnal sleep. The studies discussed previously involved a single continuous block of nighttime sleep. Many sleep researchers suggest that a minimum of 4 to 5 hr sleep be taken in an uninterrupted period to prevent impaired performance. The continuity theory of sleep postulates that continuous sleep has greater recuperative power. But if sleep occurs for smaller amounts of time during a 24-hr period, is it much less effective for enhancing mood and performance [assuming that total sleep time (TST) per 24 hr is the same]? This question has been the subject of a number of napping studies.

Hartley (28) compared task performance of a group of healthy subjects who had a single 4-hr block of sleep during 4 days with a group who had three 80-min sleep periods starting at 2310, 0530, and 1225 hr. Performance on a vigilance task administered once a day at 2100 hr was better after distributed sleep than after continuous sleep. There was a shorter period, however, between waking and testing in the distributed nap group than in the continuous sleep group (7.75 versus 16.5 hr).

Mullaney et al. (58) looked at sleep deprivation effects on a tracking task, a visual pattern memory task, and an addition task (all 3 min long). Subjects had 5 hr of sleep immediately before the start of the study, which lasted from 0500 hr of the first day to 2300 hr the next day. The group of subjects was allowed a 1-hr sleep period every 7 hr (6 hr total). Another group was allowed to sleep for 6 hr continuously (from 2300 to 0500 hr). There was only a nonsignificant trend for subjects who had continuous sleep to perform better during the latter half of the 42-hr study than those who had 1-hr sleep periods. This paper does not report the sleep efficiency (actual time asleep as a percentage of the time available) in the two conditions.

Haslam (32) compared the effects of a 4-hr early morning nap (0200–0600 hr) with four 1-hr naps (0500–0600, 1000–1200, 1700–1800, and 2300–2400 hr). Subjects were awake for 23 hr before the 4-hr sleep or the first nap. Performance tests included 15-min cancellation, addition, and logical reasoning tasks, separated by 1-min breaks. Sleep log data showed that the subjects with four 1-hr naps slept less than those in the 4-hr continuous sleep group. Despite this, there were no significant differences in task performance.

Webb (76,77) observed performance for 72 hr on many tests and measures of mood in three groups allowed one of three nap sleep regimes during three days otherwise devoid of sleep: (a) two 2-hr naps (0800–1000 hr) the mornings of days 2 and 3; (b) two 2-hr midnight naps (2200–2400 hr) after days 1 and 2; or (c) one 4-hr midnight nap (2000–2400 hr) after day 2. Examining the last test session (2400–0600 hr on the third day), he found that only four of 13 measures sensitive to sleep deprivation showed any differences between the groups, and these differences were inconsistent. One measure favored the evening nap group, one the morning group, one the 4-hr nap, and one the two 2-hr naps.

Dinges et al. (16,17) studied the effects on mood and performance of a single 2-hr nap opportunity placed at one of five times (after 6, 18, 30, 42, and 54 hr of prior wakefulness) during a 56-hr period otherwise devoid of sleep. They were especially interested in the benefit of a nap taken prior to significant sleep loss and described the problem thusly:

> temporal placement of a nap during sustained wakefulness . . . concerns the placement of a nap *between* two or more days of wakefulness: On which day is a nap most likely to yield the optimum benefits if it is the only sleep taken during 48 or more hours of wakefulness? Orne (18) . . . suggested that "prophylactic napping," which refers to napping in advance of sleep loss, may be more beneficial than napping after sleep debt accumulation, a hypothesis that seems to contradict the view that sleep cannot be "stored" in anticipation of sleep loss. The issue has not been appropriately tested in most napping studies. . . . Those studies that have attended to nap effects in advance of sleep loss (less than 24 h), or during sleep restriction, have yielded data suggesting that an early nap may be advantageous . . . , although no assessment contrasting the effects of naps taken prior to, and following, sleep loss has yet been carried out.

The results of their study showed that naps taken anytime during a sustained wakefulness protocol had clear benefits for performance, such as reaction time and mental subtraction, but not for mood and subjective sleepiness reports (16,17). The advance or prophylactic naps, after 6 and 18 hr of wakefulness, appeared to improve performance as much, if not more so, than the recovery naps, taken after 30 and 42 hr of wakefulness, despite the fact that they involved less TST and less slow wave sleep (SWS) (13).

Many factors may affect the benefits of naps taken during prolonged work. Webb (76), building on Naitoh et al. (62), identified six factors thought to be important for determining the minimum amount of sleep necessary to maintain human performance. These are (a) the length of sleep deprivation; (b) the length of the nap period; (c) the circadian placement of the nap; (d) the elapsed time between the end of the nap period and the beginning of postnap performance; (e) the circadian time of task performance, and (f) the type of performance task. Since our interest in this chapter centers on the effects of naps, only the literature related to the first four factors will be reviewed.

Length of Prior Sleep Loss and Length of Nap Periods

Woodward and Nelson (92) suggested that increased amounts of sleep are required for recovery from the effects of increased periods without sleep [see also (46)]. Approximately 8 hr of sleep are needed after a normal waking day of 16 hr. They speculated that approximately 12 hr of sleep would be required after 24 hr of continuous work, 12 to 14 hr of sleep after 36 to 48 hr of work, and 2 to 3 days of 12- to 14-hr sleep after 72 hr of work.

Morgan and Coates (55) found that, after 36 hr of work, 2 or 3 hr of rest restored performance to approximately 60% of baseline, 4 hr of rest restored 70%, and 12 hr restored 100%. Four hours of rest after 44 hr of work recovered only 40% of baseline performance. The rest periods in this study included meals and other activities as well as sleep, and the actual sleep duration was not stated.

Wilkinson (86) found that, after one night of sleep loss, vigilance and addition performance recovered to acceptable levels after 2 hr of sleep but, following two nights of sleep loss, 5 hr were needed. Rosa et al. (68) found that after 48 hr of continuous work, 4 hr of sleep were required to return reaction times to baseline, whereas after 64 hr of sleep loss, a full 8 hr of recovery sleep were necessary.

Circadian Placement of the Nap

The time when a nap is taken may influence its effects on behavior and mood. The circadian cycle is defined by the 24-hr rhythm of body core temperature, and there is an associated cycle in level of alertness. Studies have found that naps taken at various times of day and night are differentially beneficial for maintenance or recovery of performance. More benefit is gained from afternoon and evening naps than from morning naps (22,49,61,73).

Naitoh (61) showed that a 2-hr nap taken near the circadian nadir (trough nap) in alertness (0400–0600 hr) after 45 hr awake had less recuperative power for cognitive performance than a nap taken during the rising circadian phase (1200–1400 hr) after 53 hr awake (peak nap). In contrast, Dinges, et al. (17) did not observe a difference in the cognitive benefits of 2-hr naps taken near the circadian peak (1500–1700 hr) after 30 hr of wakefulness from those of naps taken near the circadian trough (0300–0500 hr) after 42 hr awake. Further studies are required to resolve the difference in cognitive performance results between the two studies. It is noteworthy, however, that both Naitoh (61) and Dinges et al. (16) observed comparable benefits for peak and trough naps on reaction time performance.

Recently, Lavie and Veler (49) showed that there were less postnap sleepiness and mood deterioration after a 2-hr nap taken between 1500 and 1700 hr than one taken between 1900 and 2100 hr. Webb (76,77) found no evidence that morning naps (0800–1000 hr) were more beneficial than evening naps (2200–2400 hr). Gillberg and Åkerstedt (24) and Gillberg (23) found no difference in morning performance between a 1-hr nap taken at either 2100 hr or 0430 hr, but their results may be confounded by large differences in time from waking to testing between the conditions.

Some of the inconsistencies among studies regarding the effects of different circadian placements of naps on performance and mood may occur because of differences in the type, frequency, and temporal placement of the behavioral measures used, as well as differences in nap lengths and in the duration of prior wakefulness.

Sleep Inertia and Nap Infrastructure

The term "sleep inertia" refers to the phenomenon of inferior task performance and/or disorientation occurring immediately after awakening from sleep relative to presleep status. It was first studied by Langdon and Hartman (47). Sleep inertia is usually so pronounced after naps during prolonged work that most investigators either do not test performance for the first 20 to 30 min after a nap or do not include the results of performance tests from this period in their analyses of nap benefits [cf. (30)]. The work described thus far is based, therefore, largely on performance measurements taken at least 30 min after naps. It remains controversial how long sleep inertia can affect performance.

Reported durations of sleep inertia have varied greatly. Tilley and Wilkinson (75) observed a marked residual awakening effect, although behavioral testing was delayed for 30 min after awakening. Earlier, Wilkinson and Stretton (89) speculated that the immediate effect of sudden awakening would dissipate in 4 min. They based this on performance at night, which was quite stable when the tests were given 4 min post-sudden awakening. The decrement measured when subjects were awakened during the night was explained as a circadian effect on performance and not as persisting sleep inertia, but it is unclear how one differentiates sleep inertia from circadian influences on performance.

Naitoh (61) and Dinges et al. (15) suggested that sleep inertia may be more severe around the circadian nadir and less severe near the circadian peak in alertness. Thus, the immediate recuperative benefits of nap sleep may be masked by sleep inertia, and the dysphoric mood caused by sleep inertia may persist longer than the negative effects on performance, which may explain why mood is often not enhanced by naps during prolonged wake-fulness (17). Labuc (43–45) has suggested ways to shorten sleep inertia, but their effectiveness will probably depend on sleep stage at the time of awak-ening and sleep infrastructure during the nap.

The infrastructure of a nap refers to the organization of sleep stages during the nap. The first 4 hr of a normal 8-hr sleep period is predominantly SWS, whereas the second 4 hr is predominantly rapid eye movement (REM) and stage 2 sleep. The minimum nocturnal sleep duration for performance main-tenance is between 4 and 5 hr (21,38), which suggests that SWS may play an important role in nap benefits.

However, nap infrastructure is not a miniature version of the pattern seen during the normal 8 hr of nocturnal sleep. It varies with nap duration, time of day, and length of prior wakefulness. Up to a point, the shorter the nap is, the higher the percentage of SWS. Although morning naps contain a higher percentage of REM sleep than naps at any other time of day, this circadian REM tendency can be overridden during prolonged wakefulness; for example, Dinges (13) found that the amount of stage 4 sleep obtained in a 2-hr nap after 30 hr of wakefulness was equivalent to the total stage 4 sleep usually obtained during a typical 7-hr night sleep. Interestingly, stage 4 sleep during the 2-hr nap in the study by Dinges (13) increased up to 30 hr of prior wakefulness, but not more. Thus, within the first 2 days of sleep loss, the longer the preceding wakefulness is, the greater the SWS percentage in a 2-hr nap.

The amount of SWS in a nap may contribute to the severity of sleep inertia. Dinges et al. (15) measured mental arithmetic upon sudden awakening from 2-hr naps during 56 hr of prolonged work. They found that subjects with more SWS had the worst performance upon awakening. Other factors also appear to contribute to the severity of the decrement, however, including sleep stage at awakening (19,20,78), length of NREM sleep (14), time of day, and prior sleep loss (68). In general, it appears that any factor associated with increased sleep depth increases the severity of sleep inertia at awak-ening (15; see Chapter 9). What remains unclear is how much time is required postnap for the negative behavioral effects of sleep inertia to give way to the positive effects of the nap.

Although NREM sleep in general and SWS in particular may be associated with the severity of sleep inertia upon awakening, there is no compelling evidence that any one sleep stage is associated with the recuperative value of sleep (as assessed well after the sleep period). Bonnet (8) found that the total amount of SWS plus REM sleep during a night of disrupted sleep pre-dicted scores on addition, vigilance, and simple reaction time tests given on the following morning. Experimental findings (42,52,64) obtained at the Naval Health Research Center (NHRC) indicate that sleep stages 2, 3, 4, and REM sleep are equally recuperative.

Ultrashort Sleep

Ultrashort sleep schedules typically involve repeated sleep periods that last less than 2 hr in duration (Broughton, personal communication). Data pertinent to the concept are discussed elsewhere in this volume, and the literature regarding sleep infrastructure observed in laboratory studies is reviewed, along with Stampi's (71,72) studies of ultrashort sleep in transatlantic yacht races (see also Chapter 8). The aspect of ultrashort sleep studies that is of interest here concerns the effects such schedules have on performance and mood.

Laboratory studies on ultrashort sleep include "short-day" experiments and disrupted sleep studies. An early short-day study, conducted by Weitzman et al. (81), involved a sleep/wake schedule of 1 hr sleep and 2 hr wakefulness (i.e., a 3-hr day) for 10 calendar days. On such a regime, subjects reduced TST from 7 hr per 24 hr during baseline sleep to approximately 4.5 hr per 24 hr, even though they were given opportunities for 8 hr of sleep each 24-hr period. No performance and mood data were available in this report.

Carskadon and Dement (10,11) imposed a sleep/wake schedule of 30 min sleep and 60 min wakefulness (90-min day) for 5.3 calendar days. TST was reduced from approximately 8 to 5 hr per day. Sleepiness, as measured by the Stanford sleepiness scale (35), increased significantly during the first day of the study but decreased to near baseline during the next 4 days.

Lubin et al. (51) reported on the effects of a sleep/wake schedule of 60/160 min followed for 40 hr. Subjects on this schedule showed significant impairment in short-term memory and sleepiness only near the end of the 40-hr period. Moses et al. (57) reported on sleep efficiency and percentage of sleep stages in these subjects (51) along with those of Weitzman et al. (81) and Carskadon and Dement (10). The ultrashort sleep schedules decreased sleep efficiency from more than 90% at baseline to less than 50% in all three studies. Lavie (48) has extensively investigated ultrashort sleep/wake schedules of less than 1 hr from the perspective of "sleepability" (see Chapter 6).

Experiments involving disruption of nocturnal sleep also have some relevance to the issue of ultrashort sleep because they provide an estimate of how frequently nocturnal sleep can be disrupted without affecting daytime functioning. Bonnet (8) reported an experiment lasting 2 consecutive nights, in which groups of subjects were awakened repeatedly: 1 min after each onset of stage 2 or REM sleep; 10 min after each onset; or 2.5 hr after each onset. Another group remained awake for the full 64 hr of the study. Polygraphic sleep recordings were obtained, and subjects completed a 30-min addition test, a 30-min vigilance test, and a 10-min simple reaction time test each morning. As expected, the total sleep loss condition produced the greatest performance decrements, followed closely by the 1-min sleep condition;

performance the morning after the second night was comparable. The 10-min sleep subjects did somewhat better, and the 2.5-hr sleep condition produced the least decrement. Bonnet (8) concluded that "the data were most parsimoniously explained by the Sleep Continuity Theory—i.e., that periods of uninterrupted sleep in excess of 10 min are required for sleep to be restorative."

MaGee et al. (53) reported a similar study that involved 1-min and 4-min disruptions of nocturnal sleep, with subsequent daytime sleepiness measured by the multiple sleep latency test (MSLT). The 4-min condition resulted in only small reductions in SWS and REM sleep and no increase in daytime sleepiness, whereas the 1-min disruption markedly altered sleep and increased daytime sleepiness. Like Bonnet (8), the authors concluded that the results supported the sleep continuity hypothesis.

In some of the only field trials of ultrashort sleep conducted to date, Stampi (71,72) (see Chapter 8) found significant correlations between mean sleep length and yacht race performance: sailors with the shortest sleep episodes and least amount of TST finished highest in the standings. These, together with the laboratory results reviewed in this section, suggest that periodic naps and short sleeps, either alone or distributed throughout a 24-hr day, can enhance aspects of performance and mood.

TWO APPROACHES TO THE STUDY OF NAPPING DURING PROLONGED WORK

There are problems with generalizing laboratory studies on the benefits of naps during prolonged work periods to actual field settings, such as those involving military operations. For example, performance degradation appears to be highly dependent on the types of tasks to be carried out (89). The tasks used in laboratory studies are often not representative of operational tasks. Moreover, laboratory testing is usually infrequent, varying from every hour or two (2,17,54–56) to only once per day (29,65). Performance capabilities may be overestimated if based on short-term, high-energy expenditures that could not be maintained for more prolonged periods (27,39,54,55,60). Intense or long-duration tasks are sensitive to sleep loss, but repetitive situations used in the laboratory may cause exaggerated decrements owing to monotony. Operational situations are usually much more complex than laboratory scenarios and contain many factors that can alter performance. The interactions of sleep loss with other environmental or situational stressors are poorly understood (40).

Two research programs have been undertaken to deal with these problems and increase the generality of the findings derived from studies of napping during prolonged work. We now review the makeup and results from these programs.

Defence and Civil Institute of Environmental Medicine Research

The research program of the Defence and Civil Institute of Environmental Medicine (DCIEM) was designed to determine performance limits and biological changes with sleep loss in military personnel engaged in command and control (C2) operations. The research environment provides a continuous high-demand battery of sensitive cognitive measures of meaningful C2-type performance. The laboratory is self-contained and can accommodate personnel for extended periods. Subjects work in individual test rooms. Closed-circuit television is used to monitor the subjects from the experimenters' control area, and slave monitors display the information on each subject's terminal screen. Continuous electroencephalographic (EEG), electrocardiographic (ECG), and other physiological responses are recorded on ambulatory cassette recorders. A computer presents the stimuli, collects the responses, and stores the data.

Protocol and Tasks

Participants in the napping and prolonged-work studies at DCIEM have included both male and female university students, as well as many different military personnel: young enlisted men, young officers, and older (35–40 years) commissioned and noncommissioned officers. On day 1 of a typical 5-day experiment, subjects are briefed on the scenario and military concepts and terminology are explained. Subjects are trained extensively on all tasks. In the evening they relax, are prepared for physiological recordings, and are allowed to sleep for 8 hr. They are awakened between 0600 and 0800 hr the next day and the scenario begins approximately 1 hr later.

The protocol simulates a brigade-level command post in which subjects assume the role of duty operations officers who have responsibility for monitoring a communications network, updating tactical maps, reading messages, and answering questions. The messages require subjects to perform a number of tasks. They must identify the locations of various units (using map grid references), describe units' activities (current or intended), select the most appropriate unit for specific tasks, decode equipment resources, and estimate travel distances and times of arrival. Most questions require short phrases to be typed on computer keyboards. Some require the scenario map to be updated. Others request that summaries be handwritten and manually filed. Previously processed messages cannot be retrieved from the computer, so the manually filed information is necessary for answering questions in later messages.

A variety of short cognitive tests that have been found to be as sensitive to sleep loss effects as longer duration tasks (34) are incorporated into the protocol, including a variant of the four-choice serial reaction time task described by Wilkinson and Houghton (88), an encoding/decoding task similar to that reported by Haslam (32), a continuous subtraction task adapted from Cook et al. (12), and a logical reasoning task devised by Baddeley (7). Some of these tasks were slightly adapted for the military scenario. The following self-report measures are also collected: the Stanford sleepiness scale (35), the School of Aerospace Medicine subjective fatigue checklist (26), and the NHRC mood scale (41). More detailed information about tasks and comparisons with results from other studies is given elsewhere (4–6,33,34).

Experimental protocols usually involve approximately 64 hr of activity on the message-processing scenario, followed by an 8-hr recovery sleep (during the same hours as the baseline sleep) and 6 hr of recovery testing the next morning. The studies follow a general design in which several 6-hr blocks of identical cognitive tasks are presented. Only the content of the military messages changes. Figure 1 illustrates the design of a study investigating the effects of a 2-hr nap opportunity between 2200 and 2400 hr, after 40 hr of sleep loss. Figure 2 shows the activities within the 11 task blocks. In this experiment there are three 2-hr work sessions per block separated by 15-min breaks, except in block 7 when a nap is substituted for the second work

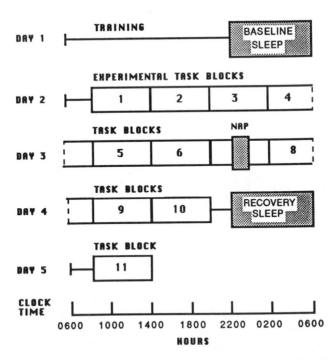

FIG. 1. Outline of design for an experimental investigation of the effects of a 2-hr nap between 2200 and 2400 hr, during 60 hr of prolonged work.

6 - HOUR TASK BLOCK

START TIME (Real)	START TIME (Elapsed)	TASKS	
08:00	**00:00**	**Scales & Battery**	
08:15	00:15	Decode (Normal)	
08:20	00:20	Messages	
08:30	00:30	Decode (Motivated)	
08:35	00:35	Messages	
08:45	**00:45**	**Scales & Battery**	**SESSION 1**
09:00	01:00	Decode (Group)	
09:05	01:05	Syllogisms	
09:20	01:20	Messages	
09:30	01:30	Decode (Normal)	
09:35	01:35	Missile Defence	
09:40	01:40	**** Break ****	
10:00	**02:00**	**Scales & Battery**	
10:15	02:15	Serial	
10:25	02:25	Messages	
10:35	02:35	Subtraction	**SESSION 2**
10:40	02:40	Messages	
10:55	**02:55**	**Scales & Battery**	
11:10	03:10	Messages	
11:20	03:20	Logical Reasoning	
11:30	03:30	Missile Defence	
11:35	03:35	**** Break ****	
12:00	**04:00**	**Scales & Battery**	
12:25	04:15	Memory (Training)	
12:35	04:25	Messages	
12:35	04:35	Digit Span	**SESSION 3**
12:45	04:45	Messages	
12:55	**04:55**	**Scales & Battery**	
13:10	05:10	Messages	
13:20	05:20	Plotting & Memory (Recall)	
13:30	05:30	Missile Defence	
13:35	05:35	**** Break ****	

FIG. 2. Experimental activities and their temporal occurrence during a typical 6-hr task block of a prolonged work scenario.

session. The experimental tasks range in duration from 5 to 15 min. A 15-min "scales and battery" package, including the four previously mentioned cognitive tasks, occurs at the beginning of each 2-hr work session and at the midpoint of each session (i.e., 1 hr later).

Results

DCIEM studies found greater decrements in performance after sleep loss combined with intensive mental work than previous studies with less emphasis on cognitive demands. Decrements of more than 30% occurred after 18 hr on duty, whereas decrements of more than 60% occurred after 42 hr. Physical fitness level appeared to be unrelated to degree of degradation in cognitive function. Exercising one third of the time at approximately 35%

of maximal oxygen uptake level did not affect cognitive functioning nor did the injection of brief (30 min) periods of strenuous exercise. Reducing cognitive work load during the day after the first night of sleep loss did not alter performance impairment during the low work load period or the subsequent night.

Because the experiments showed that performance is reduced to less than 40% of baseline after 2 nights of sleep loss and continuous work, the study outlined in Fig. 1 was performed. Subjects worked continuously for 40 hr and then received a 2-hr nap (2200–2400 hr), prior to the expected decline in performance. Subjects did not expect the nap and were not informed as to its duration. Figure 3 shows performance scores obtained on a serial reaction time task (presented at the beginning of each work session) 6 hr before and 4 hr after the 2-hr nap. The worst performance occurred immediately after awakening from the nap (i.e., behavioral sleep inertia). Performance recovered up to the level observed after 1 night of sleep deprivation, approximately 2 hr after the nap. That is, the 2-hr midnight nap allowed subjects to maintain performance during the second night without sleep at the premidnight level but was not long enough to return subjects to baseline level. These results are typical of most of the tasks studied.

A similar study explored the recuperative power (i.e., the ability to revive already degraded performance) of a 2-hr nap. In this study, the nap (again unexpected) was placed at the trough (0400–0600 hr) of the circadian cycle

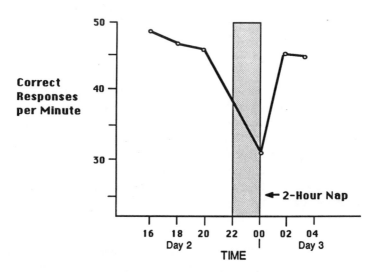

FIG. 3. Mean number of correct responses on a 2-min serial reaction time task. Data were collected after rest breaks at three places: prior to a nap, immediately after the nap, and after the next two rest breaks. The nap occurred after 40 hr of sleep loss between 2200 and 2400 hr.

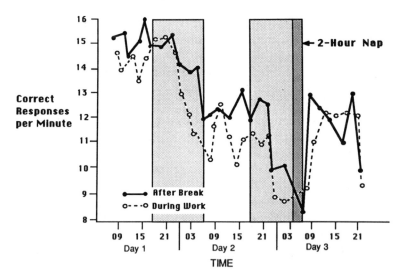

FIG. 4. Number of correct responses on a 2-min logical reasoning task given every hour during 60 hr of prolonged work. The results are divided into trials occurring after a 15-min break and trials occurring 1 hr into a 2-hr work session. *Shaded bands* signify night. Note the nap between 0400 and 0600 hr, after 46 hr of wakefulness, and its recuperative effect on performance.

after approximately 46 hr of wakefulness, several hours after the usual large decline in performance observed during the second night of sleep loss. Figure 4 shows the results for performance on a 2-min logical reasoning task that was presented once per hour. Again, task performance was worst just after awakening and subsequently improved to the level present prior to the second night of sleep loss.

As noted previously, sleep inertia can be missed if performance is not tested soon after a nap. The logical reasoning scores show improvements well within 1 hr. Figure 5 shows data from a 10-min version of the logical reasoning task that was presented only once every 6 hr. The relevance of sleep inertia depends on operational characteristics. Infantry personnel generally have a long enough period after waking to shake off sleep inertia, particularly if they use the simple effective means suggested by Labuc (43–45) for overcoming it. In contrast, C2 personnel, dozing in their workplace, might be awakened and required to work immediately while still in a state of sleep inertia.

Studies at DCIEM found similar degrees of sleep inertia occurring after a midnight nap as following an early morning nap, which is evidence against sleep inertia being primarily circadian controlled. EEG observations (see Fig. 6) revealed that the periods of sleep inertia were characterized by stage 1-like EEGs (66,67).

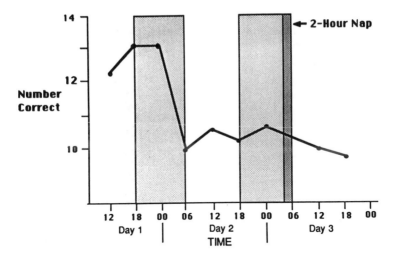

FIG. 5. Number of correct responses on a 10-min logical reasoning task given once every 6 hr during a 60-hr prolonged work experiment. A nap occurred between 0400 and 0600 hr after 46 hr of wakefulness.

NHRC Research

The NHRC laboratory can accommodate experiments on two subjects simultaneously. It has EEG, ECG, and other physiological monitoring equipment. A treadmill is used to experimentally investigate walking at various speeds and grades. The protocols provide continuous high physical and cognitive demands. They simulate reconnaissance operations in which Marines perform search and locate tasks while walking through rugged terrain. A realistic account of Force Reconnaissance Units is given by West (82).

Protocol and Tasks

During the period 1979 to 1986, seven 5-day studies of sustained operations were conducted at NHRC. In studies 1, 5, and 6, listed in Tables 1 and 2, an attempt was made to replicate the findings of Morgan and Coates (55), who sought to determine whether performance was influenced by the starting time (morning, afternoon, or late evening) of sustained operations. Their subjects started prolonged work periods at 0600, 1400, or 2200 hr. Performance remained near 100% of the baseline level for 35 hr if the continuous work episode started at 1400 hr. Performance declined significantly when the work period started at 0600 hr (between 0400 and 0800 hr). It also deteriorated precipitously from 2000 hr onward of the second day in the group starting at 2200 hr, with the lowest level of performance between 0400 and 0600 hr. These findings are important for military planners since military

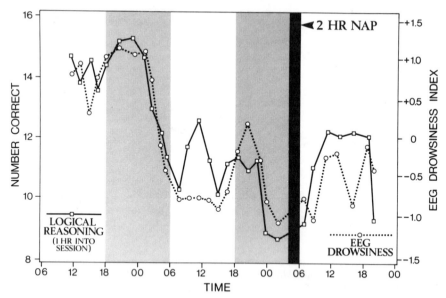

FIG. 6. Logical reasoning task results plotted against an index of EEG drowsiness. Data for the EEG drowsiness index were collected during 4-min eyes closed relaxation periods given every hr during sleep deprivation. The index was calculated by first quantifying (bipolar C_3-O_1) EEG recorded during each 4-min period using period analysis. Each period was then divided into 24 10-sec epochs on which power estimates were calculated for delta (0.5–4 Hz), theta (4–8 Hz), and alpha (8–12 Hz). Mean values were obtained for each bandwidth by collapsing across the 24 epochs. A drowsiness index was calculated as follows: $\{(AP/AP_b) - [(TP\text{-}DP)/(TP_b\text{-}DP_b)]\} + 1$, where AP, TP, and DP refer to alpha, theta, and delta power, respectively, and AP_b, TP_b and DP_b are mean baseline alpha, theta, and delta power. Baseline values were calculated by averaging the EEG for each bandwidth for the first four relaxation periods in the experiment when the subjects were fresh and produced maximal alpha. The drowsiness index will vary from approximately 1 to a negative number, the value of which will depend on the depth of slow wave sleep (67).

missions can involve prolonged work periods starting at any time of day. Hence we sought to replicate the results and investigate the effects on them of exercise and naps placed at different circadian phases.

In study 1, subjects took a 3-hr early-morning nap (0400–0700 hr); in study 5, a 3-hr premidnight nap (2000–2300 hr); and in study 6, a 3-hr late-morning nap (0900–1200 hr).

On day 1 of study 1 (morning nap) the volunteer Marines were given exercise tests to determine maximal oxygen consumption (VO₂ max). During the remainder of day 1, the subjects were trained in taking a psychological assessment battery developed at NHRC. This includes various tests sensitive to sleep loss and fatigue: (a) a simple, unprepared reaction time task modeled after Lisper and Kjellberg (50), (b) an alphanumeric visual vigilance task, (c) a four-choice serial reaction time task as described by Wilkinson and Houghton (88), (d) a logical reasoning task (7), (e) an auditory word

TABLE 1. *Characteristics of NHRC studies*

Length of nap	Exercise (E)[a] Control (C)	Mission start times		
		0800 hr	1200 hr	0000 hr
0 hr	E (30%)	Study 2		
	E (40%)	Study 4	XX[b]	XX
	C	Study 2		
	C	Study 4		
3 hr	E (30%)	Study 1	Study 6	Study 5
	C	Study 1	Study 6	Study 5
4 hr	E (40%)	Study 3	XX	XX
	C	Study 3		
8 hr	C	Study 7	XX	XX

[a] Physical work load set to be at percentage of max O_2.
[b] No study corresponding to this cell was conducted because of time constraint.

memory test (1,90), and (f) a key-tapping task as described in Friedmann et al. (21). The test battery also measures subjective ratings of mood, fatigue, and physical exertion with the NHRC mood questionnaire (41), the School of Aerospace Medicine subjective fatigue checklist (26), and the rating of perceived exertion (9).

Baseline data were collected on day 2, and the experiment began at 0800 hr on day 3. One member of each pair of subjects was randomly assigned to an exercise group and the other to a nonexercise group. During the first half hour of each hourly session, the exercise subjects performed the visual vigilance task while walking on the treadmill at a speed that kept oxygen consumption at approximately 30% of VO_2 max. Nonexercise subjects performed the same visual vigilance test seated in front of a CRT monitor.

The first 20-hr continuous work (CW1) episode ended at 0400 hr of day 4, when a 3-hr nap was allowed. Subjects were awakened at 0700 hr, given breakfast, and started on the second 20-hr CW period (CW2) at 0800 hr. CW2 ended at 0300 hr of day 5. Subjects were given dinner and then allowed

TABLE 2. *Protocol schedules for three NHRC studies involving 3-hr naps*

Study	Mission start	Orientation	Adapt. sleep	Baseline work	Baseline sleep	CW1[a]	Nap	CW2[b]	Recovery
1	Morning	08–22[c]	23–07	08–22	23–07	08–03	04–07	08–03	04–12
6	Noon	08–12		13–03	04–12	13–08	09–12	13–08	11–19
5	Midnight	08–19	20–04	05–14	15–23	00–19	20–23	00–19	20–04

[a] 20-hr continuous work episode 1.
[b] 20-hr continuous work episode 2.
[c] Time duration (hour of the day) of each phase of the studies.

to sleep for 8 hr (0400–1200 hr). The top line of Table 2 shows the data collection schedule.

The data collection protocol of study 5 (premidnight nap) was identical to study 1, except that the work schedule was shifted to start at midnight (Table 2). The data collection protocol for study 6 was also similar (Table 2). Here the schedule was shifted to start at 1300 hr by having the subjects work continuously from early morning day 1 until 0300 hr day 2, followed by an 8-hr sleep (0400–1200 hr).

Results

Table 3 shows the results of studies 1, 5, and 6. Exercising dramatically increased the degrading effects of 20 hr of prolonged work. Looking at the 10% slowest simple unprepared reaction times in each session, there were no between-nap group differences during the early sessions of CW2, suggesting that all naps were beneficial. However, during the last four sessions of CW2, the reaction times of the late-morning nap group slowed much more than those of the other two nap groups. Similar results were found for the 10% slowest reaction times and the percentage of correct responses in the four-choice task. In contrast, the early-morning nap group got a lower percentage of correct detections in comparison with the late-morning and premidnight nap groups on the alphanumeric visual vigilance task. They also showed the lowest percentage-correct recall in the word memory test during CW2.

The NHRC positive mood scale results, tapping task performance, and ratings of perceived exertion during exercise were not affected by the timing of the 3-hr nap. However, the subjects in the late-morning nap group had the greatest increase in the NHRC negative scale scores near the end of CW2 and the highest fatigue scale scores near the end of both CW1 and CW2.

In general, all nap times showed similar benefits soon after the first night of sleep deprivation, but the late-morning nap group showed more deterioration of mood and performance toward the end of the second day, even though the results of sleep stage analysis showed that subjects in this group slept normally and had sleep efficiency similar to the other groups. The last periods of CW1 and CW2 for the late-morning nap group fell during the circadian trough period (0400–0700 hr). As noted, this group had higher fatigue scores even at the end of CW1 before the nap, indicating there were effects from the jet-lag-like shift in their work schedule. This group's degradation in task performance at the end of CW2 may be secondary to direct circadian effects or to an interaction of these effects with time on the job and circadian timing of the nap.

TABLE 3. *Behavior before and after a 3-hr nap in three different studies of prolonged work with exercise*

	Study 1: early-morning nap			Study 6: late-morning nap			Study 5: premidnight nap		
		Post			Post			Post	
	Pre[a]	Proximal[b]	Distal[c]	Pre	Proximal	Distal[c]	Pre	Proximal	Distal[c]
Simple RT	817	1036	1136	1545	1038	1521	685	832	1071
(10% slow msec)	(417)	(674)	(589)	(695)	(617)	(848)	(673)	(596)	(765)
Four-choice RT	1487	1540	1568	2049	1704	2224	1417	1502	1766
(10% slow msec)	(450)	(404)	(483)	(505)	(513)	(499)	(422)	(206)	(496)
Tapping response	777	914	854	939	804	831	854	790	877
(10% slow msec)	(330)	(332)	(458)	(375)	(182)	(343)	(380)	(335)	(396)
Four-choice RT	82.6	85.3	77.8	58.4	78.2	48.2	87.7	90.7	75.4
(% correct)	(20.3)	(19.8)	(26.6)	(17.9)	(21.6)	(24.9)	(18.8)	(12.3)	(28.1)
Visual vigilance	78.7	77.1	68.6	77.8	85.9	76.7	92.4	88.2	88.9
(% correct)	(16.7)	(21.0)	(25.9)	(20.0)	(13.0)	(22.9)	(5.6)	(16.1)	(14.9)
Logical reasoning	87.6	92.0	89.9	71.8	85.4	75.0	89.7	92.9	93.1
(% correct)	(11.2)	(4.9)	(8.3)	(26.8)	(19.1)	(17.5)	(12.0)	(9.5)	(8.6)
Word memory	49.7	45.8	63.0	53.3	56.5	53.3	53.1	57.1	53.9
(% correct recall)	(11.2)	(20.9)	(11.9)	(16.9)	(22.8)	(18.8)	(18.8)	(21.1)	(19.8)
NHRC mood (negative)	6.0	6.1	7.2	12.7	4.9	12.0	6.4	7.6	6.7
	(3.9)	(5.0)	(5.7)	(7.0)	(5.4)	(6.4)	(5.3)	(8.1)	(4.4)
SAM fatigue score	9.6	8.7	8.5	12.6	9.0	13.8	10.6	9.5	11.7
	(4.3)	(4.0)	(5.7)	(5.8)	(4.3)	(3.3)	(4.4)	(3.2)	(5.9)
Borg's RPE score	10.8	10.6	11.9	12.6	10.0	12.8	11.1	10.4	11.6
	(1.6)	(2.2)	(2.1)	(3.1)	(2.1)	(2.6)	(2.4)	(4.0)	(3.0)

RT, reaction time; SAM, School of Aerospace Medicine.

[a] Mean (SD) of data taken 1–5 hr before nap during the first continuous work episode.

[b] Data gathered 1–4 hr after nap.

[c] Data gathered 15–19 hr after nap.

CONCLUSIONS AND FUTURE RESEARCH

Studies were reviewed showing that subjects (regardless of their usual sleep duration) need 4 to 5 hr of sleep per night to maintain performance levels. Naps taken during prolonged work periods can prevent or reduce sleep decrements in mood and performance and recuperate subjects from such decrements after they have occurred. The length of nap required depends on the duration of sleep loss involved and type of task to be done. There is evidence that sleep fragmented into two 2-hr naps can be as beneficial as one 4-hr sleep. Moreover, at least one study has reported that a nap before prolonged work can be as beneficial as one taken during or after such work. But how long does the nap need to be at any time of day to be beneficial?

One-minute nocturnal naps do not refresh subjects, even though they sleep 4.5 hr per night. Performance following 2 nights of such ultrashort sleep is no better than with no sleep at all. The duration of ultrashort sleep required for recuperative effects needs to be determined. How short can a nap period be before it is no longer beneficial for alertness and performance? This question continues to require research. Studies designed to answer it may also help elucidate the endocrinological and physiological factors involved in the recuperative effects of sleep.

In prolonged work scenarios, especially military operations, opportunities for sustained uninterrupted sleep generally do not exist. The only alternative to napping is prolonged wakefulness. The problem is that napping is a double-edged sword; its beneficial effects must be balanced against the negative effects of sleep inertia, which, although transient, are nevertheless important in some operational situations. For example, if sleep inertia of 5 min or longer is expected to occur, then fighter pilots must not be allowed to doze off in the cockpit of a jet below a carrier deck (3).

Sleep inertia was previously vaguely defined as a lowered psychological state following sudden awakening. We have presented EEG data collected during a period of behavioral sleep inertia that contains components of stage 1 sleep. This suggests that subjects may not be fully awake, according to physiological criteria during periods of sleep inertia, which indicates the value of polygraphic monitoring of subjects throughout behavioral sessions.

Results from studies of naps during prolonged wakefulness have not always been consistent, possibly because of interaction with, and/or masking from, direct circadian effects on performance or sleep inertia. It appears that sleep inertia severity relates to many factors including prior hours of wakefulness, duration of nap, and the time of day the nap is taken.

Another area where research is needed concerns the factors that increase the debilitating effects of sleep loss and those that increase or diminish the beneficial effects of naps on performance and mood during prolonged work. In this regard, increased continuous performance demands and cognitive load clearly seem to potentiate sleep-loss effects. The effects of exercise are less clear. The NHRC studies found that including exercise in the prolonged work periods actually increased performance deterioration, whereas studies at DCIEM found neither positive nor negative exercise effects.

Factors that may increase the benefits of naps, such as the development of polyphasic sleep/wake schedules with frequent naps rather than a single sleep period per 24 hr, have only begun to be investigated. The promise of this kind of research is that a polyphasic sleep schedule could be developed that would serve to protect the person undergoing prolonged work from serious performance deterioration, such as that seen during circadian troughs in alertness. This kind of research focuses on the value of sleep management for optimum performance capability. An alternative and complementary approach involves a search for techniques that enhance waking performance during prolonged work.

Finally, because of current limitations in predicting how a given sleep pattern will benefit individuals, drugs have been recommended for use in prolonged work conditions for the purpose of creating a proper level of activation; sedative drugs have been suggested for promoting sleep and stimulant drugs for maintaining wakefulness. However, the use of drugs for maintaining performance has many problems, such as variability in required dosage and rapidity and duration of effects. There may be drug side effects, hangovers, and/or adverse interactions between drugs or between drugs and stressors. Tolerance and dependence may develop after repeated drug use. A serious drawback is that, once taken, little can be done to reverse their course of action. On the other hand, some drugs produce predictable behavioral changes that can be useful in certain situations. For example, caffeine is often used to overcome sleepiness in prolonged work, and short-acting benzodiazepines are used to overcome situational insomnia. Research on drugs and natural substances should be continued to find optimal ways to use them in prolonged work scenarios. However, the first and best solution is to work out sleep/wake logistics and appropriate nap schedules to facilitate human performance. Only when such schedules fail should the use of drugs be considered.

In conclusion, performance under prolonged work may be affected by many factors in addition to sleep loss, including the factors discussed by Webb (76), as well as boredom, sleep infrastructure, sleep inertia, cognitive load, and physical exertion. Because of these possible complex interactions, more data are required before a behavioral model can be developed to predict the utility of napping. Currently, all these factors must be considered when designing schedules for improving performance through napping strategies. However, if studies continue to confirm that specific proportions of sleep stage activity are not needed to guarantee recuperative effects and that there is no circadian component to sleep inertia or to the recuperative effects of naps, sleep management during prolonged work would be much simpler.

ACKNOWLEDGMENTS

This work is supported in part by the Naval Medical Research and Development Command, Bethesda, Maryland, Department of the Navy, under research Work Unit No. 3M463764B995.AB.087-6. The views expressed in this chapter are those of the authors and do not reflect the official policy or position of the Department of the Navy, the Department of Defense, or the U.S. Government. This chapter is DCIEM Report No. 87-P-23 and NHRC Tech Report No. 87-21. The authors gratefully acknowledge the editorial assistance of Tamsin Kelly, M.D., in preparing the chapter.

The collaboration for this chapter was made possible by The Technical Cooperation Program (TTCP), Subgroup U (Behavioral Sciences), Technical Panel 4 (Human Factors in Manned System Integration). TTCP is an international organization of the defense departments of Australia, Canada, New Zealand, the United Kingdom, and the United States dedicated to the development and enhancement of a shared technology base in a number of key technology areas that include the behavioral sciences. This chapter is a product of TTCP UTP-4 in the Key Technical Area of Human Factors in Military Operations.

REFERENCES

1. Åkerstedt, T., and Gillberg, M. (1979): Effects of sleep deprivation on memory and sleep latencies in connection with repeated awakenings from sleep. *Psychophysiology*, 16:49–52.
2. Alluisi, E. A. (1969): Sustained performance. In: *Principles of Skill Acquisition*, edited by E. A. Bilodeau and I. M. Bilodeau, pp. 59–101. Academic Press, New York.
3. Angiboust, R., and Gouars, M. (1972): Tentative d'évaluation de l'efficacité operationelle du personnel de l'aeronautique militaire au cours de veilles nocturnes. In: *Aspects of Human Efficiency: Diurnal Rhythm and Loss of Sleep*, edited by W. P. Colquhoun, pp. 151–170. English Universities Press, London.
4. Angus, R. G., and Heslegrave, R. J. (1983): The effects of sleep loss and sustained mental work: Implications for command and control performance. In: *AGARD Conference Proceedings No. 338 Sustained Intensive Air Operations: Physiological and Performance Aspects*, pp. 11.1–11.20. NATO, Paris.
5. Angus, R. G., and Heslegrave, R. J. (1985): Effects of sleep loss on sustained cognitive performance during a command and control simulation. *Behav. Res. Meth. Instru. Comp.*, 17:55–67.
6. Angus, R. G., Heslegrave, R. J., and Myles, W. S. (1985): Effects of prolonged sleep deprivation, with and without chronic physical exercise, on mood and performance. *Psychophysiology*, 22:276–282.
7. Baddeley, A. D. (1968): A 3 min reasoning test based on grammatical transformation. *Psychonom. Sci.*, 10:341–342.
8. Bonnet, M. H. (1986): Performance and sleepiness as a function of frequency and placement of sleep disruption. *Psychophysiology*, 23:263–271.
9. Borg, G. (1985): *An Introduction to Borg's RPE-Scale*. Movement Publications, Ithaca, NY.
10. Carskadon, M. A., and Dement, W. C. (1975): Sleep studies on a 90 minute day. *Electroencephalogr. Clin. Neurophysiol.*, 39:145–155.
11. Carskadon, M. A., and Dement, W. C. (1977): Sleepiness and sleep state on a 90-minute schedule. *Psychophysiology*, 14:127–133.

12. Cook, M. R., Cohen, H., and Orne, M. T. (1972): Recovery from fatigue. Technical Report No. 55. U.S. Army Medical Research and Development Command, Fort Detrick, MD.
13. Dinges, D. F. (1986): Differential effects of prior wakefulness and circadian phase on nap sleep. *Electroencephalogr. Clin. Neurophysiol.*, 64:224–227.
14. Dinges, D. F., Orne, E. C., Evans, F. J., and Orne, M. T. (1981): Performance after naps in sleep-conducive and alerting environments. In: *Biological Rhythms, Sleep and Shift Work. Advances in Sleep Research, Vol. 7*, edited by L. Johnson, D. Tepas, W. P. Colquhoun, and M. Colligan, pp. 539–552. Spectrum, New York.
15. Dinges, D. F., Orne, M. T., and Orne, E. C. (1985): Assessing performance upon abrupt awakening from naps during quasi-continuous operations. *Behav. Res. Meth. Instru. Comp.*, 17:37–45.
16. Dinges, D. F., Orne, M. T., Whitehouse, W. G., and Orne, E. C. (1987): Temporal placement of a nap for alertness: Contributions of circadian phase and prior wakefulness. *Sleep*, 10:313–329.
17. Dinges, D. F., Whitehouse, W. G., Orne, E. C., and Orne, M. T. (1988): The benefits of a nap during prolonged work and wakefulness. *Work and Stress*, 2:139–153.
18. Evans, F. J., and Orne, M. T. (1976): Recovery from fatigue. Technical Report No. 65. U.S. Army Medical Research and Development Command, Fort Detrick, MD.
19. Feltin, M., and Broughton, R. J. (1968): Differential effects of arousal from slow wave versus REM sleep. *Psychophysiology*, 5:231.
20. Fort, A., and Mills, J. N. (1972): Influence of sleep, lack of sleep and circadian rhythm on short psychometric tests. In: *Aspects of Human Efficiency*, edited by W. P. Colquhoun, pp. 115–128. The English Universities Press, London.
21. Friedman, J., Globus, G., Huntley, A. B., Mullaney, D., Naitoh, P., and Johnson, L. C. (1977): Performance and mood during and after gradual sleep reduction. *Psychophysiology*, 14:245–250.
22. Gillberg, M. (1984): The effects of two alternative timings of one-hour nap on early morning performance. *Biol. Psychol.*, 19:45–54.
23. Gillberg, M. (1985): Effects of naps on performance. In: *Hours of Work*, edited by S. Folkard and T. Monk, pp. 77–86. Wiley, New York.
24. Gillberg, M., and Åkerstedt, T. (1983): Effect of 1-hour nap on performance following restricted sleep. In: *Sleep 1982*, edited by W. P. Koella, pp. 392–394. Karger, Basel.
25. Hamilton, P., Wilkinson, R. T., and Edwards, R. S. (1972): A study of four days partial sleep deprivation. In: *Aspects of Human Efficiency*, edited by W. P. Colquhoun, pp. 101–113. The English Universities Press, London.
26. Harris, D. A., Pegram, G. V., and Hartman, B. O. (1971): Performance and fatigue in experimental double-crew transport missions. *Aviat. Space Environ. Med.*, 24:980–986.
27. Harris, W., and O'Hanlon, J. F. (1972): A study of recovery function in man. U.S. Army Engineering Laboratories, Aberdeen Proving Ground Tech Memo No. 10-72. Aberdeen Proving Ground, MD.
28. Hartley, L. R. (1974): A comparison of continuous and distributed reduced sleep schedules. *Q. J. Exp. Psychol.*, 26:8–14.
29. Haslam, D. R. (1981): The military performance of soldiers in continuous operations: Exercise early call I and II. In: *Biological Rhythms, Sleep and Shift Work*, edited by L. C. Johnson, D. I. Tepas, W. P. Colquhoun, and M. J. Colligan, pp. 435–458. Spectrum, New York.
30. Haslam, D. R. (1982): Sleep loss, recovery sleep, and military performance. *Ergonomics*, 25:163–178.
31. Haslam, D. R. (1984): The military performance of soldiers in sustained operations. *Aviat. Space Environ. Med.*, 55:216–221.
32. Haslam, D. R. (1985): Sleep deprivation and naps. *Behav. Res. Meth. Instru. Comp.*, 17:46–54.
33. Heslegrave, R. J., and Angus, R. G. (1983): Sleep loss and continuous cognitive work. In: *Proceedings of the 24th Defence Research Group (NATO) Seminar on the Human as a Limiting Element in Military Systems, Vol. 1*, pp. 61–108. NATO, Paris.
34. Heslegrave, R. J., and Angus, R. G. (1985): The effects of task duration and work-session location on performance degradation induced by sleep loss and sustained cognitive work. *Behav. Res. Meth. Instru. Comp.*, 17:592–603.

35. Hoddes, E., Zarcone, V., Smythe, H., Phillips, R., and Dement, W. C. (1973): Quantification of sleepiness: A new approach. *Psychophysiology*, 10:431–436.
36. Horne, J. A. (1978): A review of the biological effects of total sleep deprivation in man. *Biol. Psychol.*, 7:55–102.
37. Horne, J. A. (1988): *Why We Sleep: The Functions of Sleep in Humans and Other Mammals*. Oxford University Press, New York.
38. Horne, J. A., and Minard, A. (1985): Sleep and sleepiness following a behaviorally "active" day. *Ergonomics*, 28:565–575.
39. Johnson, L. C. (1979): Sleep disturbances and performance. In: *Sleep, Wakefulness and Circadian Rhythm* (*NATO AGARD Lecture Series No. 105*), edited by A. N. Nicholson, pp. 8:1–8:17. Agard, Paris.
40. Johnson, L. C. (1982): Sleep deprivation and performance. In: *Biological Rhythms, Sleep and Performance*, edited by W. B. Webb, pp. 111–141. Wiley, Chichester, England.
41. Johnson, L. C., and Naitoh, P. (1974): *The Operational Consequences of Sleep Deprivation and Sleep Deficit*. NATO Advisory Group for Aerospace Research and Development, Paris.
42. Johnson, L. C., Naitoh, P., Moses, J. M., and Lubin, A. (1974): Interaction of REM deprivation and stage 4 deprivation with total sleep loss: Experiment 2. *Psychophysiology*, 11:147–159.
43. Labuc, S. (1978): A study of performance upon sudden awakening. Report APRE No. 1/78. Army Personnel Research Establishment.
44. Labuc, S. (1978): Performance upon awakening from four hours of sleep per night. Report APRE No. 7/78. Army Personnel Research Establishment.
45. Labuc, S. (1978): The effect of a one minute alerting procedure on performance after sudden awakening from sleep. Army Personnel Research Establishment.
46. Laird, D. A., and Muller, C. G. (1930): *Sleep, Why We Need It and How to Get More of It*. The John Day Company, New York.
47. Langdon, D. E., and Hartman, B. (1961): Performance upon sudden awakening. U.S. Air Force School of Aerospace Medicine, SAM Rep. 62-17. Brooks AFB, TX.
48. Lavie, P., and Scherson, A. (1981): Ultrashort sleep-waking schedules. I. Evidence of ultradian rhythmicity in 'sleepability.' *Electroencephalogr. Clin. Neurophysiol.*, 52:163–174.
49. Lavie, P., and Veler, B. Timing of naps: Effects on post nap sleepiness levels. *Work and Stress* (*in press*).
50. Lisper, H., and Kjellberg, A. (1972): Effects of a 24-hour sleep deprivation on rate of decrement in a 10-minute auditory reaction time task. *J. Exp. Psychol.*, 96:287–290.
51. Lubin, A., Hord, D. J., Tracy, M. L., and Johnson, L. C. (1976): Effects of exercise, bedrest and napping on performance decrement during 40 hours. *Psychophysiology*, 13:334–339.
52. Lubin, A., Moses, J. M., Johnson, L. C., and Naitoh, P. (1974): The recuperative effects of REM sleep and stage 4 sleep. Experiment I. *Psychophysiology*, 11:133–146.
53. MaGee, J., Harsh, J., and Badia, P. (1987): Effects of experimentally-induced sleep fragmentations on sleep and sleepiness. *Psychophysiology*, 24:528–534.
54. Morgan, B. B., Brown, B. R., and Alluisi, E. A. (1974): Effects of sustained performance of 48 hours of continuous work and sleep loss. *Hum. Factors*, 16:406–414.
55. Morgan, B. B., and Coates, C. D. (1974): Sustained performance and recovery during continuous operations. Interim Technical Rep. ITR-74-2. Old Dominion University, Norfolk, VA.
56. Morgan, B. B., Coates, G. D., Brown, B. R., and Alluisi, E. A. (1973): Effects of continuous work and sleep loss on the recovery of sustained performance. U.S. Army Technical Memo 14-73, Human Engineering Lab., Aberdeen Proving Ground, MD.
57. Moses, J. M., and Hord, D. J., Lubin, A., Johnson, L. C., and Naitoh, P. (1975): Dynamics of nap sleep during a 40 hour period. *Electroencephalogr. Clin. Neurophysiol*, 39:627–633.
58. Mullaney, D. J., Kripke, D. F., Fleck, P. A., and Johnson, L. C. (1983): Sleep loss and nap effects on sustained continuous performance. *Psychophysiology*, 20:643–651.
59. Naitoh, P. (1969): Sleep loss and its effects on performance. U.S. Navy Medical Neuropsychiatric Research Unit Technical Rep. 68-3. U.S. Navy Medical Neuropsychiatric Research Unit, San Diego.

60. Naitoh, P. (1976): Sleep deprivation in human subjects: A reappraisal. *Waking Sleeping,* 1:53–60.
61. Naitoh, P. (1981): Circadian cycles and restorative power of naps. In: *Biological Rhythms, Sleep and Shiftwork. Advances in Sleep Research, Vol. 7,* edited by L. Johnson, D. Tepas, W. P. Colquhoun and M. Colligan, pp. 553–580. Spectrum, New York.
62. Naitoh, P., Englund, C., and Ryman, D. (1982): Restorative power of naps in designing continuous work schedules. Naval Health Research Center Technical Rep. 82-25. Naval Health Research Center, San Diego.
63. Naitoh, P., Englund, C. E., and Ryman, D. H. (1986): Sleep management in sustained operations: User's guide. Naval Health Research Center Technical Rep. 86-22. Naval Health Research Center, San Diego.
64. Naitoh, P., Johnson, L. C., and Lubin, A. (1973): The effect of selective and total sleep loss on the CNV and its psychological and physiological correlates. In: *Event-Related Slow Potential of the Brain: Their Relations to Behavior,* edited by W. C. McCallum and J. R. Knott, pp. 213–218. Elsevier, Amsterdam.
65. Opstad, P. K., Ekanger, R., Nummestad, M., and Raabe, N. (1978): Performance, mood, and clinical symptoms in men exposed to prolonged, severe physical work and sleep deprivation. *Aviat. Space Environ. Med.,* 49:1065–1073.
66. Pigeau, R. A., Angus, R. G., and Heslegrave, R. J. (1987): Electrophysiological measures of mental fatigue and declining performance resulting from sleep loss. In: *Proceedings of the 29th Annual Conference of the Military Testing Association, Ottawa, 19–23 Oct.,* pp. 584–589.
67. Pigeau, R. A., Heslegrave, R. J., and Angus, R. G. (1987): Psychophysiological measures of drowsiness as estimates of mental fatigue and performance degradation during sleep deprivation. In: *AGARD Conference Proceedings No. 432, Electric and Magnetic Activity of the Central Nervous System: Research and Clinical Applications in Aerospace Medicine,* pp. 21.1–21.16. NATO, Paris.
68. Rosa, R. R., Bonnet, M. H., and Warm, J. S. (1983): Recovery of performance during sleep following sleep deprivation. *Psychophysiology,* 20:152–159.
69. Rutenfranz, J., Aschoff, J., and Mann, H. (1972): The effects of a cumulative sleep deficit, duration of preceding sleep period and body-temperature on multiple choice reaction time. In: *Aspects of Human Efficiency,* edited by P. Colquhoun, pp. 217–230. The English Universities Press, London.
70. Smith, M. (1916): A contribution to the study of fatigue. *Br. J. Psychol.,* 8:327–350.
71. Stampi, C. (1985): Ultrashort sleep-wake cycles improve performance during one-man transatlantic races. In: *Sleep '84,* edited by W. P. Koella, E. Ruther, and H. Schulz, pp. 271–272. Gustav Fischer Verlag, Stuttgart.
72. Stampi, C. (1985): Ultrashort sleep-wake cycles during single-handed transatlantic races. In: *Circadian Rhythms in the Central Nervous System,* edited by P. H. Redfern, I. C. Campbell, J. A. Davis, and K. F. Martin, pp. 229–232. VCH (Macmillan), Basingstoke, England.
73. Taub, J. M. (1979): Effects of habitual variations in napping on psychomotor performance, memory and subjective states. *Int. J. Neurosci.,* 9:97–112.
74. Taub, J. M., and Berger, R. J. (1973): Performance and mood following variations in the length and timing of sleep. *Psychophysiology,* 10:559–570.
75. Tilley, A. T., and Wilkinson, R. T. (1984): The effects of a restricted sleep regime on the composition of sleep and on performance. *Psychophysiology,* 21:406–412.
76. Webb, W. B. (1985): Experiments on extended performance: Repetition, age, and limited sleep periods. *Behav. Res. Meth. Instru. Comp.,* 17:27–36.
77. Webb, W. B. (1987): The proximal effects of two and four hour naps within extended performance without sleep. *Psychophysiology,* 24:426–429.
78. Webb, W. B., and Agnew, H. R., Jr. (1964): Reaction time and serial response efficiency on arousal from sleep. *Percept. Mot. Skills,* 18:783–784.
79. Webb, W. B., and Agnew, H. W., Jr. (1965): Effects of a restricted sleep regime. *Science,* 150:1745–1747.
80. Webb, W. B., and Agnew, H. W., Jr. (1974): The effects of a chronic limitation of sleep length. *Psychophysiology,* 11:265–274.
81. Weitzman, E., Nogeire, C., Perlow, M., et al. (1974): Effects of a prolonged 3-hour sleep

wakefulness cycle on sleep stages, plasma cortisol, growth hormone and body temperature in man. *J. Clin. Endocrinol. Metab., 38*:1018–1030.

82. West, F. G. (1967): The Indians. In: *Small Unit Action in Vietnam Summer* 1966, pp. 59–67. History and Museums Division, Headquarters, U.S. Marine Corps, Washington, DC.
83. Wilkinson, R. T. (1961): Interaction of lack of sleep with knowledge of results, repeated testing and individual differences. *J. Exp. Psychol., 62*:263–271.
84. Wilkinson, R. T. (1965): Sleep deprivation. In: *The Physiology of Human Survival*, edited by O. G. Edholm and A. L. Bachrach, pp. 399–430. Academic Press, New York.
85. Wilkinson, R. T., (1969): Sleep deprivation: Performance tests for partial and selective sleep deprivation. In: *Progress in Clinical Psychology*, edited by L. A. Abt and B. F. Riess, pp. 28–43. Grune and Stratton, New York.
86. Wilkinson, R. T. (1972): Sleep deprivation—eight questions. In: *Aspects of Human Efficiency*, edited by W. P. Colquhoun, pp. 25–30. The English Universities Press, London.
87. Wilkinson, R. T., Edwards, R. S., and Haines, E. (1966): Performance following a night of reduced sleep. *Psychonom. Sci., 5*:471–472.
88. Wilkinson, R. T., and Houghton, D. (1975): Portable four-choice reaction time test with magnetic tape memory. *Behav. Res. Meth. Instru. Comp., 7*:441–446.
89. Wilkinson, R. T., and Stretton, M. (1971): Performance after awakening at different times of night. *Psychonom. Sci., 23*:283–285.
90. Williams, H. L., Gieseking, C. F., and Lubin, A. (1966): Some effects of sleep loss on memory. *Percept. Mot. Skills, 23*:1287–1293.
91. Williams, H. L., Lubin, A., and Goodnow, J. J. (1959): Impaired performance with acute sleep loss. *Psychol. Monogr., 73*(14).
92. Woodward, D., and Nelson, P. D. (1974): *A User Oriented Review of the Literature on the Effects of Sleep Loss, Work-Rest Schedules, and Recovery on Performance.* Office of Naval Research, Arlington, VA.

Sleep and Alertness: Chronobiological, Behavioral, and Medical Aspects of Napping, edited by
D. F. Dinges and R. J. Broughton.
Raven Press, Ltd., New York © 1989.

12

Cultural Perspectives on Napping and the Siesta

*Wilse B. Webb and †David F. Dinges

Department of Psychology, University of Florida, Gainesville, Florida 32611, USA, and †Unit for Experimental Psychiatry, The Institute of Pennsylvania Hospital and University of Pennsylvania, Philadelphia, Pennsylvania 19139-2798, USA

Daytime napping is often associated with certain cultures, as is evident in the use of the phrase "siesta culture." Despite recognition of differences in napping behavior, the cultural/epidemiological literature on napping and daytime sleep is limited. Four studies on napping in siesta cultures have been reported (1,5,6,8). These are reviewed here with an eye toward the extent to which cultural differences contribute to napping or the lack thereof (see also Chapter 9 for a review of napping in adults). In addition, however, a broadened assessment of napping across cultures was undertaken through an examination of the Human Relations Area Files. These data are described and presented first.

THE HUMAN RELATIONS AREA FILES AND NAPS

The Human Relations Area Files were begun as the Cross Cultural Survey by Yale University in 1937. The purpose was to assemble and classify basic information on a sample of the peoples of the world. The procedure was to excerpt from the world literature, translate into English where necessary, descriptive statements about the cultures of the world and to classify and code these statements. These statements formed a bibliographic unit that identified the culture, the source, the date of the field statement, the date of publication, and a code category of the statement. Two categories pertinent to napping were included in these codes.

Under the general code for "Living Standards" was the category "Sleeping," which coded statements about hours of sleeping, postures in sleeping, segregation in sleeping, bedding, ideas about sleep and sleepiness, presence of naps and siestas, etc. In reviewing the files, it became apparent that the category "Daily Routines" also included statements about sleep and napping in relation to successions of activities throughout a typical day, times of arising and retiring, hours of work and relaxation, daily chores, longer rhythms, etc.

The Cross Cultural Survey has continued to the present time. During World War II, under the Navy Department, the file was focused on extending the materials on the Japanese-held Pacific Islands. Similarly, from 1946 to 1947, the University of Nebraska conducted a special effort on the Plains Indians. From 1948 to 1949, in collaboration with the Social Sciences Research Council and the Carnegie Foundation, an inter-university organization called the Human Relations Area File, Inc., was formed. The present file is microfilmed with approximately 225 participating universities. These files are indexed in two primary sources (2,3).

Currently, the file contains coded descriptive statements on more than 1,700 cultural units. They extend in time from the prehistoric to the present and are grouped under eight major land areas. Table 1 displays an illustrative citation for Nigeria, from the African grouping.

Each descriptive statement about a culture is coded under one or more of approximately 500 classifying categories. As noted previously, two of these categories, Sleep and Daily Routines, provide specific reference to naps and napping. The entire file of these two categories was reviewed. Eighty-seven items were found that made specific reference to sleeping during the day. These references could be sorted into three groupings: (a) indications of regular "siesta" patterns; (b) reports of daytime sleeping that did not appear to be regular; and (c) negative descriptions of daytime sleep. These are presented in three separate tables. Approximately one half of the citations were too general to permit classification.

Tables 2 through 6 present citations of siesta cultures. Grouped under the major land areas, each citation presents the country, any specialized unit

TABLE 1. *Excerpt from the Human Relations Area File on Nigeria[a]*

FF1 Nigeria	General data on geography, demography, and indigenous people and cultures.
FF2 Colonial Nigeria	General and specific data on colonial rule.
FF4 Ada	Specific data on the Ada, including the Abam and Aro, who are a culturally distinctive branch of the Ibo (FF26).
FF5 Afo	Specific data on the Afo tribe.
FF6 Angas	Specific data on the Angas tribe, including the related Ankwe, Bwol, Dmuk, Goram, Gurkha, Miriam, Montoil, Ron (Baron, Boram), and Sura.

[a] The entry extends through 65 tribal groups within Nigeria (i.e., FF65).

TABLE 2. *The siesta cultures of Asia*

Country (subgroup)	Date	Setting	Time (hr)	Comments[a]
Afghanistan	1943	Village		1
Burma	1951	Agricultural	1200–1400	
Cambodia	1938	Agricultural	1200–1500	
India (Bhil)	1954	Agricultural	1200–1400	
India (Calcutta)	1978	Urban	?–1600	2
India (Gujarati)	1928	Agricultural	1300–1500	
India (Santal)	1939	Agricultural	1200–1400	
India (Tamil)	1971	Weaving	1200–?	
Korea	1911	Agricultural		3
Malaya (Semang)	1939	Village	?	
Manchuria	1956	Agricultural		4
Taiwan (Hokkien)	1970	Agricultural		5
Thailand	1898	Urban		6
Tibet	1940	Agricultural	1400–1500	
Vietnam	1926	Urban	1100–1400	7

[a] Quotations and sources may be located in the Human Relations File under "sleeping" and for culture as indexed in Murdoch (2):

1: "like other races close to nature in hot lands, [they] respect the sun. . . ."
2: "Those who are most fortunate . . . take siestas in the afternoon."
3: "About nine o'clock . . . he eats, and after having a smoke he lies down for a nap. . . . At noon he . . . takes another nap. . . ."
4: "When summer comes. . . . After lunch naps are a habit . . . ."
5: "most field workers [in hot weather] have a nap after lunch but in the winter with short days, the workers return to the fields within an hour."
6: "during the broiling heat . . . officialdom sleeps and rests."
7: "all shops except those of the heat impervious Orientals, closed shop. . . ."

within the country, the date of the field observation, the setting, the time of the siesta when cited, and any other relevant comments.

Table 7 presents citations about daytime sleep that do not appear to reflect siesta patterns of sleep. This table is arranged in the same land mass order as Tables 2 through 6 and cites the country and subunit, the date of observation, and the pertinent excerpt.

Table 8 presents the few statements indicating that daytime sleep tended not to be a part of a culture. As in other tables, these are arranged by land mass, citation of country and subculture, date of observation, and relevant quotation.

Although interesting, these data cannot be treated actuarially. Moreover, they may not represent the extent to which daytime napping occurs in all cultures. The data are the asystematic observations of intrepid explorers and selective anthropologists and heavily weighted toward undeveloped cultural groups. For example, 24% of the index volume pages citing specific cultural units is from Africa, whereas 2% is from Russia (2). The 14% from North America is dominated by early Indian cultures. Furthermore, each observation was from the unique perspective of the observer. Although a file

TABLE 3. *The siesta cultures of Africa*

Country (subgroup)	Date	Setting	Time (hr)	Comments[a]
Kenya (Masai)	1908	Cattle	1200–?	1
Kenya (Nairobi)	1967	Urban	Afternoon	2
Nigeria (Ibo-West)	1920	Agricultural	Afternoon	
Nigeria (Nupe)	1922	Village	1300–1500	
Sierra Leone (Mende)	1945	Village	1200–1500	
South Africa (Bushman)	1934	Nomad	1100–1600	3
Uganda (Ganda)	1911	?	1200–1400	4
West Africa (Dogon)	1935	Agricultural	Midday	
West Africa (Mossi)	1908	Agricultural	1200–14/1500	
Zaire (Azande)	1953	Agricultural	1100–1500	
Zaire (Pigmy)	1958	Nomad	Midday	5

[a] Quotations and sources may be located in the Human Relations File under "sleeping" and for culture as indexed in Murdoch (2):
  1: "After the meal one takes a nap, plays a little game or picks up some manual work. . . ."
  2: "families tend to take naps more often. . . ."
  3: "From eleven to four the camp is deserted, and even the old people sleep somewhere in the forest. The hunters and the women . . . also take a long midday rest during the hot afternoon hours."
  4: "Most of the better class went to rest at noon for two hours. . . ."
  5: "the Pygmies have learned from the animals around them to doze with one eye open and a sleep midday camp can become filled in a minute with shouts and yells. . . ."

TABLE 4. *The siesta cultures of Middle East*

Country (subgroup)	Date	Setting	Time (hr)	Comments[a]
Egypt (Fellahin)	1878	Urban	Midday	1
Ethiopia (Amahara)	1961	Village	1200–?	
Iran	1910	Urban	?	2
Oman	1945	Urban	1200–1600	3
Senegal	1955	Urban	1200–?	
Syria (Rwala)	1928	?	Midday	4

[a] Quotations and sources may be located in the Human Relations File under "sleeping" and for culture as indexed in Murdoch (2):
  1: "He takes care not to make his midday sleep too short, and he lies down in his house or his shop, in the cafe, or in any shady spot; at this time the streets and markets become deserted."
  2: "siestas—a habit indulged in by the lowest as well as the highest, and which it is a crime to disturb. . . ."
  3: "[siesta] a custom bequeathed by their Moorish cousins. . . ."
  4: "Sleep during the hot midday, kajjalno . . . is preferred to night sleep. . . ."

TABLE 5. *The siesta cultures of Europe and Oceania*

Country (subgroup)	Date	Setting	Time (hr)	Comments[a]
Greece	1957	Village	Afternoon	
Greece (Athens)	1957	Urban	Midday	1
Yugoslavia	1939	Agricultural	1300–1400	
Australia (Murngin)	1925	Nomad	Midday	2
Java	1959	Village	1200–?	3
Micronesia (Marshalls)	1947	Village	Midday	
Papua (Orokaiva)	1925	Village	1200–1400	4
Philippines	1952		Midday	5
Samoa	1948	Village	Midday	6

[a] Quotations and sources may be located in the Human Relations File under "sleeping" and for culture as indexed in Murdoch (2):
1: "here, too, the midday siesta is rigorously observed. . . ."
2: "The midday interlude is firmly established throughout the country. . . ."
3: "If the day is warm there will be . . . almost universal somnolence . . . for an hour or two. . . ."
4: "The scorching midday sun forces people to stop working . . . to take a rest . . . or sleep. . . ."
5: "allowing the nomad six hours of sleep at night in addition to that during the heat of the day. . . ."
6: "Samoans feel in regard to the hottest part of the day as did the discerning person who wrote about mad dogs and Englishmen. . . ."

TABLE 6. *The siesta cultures of North and South America*

Country (subgroup)	Date	Setting	Time (hr)	Comments[a]
Mexico (Seri)	1955	?	Midday	
Yucatan (Yucatac)	1936	Village	1400–1600	
Amazonia (Yanoama)	1968	Agricultural	1100–1400	
Brazil (Tupinambs)	1864	Agricultural	1000–?	1
Colombia (Kogi)	1949	Village	1200–1400	2
Haiti	1966	Agricultural	1200–1400	
Jamaica	1954	Village	1200–1400	
Mato Grosso (Traumai)	1938	Village	1100–1600	3
Paraguay (Guana)	1947	Village	Midday	4
Paraguay (Guana)	1947	Agricultural	1200–1400	5

[a] Quotations and sources may be located in the Human Relations File under "sleeping" and for culture as indexed in Murdoch (2):
1: "When the sun is strong . . . about ten, they leave and they go to eat and sleep. At two in the afternoon, when the heat decreases, they return to work. . . ."
2: "From 12 to 2 in the afternoon, more or less, the men rest or mend some object while the women cook. . . ."
3: "As a rule they ate upon returning and then stayed in the village the rest of the day, working at various crafts as the spirits moved them. Actually a great deal of time was spent loafing and sleeping."
4: "After eating both men and women lie in hammocks to rest. . . ."
5: "when the sun reaches the zenith, the men return to the house . . . after eating, both men and women lie in the hammocks to rest. . . ."

TABLE 7. *Cultures with irregular and nonsiesta daytime sleep*

Area	Country (subgroup)	Date	Comment[a]
Asia	Okinawa	1955	"Midsummer . . . is one of the busiest agricultural seasons, but exhaustion causes many people to take naps during the noon hour."
	Okinawa	1953	"Much of the fisherman's work at sea is at night. . . . For this reason, there is considerable day time sleeping among them."
	Japan (Ainu)	1924	"While weir fishing every night, they took a nap and rested by turns in the hut built by the weir."
	Korea	1891	"The Koreans are most irregular in their habits, for, slumbering as they do at all hours of the day, they feel sleepless at night, and are compelled as a consequence to sit up."
Africa	Nigeria (Tiv)	1952	"Adults nap during the day in snatches of leisure or when they feel like it. . . ."
Middle East	Somaliland (Negroid)	1880	"As regards sleep, they do not seem to be tied to certain times. . . . They make concessions to . . . sleep whenever the opportunity presents itself [during caravan journeys]."
Europe	Bulgaria	1949	"in spring and supper . . . the tired workers had been at their task since sun up. In the middle of the morning, at noon, and again in the midafternoon, the toiler stopped long enough for food . . . and a little sleep."
	Malta	1956	"All this means is an early daily start [2 to 4 am] . . . stop for a light breakfast. Lunch is at 11 am or noon followed by a siesta of some hours, depending on the season and pressure of work. . . ."
Oceania	Polynesia (Tikopia Island)	1929	"Tikopia are used to sleeping at odd hours during the day and have not that traditional association of daylight with working period and night with sleep period. . . . Their attitude toward sleep is governed greatly by night fishing. . . . The practice is to lie down and cover oneself up because one is sleepy and not because it is time to go to bed."

TABLE 7. (*continued*)

Area	Country (subgroup)	Date	Comment[a]
	New Zealand (Pukapuka Island)	1934	"The Pukapukan distinguish . . . more than 35 kinds of sleep, according to position, movements during sleep and soundness . . . The amount of sleep varies from 4 to 8 hours each night, sometimes with an hour or more of sleep during the day."
	Micronesia (Woleai)	1948	"Unless there is something special . . . their average day has no special commitment. Usually they go to the canoe house, where they may make rope, or . . . just sit around . . . or perhaps take a nap."
	Truk	1920	"the natives eat whenever they please and also sleep whenever they please. The time of day does not matter."
North America	Alaska (Tlingit)	1914	"The Tlinget sleep whenever he is inclined to do so. We have found them in bed at all hours of the day. . . ."
	Pacific NW (Salish Indians)	1930	"Should anyone decide to take a nap . . . he would go to sleep right in or near the tipi. . . ."
South America	Venezuela (Warao Indians)	1943	"Guaros are light sleepers. . . . During the day they can easily make up for nighttime interruptions . . . since neither business nor work is customarily overwhelming for them."
	Patagonia (Yahgan)		"Yahgan sleep together as a group and frequently interrupt each others sleep . . . a tired person will lie down for a nap anywhere and at anytime. . . ."
	Argentina (Toba)		"the hunters would drift back to the camp. . . . During what remained of the day the Indians would sit about resting and doing odd jobs, especially attending to their weapons. . . ."

TABLE 8. *Cultures with little or no daytime sleep*

Area	Country (subgroup)	Date	Comment[a]
Asia	Sino/Tibet	1936	"the tent nomads . . . get up at the first hour of dawn. . . . They only occasionally take a nap in the middle of the day."
	Tibet	1920	"It is a general Tibetan belief that it is bad for people to sleep in the daytime, for it is thought that doing so, especially during summer induces fever. . . ."
	India (East Punjab)	1934	"And you never sleep by day? For a half-hour—only when it is hot."
	Japan Islands (Matsunagi)	1951	No naps in typical daily log.
	(Niike)	1957	No naps in typical daily log.
Africa	Nigeria (Southern Ibo)	1934	"They are not quite such early risers as those of the North, but also they take no rest during the day. . . ."
	Zaire (Azande)	1924	"It is not customary to sleep in the day-time."
North America	Yucatan (Maya)	1941	"From ten to three the plaza takes on a quite, lazy appearance, as if all the villagers were asleep. . . . This picture may leave . . . the idea [of] what is supposed to be a universal practice in warm climates, the Spanish siesta, but this is not so. The men are at work in their cornfields away from town. Very rarely do the Maya nap during the day . . . the women are constantly busy with cooking, sewing, mending, or taking care of children."
	Southwest United States (Zuni)	1953	"She remarks that one should not take a nap but that she took a nap yesterday afternoon."
	(Navaho)	?	"A man lying down during the daytime is a lazy man in the opinion of the Navaho."

statement about daytime sleep may be taken as a factual observation, the absence of a statement may indicate a number of possibilities including disinterest in daytime napping behavior on the part of the observer, a failure to observe whether people slept during the daytime, or an observation that daytime sleep was absent that was not so noted in the comments. The tables list, therefore, only those instances in which a statement about daytime sleep was explicitly placed in the file. Countries and regions where siesta or midday napping are commonly reported (e.g., Northern Italy, China) are not included if this activity was not mentioned in the file. Thus, if a country or cultural subgroup is not listed in the tables (for siesta, irregular daytime napping, or no napping), it does not mean that daytime sleep does not occur, but rather only that no firm conclusion can be drawn from the file regarding the incidence and prevalence of napping.

Nevertheless, some general impressions of daytime sleep emerge from the data in the tables. Particularly in Asia (Table 2) and where observed in Africa (Table 3), naps or siestas were noted as being quite common in agricultural and rural settings. The timing of the daytime sleep was remarkably consistent across cultures. For example, of 96 instances in Tables 2 through 6 involving mention of specific hours of the day when naps were observed, the hours between noon and 1400 were mentioned 76% of the time, with 1500 and 1600 hr mentioned 17% of the time, and the remaining 7% of the time involved the hours before noon; no file entry contained comments about regular daily sleep after 1600 hr. The term "afternoon" was used to describe the regular nap of three cultures, and the term "midday" was used with 11 cultures. If these terms are taken to refer to the hours between 1200 and 1600, it is fair to state that virtually all siesta cultures described in the file have preferred times for napping between 1200 and 1600 hr. Less information was available about the duration of naps than their timing, but, in general, siesta or nap times appeared to last approximately 2 hr. Comments regarding the prevalence of napping often pertained to the heat and the sun.

Variations in these standard patterns were noted, however. Seasons were prominent factors, with siestas most common in the summer months. There was further evidence, particularly in urban settings, that naps were related to social class status; the higher classes appeared to more consistently maintain the nap pattern.

Irregular daytime sleep (Table 7) and prohibitions against daytime naps (Table 8) tended to be associated with less agriculturally centered cultures: nomadic groups; fishing cultures, where night fishing was a part of the culture (Oceanic and island groups); and herdsmen and hunter groups with less settled village patterns such as those reported from Africa and South America. In only a few cultures were there observations of daytime sleep being proscribed.

There was a clear geographic factor associated with comments in the file about napping. The farthest siesta culture from the equator was the Manchurian citation at approximately 45° north latitude. With the exception of Alaska, virtually all siesta cultures (Tables 2–6) and cultures with irregular napping (Table 7) reported in the files are located between latitudes of 45° north and 45° south, which contains most of the continental United States, South America, and Asia; all of Africa, Australia, the Middle East and

Oceania; and the southern portions of Europe. More important, of those cultures showing a strong siesta pattern (see Tables 2–6), 63% are between latitudes of 20° north and 20° south; 89% are between 30° north and 30° south. Beyond this geographic band, the siesta culture (even in this sample, which emphasizes nonindustrial cultures) is seldom reported.

Equally striking from this world view is the fact that despite siesta cultures being widespread in tropical climates, it is likely that even here they are not so prevalent as to be a universal phenomenon. Although the status of "nonreports" is problematic, the vast number of cultures in which this obvious daily behavior was not reported is impressive.

FOUR STUDIES OF NAPPING

Among the four studies that have been reported of napping in contemporary siesta cultures, three provide data on the nature of naps in three traditional siesta culture settings: Italy (1), Greece (6), and Mexico (8). The fourth explores a sociological theory about cultural development using siesta data from Peru and Belgium (5).

The study by Lugaresi and colleagues (1) uses data from interviews of 5,713 subjects who were a representative sample of the population of the state of San Marino in east central Italy. Approximately one half of the subjects resided in an urban setting and the other half lived in the countryside. Their ages ranged from 3 to 94 years. The study by Soldatos and colleagues (6) was drawn from a cross-sectional mental health survey of 1,061 persons between the ages of 19 and 65 years living in Byronas, a suburb of Athens, Greece. The Taub study (8) used a questionnaire given individually to a representative sample (age range 20–89 years) of the residents in Hermosilla, Mexico.

The data from Italians (1) are complicated by two factors: (a) one third of the sample was less than 20 years of age and (b) the results were subdivided into "good" and "poor" sleepers (the latter was defined by responses of "always or almost always sleep poorly"). The group designated as poor sleepers constituted 13% of the sample and had more female (17%) than male subjects (10%), and more older persons.

Among the group designated as good sleepers, 8% reported napping every afternoon and 36% napped "occasionally." "Habitual nappers" (presumably the group napping every afternoon) were more common among children under 10 and persons over 65 years of age (36%). The poor sleepers habitually napped in the afternoon significantly more often than did the good sleepers (18% versus 8%, $p < 0.01$).

The studies in Greece and Mexico used a comparable age sampling that did not include persons under 19 years of age. The study by Soldatos et al. (6) of Greeks reported that 42% of their sample indicated that they napped four or more times per week with 15% napping daily. The study of Mexicans by Taub (8) reported a significantly higher percentage of napping four or more times per week (76%) "during at least one season per year." It was noted, however, that there were seasonal differences. Fifty percent of those who napped (37% of the total sample) did so year-round, whereas another 29% did so only during the summer.

The studies in Greece and Mexico both observed increases in napping as a function of age. No differences were found as a function of gender in the Mexican sample, but the Greek sample showed some limited evidence of higher levels of napping among the men. The mean duration of napping in the Greek sample was 79 min (SD = 38 min), with women taking significantly shorter naps. No relationship between age and duration of nap was found. The mean duration of napping in the Mexican sample was identical to that of the Greek sample (79 min), but no gender differences were noted. There was a decrease in nap duration in the Mexican sample from 20 to 50 years of age, and an increase thereafter, which was statistically significant. In Taub's (8) study of Mexicans, there was a significant, albeit low, positive partial correlation ($r = 0.27$) between nap duration and the duration of nighttime sleep (with age partialed out).

In general, the results of studies of the traditional siesta cultures of Italy, Greece, and Mexico reflect similar and somewhat surprising characteristics. Only a limited subgroup of the samples actually napped daily. These daily nappers were 8% of the good sleepers in Italy and 15% of those in the Greek sample. Using four naps per week as a criterion for habitual napping, only 42% of the Greek sample and 37% of the Mexican sample followed this pattern (year-round in the case of the latter), although napping was more frequent in the Mexican sample.

In both the Greek and Mexican studies, naps average 1.31 hr in length, but there were large standard deviations. Nevertheless, this average strikingly approximates the figures reported for daytime naps of younger children (see Chapter 3) and healthy young adults in the United States (see Chapter 9). In all studies, napping increased in frequency with age; gender differences were not reliable across samples. The Italian study found evidence for more napping among poor sleepers. The Mexican study reported a small positive correlation between nighttime sleep and nap durations.

A sociological study, which used napping as an index of time utilization, may help to interpret the limited level of napping reported in the three survey studies and may indicate the role of industrialization on napping. The Belgian sociologist, Rudolph Rezsohazy, has been concerned with the role of sociocultural factors in cultural and economic development and published his views in a 1970 book (4). From the standpoint of napping, Rezsohazy (5) has gathered some particularly interesting comparative data from a highly industrialized setting (i.e., Brussels and the Belgian village of Nodebais) and a "transitional" setting (i.e., Lima-Calleo, Peru). In this study, the daytime rest or siesta was a primary index relative to time usage or time "value."

Rezsohazy (5) hypothesized that "the industrial societies value time . . . so that it considers its flow without activities, during rest, as not utilized, as a loss." In traditional (underdeveloped) countries there is "low value of time." He used "the quantity of time devoted to rest and to siesta as a revealing index [of these tendencies]." Rezsohazy (5) presented a number of figures comparing the "rest/siesta" levels of Belgium and Peru along "developmental dimensions." Figures 1 and 2 present two of these comparisons.

Figure 1 compares the hours devoted to daily rest and/or siesta in Belgium (B) and Peru (P) relative to educational levels. Figure 2 makes the same comparisons relative to occupational status. The data are drawn from time budget data (i.e., complete single-day, detailed analyses of the activities of large samples). It is clear that for the Belgians, the amount of daily rest and/ or siesta is low and generally unrelated to either educational or occupational

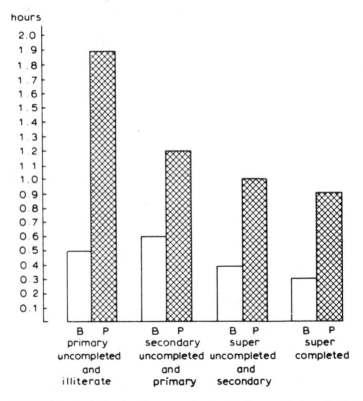

FIG. 1. Rest and siesta during the day according to the degree of education in Belgium (B) and Peru (P). (From ref. 5.)

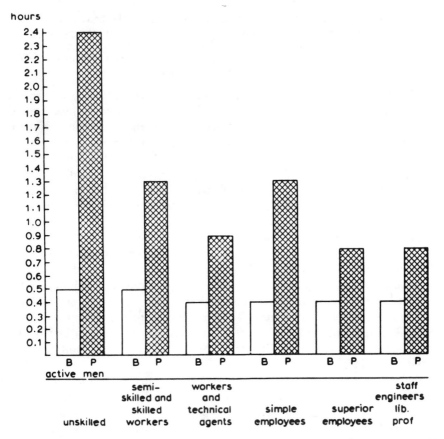

FIG. 2. Rest and siesta during the day according to occupational status in Belgium (B) and Peru (P). (From ref. 5.)

status. In the developing culture of the Peruvians, however, more time is devoted overall to daytime rest and/or naps, regardless of educational or occupational level, and the amount of time is clearly influenced by cultural status indices. Rezsohazy (5) summarized the data thusly:

the quantity of time devoted to rest and siesta during the day is two to five times greater in the Peruvian site than in the Belgian sites. . . . The observed quantity reveals no notable oscillation in Belgium according to different indices of development. . . . But in Peru . . . the indices of development bring out notable differences: the rural, the unskilled workers, the illiterate or those whose education was not above the primary level, take the most rest . . .

Rezsohazy (5) concluded that

progressing toward the transitional society and, from there toward the industrial . . . , society arrives at a point where the problems . . . require the quantification of time, its mastery, and its saving . . . mechanisms of social control . . .

are put to use. One of the first victims of this change is the daily rest period, *la siesta*. It is interesting to see how the siesta is in process of disappearing in Spain parallel to industrialization. In Chile . . . the siesta has recently been suppressed by decree. When development has reached a level of generalized well being for all people then perhaps man will have to relearn how to rest.

Although Rezsohazy's (5) evaluation illustrates that cultural development, such as that brought on by industrialization and increasing numbers of educated and skilled members in a society, can influence the time devoted to daytime rest and napping, it does not consider the influence on rest and napping of changes in nocturnal sleep. To the extent that daytime napping reflects a compensation for increased sleep pressure or unsatisfied sleep need, cultures that sleep more at night would be expected to have less daytime napping. Although Peru is a siesta culture (and much closer to the equator than Belgium), Belgians apparently place a high value on obtaining sufficient nocturnal sleep. Data for this point derive from the Multinational Comparative Time Budget Research Project, which was coordinated by Alexander Szalai (7) and formed the basis of his book on more than 30,000 time budgets, covering 24-hr periods of adults in urban and suburban cities in 13 countries. With the exception of Peru, the sites were industrialized and in the northern hemisphere. They included Belgium, Bulgaria, Czechoslovakia, the Federal Republic of Germany, France, German Democratic Republic, Hungary, Poland, the United States, U.S.S.R., and both rural and urban areas in Yugoslavia.

Webb (10) analyzed the sleep reported in the surveys of the participants from these 12 industrialized areas. Of particular interest is the fact that Belgium had the highest proportion of respondents reporting 7 hr or more of sleep each day (88%); this included night and "incidental" day sleep, but not naps, which were counted in the category of "nap or rest." Seven countries (Czechoslovakia, German Democratic Republic, Hungary, Poland, the United States, U.S.S.R., and Yugoslavia) had three times as many people (i.e., 32%–37%) reporting that they obtained less than 7 hr sleep, compared with Belgians (11%). Belgians not only had the highest percentages in the 7- to 8.9-hr zone (70%) but also the third highest proportion in the 9- to 10-hr zone (18%). Although report bias may account for some of these cultural differences, it appears that Belgians (and the French as well) place a relatively high priority on obtaining adequate sleep. Since napping was a separate category and only "incidental" day sleep was included with nocturnal sleep, it is reasonable to assume that Belgians value obtaining adequate nocturnal sleep in particular. Extrapolating from the data presented by Webb (10), the 2,077 Belgian respondents averaged as much as 35 min more sleep per 24-hr period than the 1,243 U.S. respondents and 45 min more than the 2,891 respondents from the U.S.S.R.

If Belgians obtain more sleep (nocturnal) than other cultures, then they may have less need for compensatory daytime naps to replace missed nighttime sleep and hence less daytime napping. As discussed elsewhere, sleep need can be a significant factor in producing daytime napping among adults (see Chapters 4, 9, and 10). Consequently, the lesser daytime rest and/or napping found in Belgium relative to that found in Peru may reflect as much the role of cultural attitude toward nocturnal sleep or the ability to obtain adequate nocturnal sleep (e.g., undisturbed by tropical heat) as the role of industrialization and education. Certainly the problem highlights the need to consider napping, even in tropical siesta cultures, relative to nocturnal sleep and the entire 24-hr period, and not merely in isolation. Unfortunately, there are virtually no data available on the relationship between nocturnal sleep and daytime napping across cultures.

DISCUSSION

Sociocultural factors may potentially facilitate or inhibit behavioral responses. Arguably, there is an endogenous circadian tendency for a daytime sleep period (or nap) in the afternoon. In some cultures this tendency is expressed, in others it is not. The data in this chapter suggest that some cultural factors may be facilitative and some inhibitory, and that these may influence the opportunity for napping either directly or indirectly.

Clearly the most facilitative factor appears to be a tropical climate (the tropics are defined as those areas between latitudes 23.27° north and south latitude of the equator). At least in terms of cross-cultural files recorded to date, habitual napping by large segments of a society is only approximately half as common outside the tropical latitudes. Within this regional zone, on the basis of commentary, the specific facilitator appears to be the heat of the midday period or afternoon sun. The force of this factor, per se, is apparent in the seasonal tendencies of naps, as they are often reported to be more common during the summer or high temperature periods. What is less certain is the extent to which daytime napping in tropical climates is owing primarily to the torrid heat of midday (9) and/or to the overall disruptive effect of heat (and humidity) on sleep, including nocturnal sleep, which in turn leads to a tendency toward compensatory daytime napping.

A second facilitative factor, particularly in underdeveloped cultures, may be a geographically stable setting, in which work patterns can follow a regular daily routine, such as that found in agriculturally centered societies in village settings. In such settings, a daily rest, siesta, or nap period appears to be easily managed.

Contraindicative factors, even in tropical climates, include nomadic lifestyles, such as those of the hunter or herdsman cultures and cultures whose economy and food gathering involve nondaytime sleep periods. When daytime napping does occur in these groups, especially among the latter, it appears to be in the form of compensatory (replacement) daytime sleep for lost nighttime sleep.

Custom clearly plays a facilitative role, as may be seen in the studies reporting on four contemporary sleep cultures: Italy, Greece, Mexico, and Peru. However, it should be noted that, even in the facilitative cultures, daytime napping is not a ubiquitous activity. Only approximately 10% (Italy and Greece) to 35% (Mexico) of the individuals took daily naps and only approximately 50% took naps four times a week year-round.

Some part of the reduced response in these cultures may be attributable to modern industrialization. Rezsohazy's (5) data and arguments support the concept that increased development of a culture resulted in time being viewed as valuable and with rest and siestas being viewed as losses. No less may have taken place during the industrial revolution in the United States, when Thomas Edison (an inveterate napper) invented artificial light with the (vain) hope that it would reduce the need for sleep. The emphasis on the notion "time is money" that appears to come with industrialization apparently results in an active cultural suppression of siestas and naps. In terms of daytime napping, the definitive test of this hypothesis would require a study of the incidence and prevalence of napping within a siesta culture from pre- to post-industrialization.

Cultural differences in napping behavior also appear to exist, however, independently of industrialization status. There are not only differences in what is considered acceptable sleep/wake activity during daylight hours but also in what is considered normal nocturnal sleep. That such differences may exist among industrialized countries of the northern hemisphere, for example, is suggested in the sleep data from the Multinational Comparative Time Budget Research Project (10). To the extent that daytime napping is influenced by inadequate nocturnal sleep, such differences can be expected to affect napping behavior. Whether these differences actually result in significant differences in total sleep obtained per 24 hr among cultures, and possibly also in the degree to which societal members are chronically sleepy during the daytime, is unknown.

The near lack of siesta cultures outside the tropics also cannot be simply ascribed to industrial development. Even in the Human Relations File, which emphasizes traditional cultures, the low rate of regular napping (Tables 2–6) outside the tropical regions is striking. Although nonreporting must certainly contribute to this, it also suggests that tropical (and subtropical) climates, and the difficulties they can create for physical activity during the afternoon, are powerful factors in the expression of napping. Aside from midday heat forcing a suspension of physical activity, and hence permitting a passive expression of sleep, a tropical climate can make nocturnal sleep difficult. It is not impossible, therefore, that another way in which tropical climates contribute to the likelihood of napping is through a disruption (i.e., shortening or fragmentation) of nocturnal sleep, which in turn results in daytime sleepiness and compensatory napping.

It is also interesting that virtually all siesta cultures noted in the Human

Relations File and those observed to engage in regular napping are between latitudes 45° north and 45° south of the equator (see Tables 2–6). Approximately 85% of the nearly 5 billion people on the earth live in this zone (Population Reference Bureau, Inc., personal communication). With so much of the population between the latitudes in which regular napping occurs, it seems likely that on any given day, a substantial, albeit unestimable, minority of the earth's inhabitants take a daytime nap, primarily in the afternoon hours. Within this 45° zone the photoperiod is highly predictable, never varying by more than a 2-to-1 ratio at the extremes (i.e., light:dark ratio of 2:1 in summer and 1:2 in winter). The timing of light and other environmental zeitgebers associated with it are, therefore, reliable features of this zone of regular napping on the planet.

Unlike cultures with siesta or regular napping patterns, cultures with irregular daytime sleep (see Table 7) appear to be the result of increased environmental (or occupational) variability. For example, an overall asystematic sleep/wake pattern is evident in some circumpolar cultures (Table 7). The Human Relations File contains the following observations, made in 1936, of the circumpolar Samoyed nation of Northern Siberia:

> With the coming of the hunting season the usual order of the day is completely shattered . . . the life of one Naganasan family is typical . . . the head of the family . . . continues to fish until two or three o'clock in the morning. Two grown sons constantly lack sufficient sleep, since they take turns watching the domesticated reindeer at night. After a short sleep, lasting no more that three hours, [they] hunt wild reindeer or take part in the collective hunt for wild geese with nets . . . the women dress game. . . . In the evenings they have to repair and dry hunting clothes . . . and they, too, sleep no more that two or three hours [RU4 File (sleep) (2)].

The Chukchee are also a Siberian circumpolar tribe. From observations made from 1919 to 1921 the following notes were made:

> As the nights were growing lighter the irregular habits of the Chukchi became almost intolerable [by] their complete disregard for all division of time. . . . One day they would sleep till noon, and would not crawl in their skins until midnight, get up and eat and go to sleep again in the morning. Day or night were ideas which no longer existed. . . . [RY2 File (sleep) (2)].

A third citation refers to the Alaskan Aleut Indian of the Pribilov Islands: "One of the peculiarities of this group [was] that at any and all hours of the night during the summer season. . . . A number of individuals are always up and around during the entire night and day" [NA6 File (sleep) (2)].

These typical observations of highly irregular sleep in circumpolar groups, compared with the more regular dawn-to-dusk patterns of tropical, subtropical, and temperate climate cultures, point to a correlative, if not causal, relationship associated with napping. Namely, the more stable the light/dark zeitgebers, the greater the likelihood of stable sleep/wake patterns and siestas. This appears to hold from the highly reliable 12-hr light/12-hr dark ratio at the equator to the highly variable photoperiod at circumpolar regions. Moreover, it is particularly true of traditional cultures in which artificial time devices are not readily available or widely used. Unfortunately, it is not based on the firmest of data and serves only to highlight the need for more systematic cross-cultural research on sleep/wake patterns as a function of geophysical, sociocultural, and biobehavioral parameters.

SUMMARY

Cultural differences in napping and siestas are apparent. Siesta cultures are associated with tropical regions. This may be a function of the regularity of zeitgebers, the stimulus control of midday and seasonal heat, and/or the effect of heat on nocturnal sleep. Within such a setting, a homogeneous culture with stabilized domestic units seems to be necessary. Factors that may inhibit napping are nomadic and unstable domestic settings and irregular food supplies such as occur in fishing and hunting cultures. In our increasingly industrialized societies, the requirements of stable and efficient production serve as strong inhibitors. It is interesting to note that both high irregularity and high regularity may be inhibitory.

Of particular import relative to these cultural considerations is that siesta cultures worldwide, even in older and traditional settings, appear to be the exception rather than the rule. Further, even within contemporary facilitative settings, such as Italy, Greece, and Mexico, all of which are pressing for industrial development, the strength of the nap pattern is remarkably limited. These observations suggest a relatively low basal endogenous tendency, although its presence may be evidenced by the remarkably stable timing of siestas in early to midafternoon. Given the cultural relativity of napping, it remains a challenge to comprehend and interpret the differences in napping behavior within and between persons of a given culture. There is a need to acquire more systematic data on cross-cultural variability in sleep/wake patterning in general and the factors that produce daytime sleep and sleepiness, in particular.

ACKNOWLEDGMENTS

The review and substantive evaluation on which this chapter is based were supported in part by grants MH-19156 and MH-44193 from the National Institute of Mental Health, U.S. Public Health Service, and in part by a grant from the Institute for Experimental Psychiatry Research Foundation.

REFERENCES

1. Lugaresi, E., Cirignotta, F., Zucconi, M., Mondini, S., Lenzi, P. L., and Coccagna, G. (1983): Good and poor sleepers: An epidemiological survey of the San Marino population.

In: *Sleep/Wake Disorders: Natural History, Epidemiology and Long Term Evolution*, edited by C. Guilleminault and E. Lugaresi, pp. 1–12. Raven Press, New York.

2. Murdock, G. P. (1972): *Outline of World Cultures*, 4th ed., rev. Human Relations File, New Haven, CT.
3. Murdoch, G. P., Ford, C. S., Hudson, A. E., Kennedy, R., Simmons, L. W., and Whiting, J. W. M. (1971): *Outline of Cultural Materials*, 4th ed., rev. Human Relations Files, New Haven, CT.
4. Rezsohazy, R. (1970): *Temps Social et Developpement*. La Renaissance du Livre, Brussels.
5. Rezsohazy, R. (1972): The methodological aspects of a study about the social notion of time in relation to economic development. In: *The Use of Time*, edited by A. Szalai, pp. 449–465. Mouton, The Hague/Paris.
6. Soldatos, C. R., Madianos, M. G., and Vlachonikolis, I. G. (1983): Early afternoon napping: A fading Greek habit. In: *Sleep 1982*, edited by W. P. Koella, pp. 202–205. Karger, Basel.
7. Szalai, A. (1972): *The Use of Time*. Mouton, The Hague/Paris.
8. Taub, J. (1971): The sleep-wakefulness cycle in Mexican adults. *J. Cross-Cultural Psychol.*, 44:353–362.
9. Webb, W. B. (1978): Sleep and naps. *Specul. Sci. Technol.*, 1:313–318.
10. Webb, W. B. (1985): Sleep in industrialized settings in the northern hemisphere. *Psychol. Rep.*, 57:591–598.

Sleep and Alertness: Chronobiological, Behavioral, and Medical Aspects of Napping, edited by D. F. Dinges and R. J. Broughton. Raven Press, Ltd., New York © 1989.

13

Sleep Attacks, Naps, and Sleepiness in Medical Sleep Disorders

Roger J. Broughton

Division of Neurology, University of Ottawa and Ottawa General Hospital, Ottawa, Ontario, Canada K1H 8L6

For decades the existence of more or less involuntary napping (as so-called "sleep attacks") has been recognized as a characteristic symptom of a variety of sleep disorders. These have included narcolepsy, sleep apnea, the symptomatic hypersomnias, and idiopathic hypersomnia.

In his original description of narcolepsy, Gélineau (58) noted the coexistence of *les accès du sommeil* (attacks of sleep) and cataplexy (which he called *astasia*). In combination with sleep paralysis and vivid hypnagogic hallucinations, these four symptoms later comprised the "narcolepsy tetrad" of Yoss and Daly (152). Charles Dickens (47), in *The Posthumous Papers of the Pickwick Club,* described the sleep episodes of the fat boy Joe, who showed all of the features of what came to be called the "Pickwickian syndrome" (i.e., excessive drowsiness, obesity, noisy breathing in sleep, and a flushed face suggestive of polycythemia). The daytime sleepiness and napping in the Pickwickian syndrome were shown 128 years later to be related to obstructive sleep apnea (57,77).

Similarly, around the time of the 1917 to 1924 encephalitis pandemic, von Economo (147,148) described daytime sleep attacks as a symptom in patients whose lesions predominated in the posterohypothalamic and midbrain brainstem areas, whereas lesions in the preoptic forebrain areas correlated with organic insomnia. In the 1950s and 1960s, a syndrome (often familial) of idiopathic hypersomnia characterized by daytime sleep episodes associated with abnormally prolonged and deep night sleep, often associated with sleep drunkenness upon awakening, was identified and differentiated from the numerous forms of symptomatic hypersomnia by B. Roth and colleagues (119,123).

These four conditions of narcolepsy, idiopathic central nervous system hypersomnia, symptomatic central nervous system hypersomnia, and sleep apnea remain to this day the object of the majority of studies of pathological diurnal sleep. All medical conditions identified to date that lead to daytime sleep attacks have been found to be characterized also by some degree of excessive daytime sleepiness (EDS) between attacks, which can be objectified most directly as an abnormal diurnal rapidity of falling asleep under standardized conditions.

Insomnia patients may also experience subjective daytime sleepiness. This appears, however, to be a qualitatively distinguishable trait, because such patients generally have difficulty (rather than facility) in falling asleep. This remains at least in part true during the daytime, as well as at night. Unlike hypersomnic patients, those with insomnia do not exhibit markedly shortened diurnal sleep latencies (115) and indeed these may be prolonged (139). The relative inability to fall asleep at any time appears to be a salient feature of insomnia. Insomnia patients therefore can be considered in general to show features of hyperarousal around the 24 hr, just as patients with EDS appear to be abnormally hypoaroused. Sleepiness and napping in this patient group will not be considered further, nor will naps in relation to the circadian sleep/wake schedule disorders such as in jet lag or in shift work (see Chapters 9 and 10).

These two major groups of sleep disorders have been differentiated in a useful and widely used nosology (6) into disorders of excessive sleepiness (DOES) and disorders of initiating and maintaining sleep (DIMS). This terminology was adapted in lieu of the similar group labels employed in earlier formally proposed classifications such as disorders of excessive sleep versus disorders of inadequate sleep (15) or hypersomnias versus insomnias (38). The DOES label is arguably preferable to the classical term hypersomnia (*hyper*, excessive; *somnia*, sleep), as studies using continuous 24-hr recordings of at least some DOES conditions, like narcolepsy-cataplexy, have shown that total sleep time (TST) in most patients is not increased, but rather sleep is simply redistributed in time. This has held true for both laboratory studies (12,72) and for ambulant home monitoring studies (25,26).

In this chapter, the term *sleepiness* will refer to the subjective feeling state of sleep need, *nap* will refer to the voluntary or semivoluntary choice of obtaining daytime sleep of duration less than the major nocturnal sleep period, and *sleep attack* will refer to episodes of more or less irresistible sleep that overwhelm the individual, when he or she is attempting to remain awake. It is, of course, recognized that these three concepts represent relatively artificial divisions in a continuum of sleep pressure leading to overt sleep.

APPROACHES TO THE STUDY OF PATHOLOGICAL DAYTIME SLEEP AND SLEEPINESS

Electroencephalogram

Prior to the (now one-third-century-old) discovery of rapid eye movement (REM) sleep by a student, Eugene Aserinsky, and his physiologist mentor, Nathaniel Kleitman (5), and the subsequent widespread study of sleep by modern polygraphic recording techniques, much pioneering work on the nature of naps, sleep attacks, and even sleepiness was accomplished by the traditional electroencephalogram (EEG) approach. This can best be exemplified by consideration of studies of narcolepsy; however, the same holds true in the case of other hypersomnias, especially the symptomatic varieties.

Although the eminent neurologist Kinnear Wilson (150) had considered both narcoleptic sleep attacks and cataplexy as forms of epilepsy, and indeed some authors presented EEG findings supportive of this diagnosis (42,140), it soon became evident that only physiological sleep patterns were present, as in fact had been firmly stated by Blake et al. (13) in the first EEG report of the condition. In addition to EEG features of physiological sleep, other patterns were repeatedly reported that were usually described as signs of persistent drowsiness between these sleep episodes (13,43,48,56,108,120). Early EEG studies during (sometimes induced) cataplexy were found to show patterns of wakefulness (43,56,67) and during sleep paralysis they were reported as those of either full wakefulness or light drowsiness (43,56,108,120). Subjects were shown to be able to fall asleep even during voluntary hyperventilation for EEG activation (43). Paradoxical alpha blocking during somnolence was emphasized (43,120). In the relatively rare symptomatic cases of narcolepsy-cataplexy, Roth (120) noted a greater frequency of minor EEG abnormalities.

In sum, the main contributions of traditional EEG studies were that they ruled out any epileptic mechanism for the sleep attacks and other episodic symptoms, showed that sleep attacks have the EEG features of physiological sleep, and unequivocally documented the presence of marked drowsiness between the sleep attacks.

Polysomnography

With the discovery of REM sleep many laboratories, mainly in Europe and Japan, rapidly took up the investigation of sleep/wake patterns both of night (Fig. 1) and daytime sleep in this important group of medical disorders. The propensity in narcolepsy-cataplexy toward sleep-onset REM periods (SOREMPs), a phenomenon first noted by Vogel (145), was shortly thereafter simultaneously documented in case series by three centers (71,113,142).

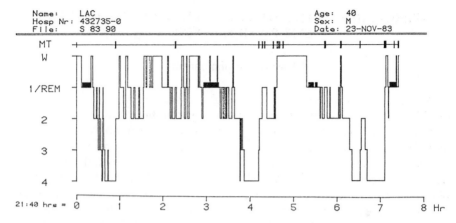

FIG. 1. Night sleep in a typical patient with narcolepsy-cataplexy. The histogram shows a sleep-onset REM period, even distribution of SWS and REM sleep across the thirds of the night, frequent awakenings and stage shifts, and increased REM fragmentation.

It is significant to note, however, that from very early on the tendency of such patients to fall into both types of sleep at night and in the daytime was widely recognized (9,39,46,55,69,123).

The presence in idiopathic hypersomnia of relatively prolonged and deep daytime sleep episodes lacking SOREMPs and containing much slow wave sleep (SWS) was described by Roth et al. (125). Daytime sleep in sleep apnea was less rigorously studied but was found to mainly consist of relatively brief repetitive episodes of light nonrapid eye movement (NREM) sleep (57,77,135).

24-Hr Recordings

Continuous polysomnographic recordings around the 24 hr to document spontaneous sleep/wake patterns in the hypersomnias were initiated by Gastaut et al. (55), Berti-Ceroni et al. (9), Passouant et al. (105,106), and Guilleminault et al. (63). Telemetric recordings allowing freedom of patients from being wired by an electrode-cable system were first reported by the Montpellier group (39,105). Ambulant monitoring of habitual 24-hr sleep patterns in the normal home environment was introduced by this author's laboratory (25,26,33,66). Others have performed 24-hr monitoring in time-free environments to investigate the nature of the biorhythmic aspects of sleep/wake states of narcoleptics under disentrained conditions (78,134). These laboratory and home studies permit documentation of spontaneous sleep/wake patterns under a variety of conditions.

Multiple Sleep Latency Test and Its Variants

A major advance in the studies of napping and sleepiness was the introduction of structured tests of sleep tendency recorded at specific times of the day. The Multiple Sleep Latency Test (MSLT) (40,116) analyzes the rapidity of onset and nature of daytime sleep under subject instructions to try to fall asleep as quickly as possible at fixed 2-hr intervals throughout the day, usually at 1000, 1200, 1400, 1600, and 1800 hr (Fig. 2). This test appears to tap directly into physiological sleep tendency (41).

EDS patients, however, must more or less continuously attempt to suppress daytime sleep, because of job requirements or other needs to perform either more or less continuously (e.g., while driving) or at unforeseeable intervals. In order to quantify the capacity to remain awake, approaches that are the inverse of the MSLT have more recently been introduced. In these tests the subjects are similarly recorded during multiple naps throughout the daytime but are instructed to try to remain awake. The procedures

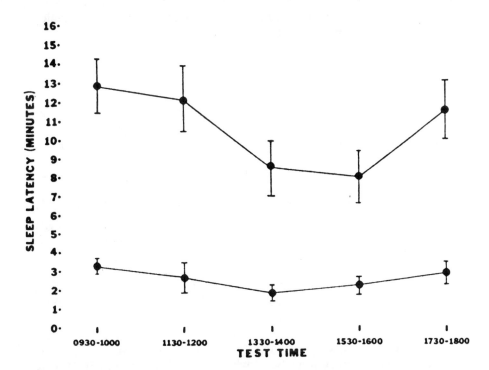

FIG. 2. Circadian pattern of MSLT mean sleep latencies every 2 hr in normal controls (*above*) and narcoleptics (*below*). Narcoleptics show significantly shorter sleep latencies for all naps. Both groups exhibit a midafternoon increase of sleepiness expressing the circasemidian pattern of sleep propensity. (From ref. 116.)

include the so-called repeated test of sustained wakefulness (RTSW) (64,65) and the maintenance of wakefulness test (MWT) (90,91) (see Chapter 4).

In both types of structured tests (of either sleep tendency or ability to suppress sleep), the emphasis has been on sleep latency as a measure of sleepiness and on the frequency of daytime SOREMPs to help diagnose narcolepsy in the absence of other known causes of direct entry into REM sleep. The latter include very irregular sleep/wake schedules in normals, endogenous depression, and the withdrawal of REM-suppressant medication.

Roth and colleagues (126,127) presented cogent arguments that the depth, as well as the rapidity, of daytime sleep represents an important measure of daytime sleep tendency. Depth is not assessed using the previous techniques. Based on a single 45-min midafternoon siesta recording, these authors introduced a weighted measure approach that considers the latency and duration for all five sleep stages and gives rise to a so-called polygraphic index of sleepiness (PIS) and a polygraphic score of sleepiness (PSS). The latter index, in particular, clearly differentiates daytime sleep of major groups of hypersomnic patients (narcolepsy-cataplexy, idiopathic hypersomnia, and sleep apnea) from controls, and indeed of narcolepsy-cataplexy from the other forms of hypersomnia.

Other Techniques

Daytime naps, and the sleepiness that precedes and succeeds them, have also been investigated by a variety of other techniques. Sleepiness has been measured by *subjective* scales, such as the Stanford sleepiness scale (73,74) and the 100-mm visual analog scale. *Pupillometry* has proved a sensitive, if technically difficult, measure (109,131,132,154). Objective *performance* tests have been used to assess the deficits from sleepiness in both uncontrolled (10,61) and controlled (143,144) studies. They have also documented the recuperative effects of naps (10,59), the presence of very brief "microsleeps"(60), and the slower waxing and waning fluctuations in alertness and performance over time (144).

Evoked potential measures were introduced as a possible rapid and sensitive means of quantifying EDS in sleep disorder patients by Broughton et al. (30). Evoked potential studies have used both simple auditory potentials evoked by tone stimuli (31) or by click stimuli (110), and so-called complex evoked potentials including those of the P300 paradigm and the contingent negative variation (CNV) paradigm (1,2,22,23). Finally, *computerized EEG* studies by power spectral analysis, which is sensitive to ultradian fluctuations of alertness in normals (88,89,99), have also been employed (Broughton et al., unpublished observations). Topographical EEG mapping employed in sleep (37,50) also has promise in this area.

All of these approaches have helped to clarify and, equally important, quantify many of the aspects of daytime sleep episodes and the intercalated daytime sleepiness of sleep disorder patients.

The main aspects of naps and sleep attacks will now be considered by diagnostic category. These aspects include, in order, the behavioral expression, the known or apparent physiological mechanisms, the chronobiological aspects, the internap (or sleep attack) status, the effects of naps, and the treatment of pathological daytime sleep and sleepiness. For almost all of these aspects, a major portion of our knowledge is based on studies of narcolepsy-cataplexy, which will therefore be considered in the greatest detail.

BEHAVIORAL ASPECTS

Narcolepsy-Cataplexy

There are surprisingly few detailed behavioral descriptions of naps and sleep attacks in any group of sleep disorder patients. Those with narcolepsy-cataplexy typically experience particularly irresistible episodes of daytime sleep, the so-called narcoleptic attack. Such sleep attacks have been described to occur while talking, eating, driving, cycling, swimming, skiing, and even during sexual intercourse and while having a painful tooth drilled (121). They are manifestly inappropriate and abnormal. As well as these truly overwhelming sleep attacks, narcoleptics also experience much less imperative sleep episodes and frequently take voluntary daytime naps. The usual duration of sleep attacks in narcoleptics is approximately 15 to 30 min, although a rare patient states that they may last 1 to 2 hr. However, as Roth (121) stated, these longer episodes more typically represent voluntary naps taken in an attempt to reduce sleepiness rather than true sleep attacks.

Clinically, sleep attacks and naps are often preceded by feelings of overwhelming sleepiness, pressure or irritation of the eyes, heaviness in the limbs, and, at times, dryness in the mouth. Such symptoms make up the complex state that is roughly quantified by subjective scales. There may also be blurring of vision, diplopia, or vertigo before falling asleep. Objective behavioral signs may include inactivity, ptosis, evident relaxation of facial musculature, and slumping of the head or body by gravitational attraction.

Such abnormally intense signs and symptoms of sleepiness, and the probability of subsequent sleep with all its manifold features, are all increased by those factors that facilitate drowsiness in normal individuals. These include a warm stuffy environment, inactivity, boredom, recent food intake, and repetitive hypnotic stimuli. Attempts to suppress sleepiness by walking around, tensing muscles, self-pinching, splashing the face with cold water, and even self-inflicted pain (such as biting the inside of the mouth) are all very common strategies in narcoleptics. At the time of sleep onset, vivid hypnagogic hallucinations, hypnagogic sleep paralysis, and a number of types of altered states of consciousness including simultaneous perception of the environment and dream imagery as a variety of "double consciousness" (19) may all occur.

Once sleep is entered, there may be quiescent or rolling eyes evident under closed or semiclosed palpebrae, slumping of the head, and regular respiration, together suggestive of passage into NREM sleep. Behavioral evidence of REM sleep, such as visible REM activity under the eyelids, irregular respiration, and facial and peripheral limb twitching, may also be observed at the start of or during an attack (16). Deep tendon jerks are depressed and may be absent, a positive Babinski may be found, and the pupils are miotic with, if in REM sleep, episodes of mydriasis during the phasic REM bursts. In short, the clinical picture is therefore one of either apparently physiological NREM or REM sleep (16).

Awakenings from sleep attacks in narcoleptics seldom show significant sleep inertia effects; most subjects in fact feel quite refreshed. They like, if possible, to nap. Awakening may sometimes be associated with hypnopompic sleep paralysis. According to Ruther et al. (129) and Roth (121), dreams are reported after only 30% to 40% of daytime sleep episodes in narcoleptics, although when awakenings are performed during recorded nocturnal REM sleep, this rises to 79% (122).

Daytime sleep episodes may also show extreme durations. These include very brief (approximately 5–30 sec) periods of ptosis, head drooping, and loss of consciousness, clearly differentiated from cataplexy and described by Ganado (53) as "momentary spells" and more recently (in association with arrest of continuous performance) as "microsleeps" by Guilleminault et al. (60). Prolonged partial sleep states also occur and appear to be the basis of the amnesic automatisms described as "fugue-like episodes" by Ganado (53) and as "automatic behavior syndrome" by Guilleminault et al. (60). Whether these are comprised of series of repetitive punctate microsleeps (60) or, more probably, of much slower waxing and waning of drowsiness with inability to sustain full alertness (144) remains speculative.

Idiopathic Hypersomnia

In idiopathic hypersomnia, the daytime sleep episodes are typically both longer and much less imperative than in narcolepsy-cataplexy (119,121). The patient usually does not fall asleep involuntarily while engaged in activities inappropriate for sleep. Patients continuously fight off sleepiness during the day and eventually feel they must lie down to nap. Daytime sleep episodes then usually last 30 to 120 min and may continue for 3 to 5 hr, rather than occurring as much briefer true sleep attacks typical of narcolepsy-cataplexy. Subjects who take shorter naps may have several a day, whereas those taking longer ones typically have only one or two per day (121).

The daytime sleep episodes usually show the clinical features of NREM sleep alone, although when prolonged for more than 50 min they may also show behavioral features of REM sleep. Patients often sleep very deeply during these daytime episodes and may be quite hard to awaken. Not infrequently, confusional arousals with sleep drunkenness will occur upon awakening from diurnal sleep, just as takes place in more than 50% of morning awakenings terminating overnight sleep (125). Patients rarely report dreaming after naps, and they often feel relatively unrefreshed even after quite prolonged daytime sleep.

Sleep Apnea Syndrome

Many but not all sleep apnea patients exhibit EDS and daytime sleep episodes. The daytime sleep episodes of sleep apnea patients are clinically more similar to those of narcolepsy than to those of idiopathic hypersomnia. They tend to be relatively brief and quite frequent and at times are subjectively quite irresistible. Their usual duration range is from brief microsleeps lasting several seconds to sleep episodes of 15 to 50 min duration, and even occasionally more prolonged periods of sleep (especially when not sleeping on the back). In obstructive sleep apnea, snoring may be a major behavioral feature of naps, just as it is of night sleep. Snoring can consist of either the rather peculiar loud inspiratory snorting, which is characteristic of the unblocking of upper airway obstruction, or a more sonorous inspiratory plus expiratory (so-called "see-saw") snoring, which indicates excessive vibration of soft tissues in the upper airway, or of both. Episodes of total cessation of all respiratory effort, indicating central sleep apnea, may also be evident behaviorally. Sleep apnea patients characteristically feel incompletely refreshed when they awaken.

Other Disorders

Daytime sleep is of interest in a large variety of other EDS disorders that are too numerous to detail here. As in idiopathic hypersomnia, the daytime sleep episodes in various types of symptomatic hypersomnia (e.g., postencephalitic, post-traumatic) tend to be prolonged, deep, and nonrestorative. Patients with so-called neurotic hypersomnia (hysterical hypersomnia) appear usually to take to their beds as an escape mechanism (121), and their daytime naps consist mainly of prolonged drowsiness rather than deep sleep.

The recurrent hypersomnias (imprecisely called periodic hypersomnias) take three main forms [cf. (11)]. The Kleine-Levin syndrome is most typically seen in male adolescents who intermittently show marked increases in TST per 24 hr associated with excessive eating (bulimia), hypersexuality, incoherent speech, irritability, confusion, and often hallucinations plus other behavioral disturbances. Patients are usually quite normal between hypersomnic periods. In menstrual hypersomnia, women may show a marked increase in daytime sleep just before, during, or immediately after menstruation. Behavioral and personality disturbances may also occur in relation to the menstrual hypersomnia. A third form of recurrent hypersomnia occurs in manic-depressive (bipolar) affective illness during the depressive periods. In all three forms, daytime sleep is considerably increased in total amount but tends to be rather light. True sleep attacks comparable with those seen in narcolepsy-cataplexy and obstructive sleep apnea are quite rare.

PATHOPHYSIOLOGY OF SLEEP ATTACKS AND NAPS

It is of interest to consider the physiology and physiopathogenesis of sleep attacks and voluntary naps in these patients. The phenomenon of EDS is discussed later.

Narcolepsy-Cataplexy

Daytime sleep in narcolepsy-cataplexy is characterized by a high frequency of SOREMPs sometimes said to be diagnostic of this condition (45). Narcoleptics do in fact generally show two or more REM-containing naps on the MSLT, and Mitler et al. (93) suggested that at least two of five MSLT naps exhibiting SOREMPs should be a diagnostic criterion for narcolepsy-cataplexy. Our laboratory has found in repeated MSLT studies of untreated patients with narcolepsy-cataplexy that there can be a considerable day-to-day variation in the number of SOREMPs, with, for example, three SOREMPs in a given patient one day and only a single one several days later. The reasons for this day-to-day variability in apparent REM pressure remain unclear. As already mentioned, SOREMPs are frequently seen in other conditions that are not discussed here, such as in normal subjects with very irregular sleep/wake habits, in endogenous depression, and in withdrawal states from REM-suppressant medication. They are even encountered quite frequently in sleep apneics (128,149). Some 50% of the daytime sleep episodes in untreated narcoleptics in fact begin with NREM rather than REM sleep. Moreover, at the onset of their illness patients who later develop full-blown narcolepsy-cataplexy may show NREM sleep attacks alone for several years as a form of monosymptomatic narcolepsy (121). As Broughton et al. (35) note, it is therefore not the ability to go more or less directly from wakefulness into REM sleep that, as some maintain, is pathognomonic of narcolepsy-cataplexy, but rather it is the presence of dissociated REM sleep to emotional stimuli during wakefulness in the form of cataplexy.

The second main feature of MSLT naps in narcolepsy-cataplexy is the extremely short latency to sleep onset. Mean sleep latency is generally less than 5 min (Fig. 2) and sometimes almost nil or essentially immediate (93,116,149). Even when narcoleptics are requested to remain awake, as in MWT and RTSW studies, sleep latencies are markedly reduced (64,90,91). The very short sleep latency in narcolepsy-cataplexy parallels the highly irresistible nature of many patients' overt sleep attacks. It appears probable that their overwhelming nature is related to the high frequency of SOREMPs, because Broughton and Aguirre (22,23) found sleep latencies to be shorter prior to REM-containing versus NREM-only MSLT naps in narcolepsy-cataplexy. A nonsignificant strong trend in the same direction is also evident in the results of Roehrs et al. (118). Moreover, it has been shown that subjective (as well as objective) sleepiness is greater prior to REM versus NREM naps (22,23). This indicates that differences exist between what may be called REM sleepiness and NREM sleepiness (20).

Most studies of the MSLT and its variants report only sleep latency and sleep-onset REM period data with little or no description of other nap sleep parameters. It therefore remains uncertain from these tests whether, for instance, REM sleep in the daytime tends to be as fragmented as it is at night (33,95) or whether overall sleep continuity is similarly disrupted as at night. The afternoon 45-min siesta studies of Roth et al. (127) show that considerable SWS (as well as REM sleep) may occur at that time.

Use of continuous 24-hr telemetric recordings, and especially ambulant home recordings (Fig. 3), provides much better documentation of spontaneous sleep patterns, although such techniques lack certain advantages relating to the structured nature of the MSLT and its variants. In particular,

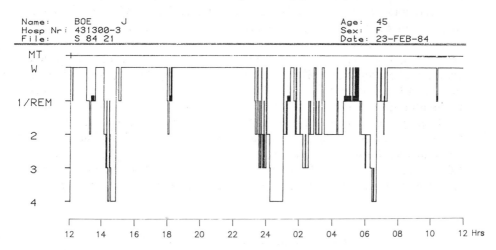

FIG. 3. A 24-hr ambulant sleep/wake recording in a 45-year-old female patient with narcolepsy-cataplexy. Recording begins at 1200 hr. There are four sleep attacks (three with SOREMPs), a 60-min midafternoon nap with SWS, and an 8-hr overnight sleep period. The latter lacks a sleep-onset REM period but shows even SWS distribution across the night, frequent awakenings, and REM sleep fragmentation.

one cannot easily measure the important variable of sleep latency. Of course, such recordings are extremely time-consuming. Laboratory studies with patients sleeping *ad libitum* in bed confirm frequent SOREMPs during semivoluntary naps and sleep attacks throughout the day. However, daytime SOREMPs in narcoleptics become much less frequent during prolonged recordings in which subjects are seated and performing (146) or going about their regular daytime activities (26). This appears to reflect both the suppressant effect of sitting or upright posture on SOREMPs, as first documented two decades ago by Hishikawa et al. (70), and the reduced daytime sleep occurring under such circumstances.

Idiopathic Hypersomnia

B. Roth and collaborators (121,125,127) reported that daytime sleep in idiopathic hypersomnia tends to be prolonged (generally lasting more than 30 min) and deep, with much SWS. The condition appears to represent an inherited (constitutional) excessively strong sleep system either involving sleep overall or involving mainly NREM sleep.

Sleep Apnea

The sleep latencies on MSLT in sleep apnea patients are exceedingly short, generally on the order of 2 to 5 min, and therefore are similar to those of narcoleptics (115,128,149). Moreover, although SOREMPs on MSLT are more frequent in narcolepsy-cataplexy, as mentioned, they are also quite common in obstructive sleep apnea (OSA). In the five-nap MSLT data of Walsh et al. (149), all of 13 narcoleptics met the two sleep-onset REM period criterion of Mitler et al. (93) for narcolepsy-cataplexy; but this was also true of 4 of 14 OSA patients investigated. Similarly, MWT studies in which patients attempt to suppress sleepiness have shown (36) frequent SOREMPs in sleep apneics (mean of 1.4 across 5 naps), although again occurring less frequently than for matched narcoleptics (mean of 2.7 across 5 naps). The spontaneous diurnal sleep patterns in sleep apneic patients have yet to be fully reported. Daytime sleep appears to be a result of marked night sleep fragmentation, repeated microarousals (128), or recurrent cerebral hypoxia in sleep.

CHRONOBIOLOGICAL ASPECTS

Consideration of the chronobiological aspects of sleep disorders raises two main issues that pertain to daytime naps or sleep attacks. The first concerns the extent to which the pathophysiology of such sleep disorders might be a consequence of an underlying primary biorhythmic disorder; Kripke (81) was the first to propose that narcolepsy-cataplexy might be considered as a disorder of biorhythmic control mechanisms. The second issue concerns the extent to which the normal temporal distribution of sleep/ wake patterns is maintained in sleep disorder patients who exhibit daytime sleep episodes. Specifically, to what extent are the following normal phenomena preserved: the circadian periodicities of the major sleep period and of REM sleep and SWS within it, the circasemidian propensity for a second major sleep period usually in the midafternoon, and the daytime ultradian variations in alertness/drowsiness?

Circadian Aspects

Fully entrained normal adult subjects who are active during the usual daytime period typically show a monophasic nocturnal sleep pattern sometimes supplemented by a daytime nap. Our considerable knowledge of human circadian rhythms is derived in large part from studies of the regular circadian recurrence of nocturnal sleep, its association and interaction with the circadian variation in central body temperature, and the different periodicities and consequent temporal dissociation (e.g., internal desynchronization) of sleep and body temperature while free-running in environments lacking zeitgebers. Although few quantitative studies of sleep-related circadian rhythms have been reported in hypersomnic patients, a number of aspects have been documented and warrant comment.

First, although the increase in the proportion of daytime to total 24-hr sleep present in all such conditions obviously indicates a broadening in the circadian acrophase of sleep (the period of greatest probability of sleep occurrence), sleep nevertheless remains predominantly nocturnal in distribution. All continuous 24-hr studies of sleep patterns in narcolepsy (9,12,25,26,33,39,95,105,106) have reported that the greatest amounts of sleep per hour remain during the nighttime (cf. Fig. 3). This is even true when daytime sleep is further greatly facilitated by keeping patients in bed much of the day, which can lead to markedly polyphasic patterns reminiscent of those of the neonate, as pointed out by Passouant et al. (105).

Just as the circadian acrophase of sleep is broadened, so, it appears, are those of individual sleep stages at night. Thus, the presence of SOREMPs (Fig. 1) leads to an apparent phase advance of the ultradian REM cycle with a more even probability of REM sleep occurring during the night. This compares with the normal peaking in the last third, as reconfirmed throughout the literature and analyzed in detail, for instance, by Endo et al. (49). Our reanalysis of the extended sleep study data of Gagnon and colleagues (51,52) has indicated that the true habitual acrophase of REM sleep is 7.5 to 8.0 hr after sleep onset. That is, it occurs at or just after the usual time of awakening

(24). Central body temperature has been found to also show a phase advance in narcolepsy (98). SWS (cf. Fig. 1) shows a distinct tendency to be distributed more evenly throughout the night in narcoleptic patients (25,26,33) than in normals.

Idiopathic hypersomnia and sleep apnea also show preservation of sleep mainly at night. Concerning the individual sleep stages, in idiopathic hypersomnia the normal circadian peaks of SWS in the first part of the night and of REM sleep in the latter part of the night appear relatively maintained (125). In obstructive sleep apnea there is such frequent total suppression of SWS that time-of-night analysis is difficult, if not (at least visually) impossible. Overnight recordings in apneics show that the greatest amounts of REM sleep are in the latter portion of the night, as occurs in normals.

Free-running studies in environments lacking time cues have been reported to date only for narcolepsy-cataplexy. Kamei et al. (78) studied three female subjects and confirmed the existence of a normal endogenous circadian sleep/wake cycle period of somewhat greater than 24 hr; they also noted the appearance of disorganization of the major sleep period with little effect on cataplexy or sleepiness. Schulz et al. (134) monitored both sleep/wake patterns and temperature in a single subject for 2 weeks and confirmed essentially normal patterns of an increase in the circadian sleep period beyond 24.0 hr, which occasionally increased to 45.2 hr (mean 32.9 hr) plus internal desynchronization with dissociation of sleep from the much shorter circadian rhythm of central body temperature (mean period 24.7 hr).

Circasemidian Aspects

More than a decade ago it was first proposed by Broughton (17) that humans also have an approximately twice-a-day biorhythmic propensity (i.e., sleep normally occurring at night plus a second period of sleep or sleepiness in the afternoon) and, moreover, that the afternoon period of sleep tendency might represent a second period of pressure for deep SWS occurring approximately 12 hr after the circadian acrophase of SWS in the first third of the night. Further evidence for this second period of sleep tendency has accumulated rapidly and is reviewed elsewhere [cf. (21)] (see Chapter 5). The secondary midafternoon increase in sleep propensity is clearly preserved in pathology.

Narcoleptics exhibit definite relative shortening of MSLT latency (cf. Fig. 2) in their 1400- and 1600-hr naps (116,117). Further detailed analysis of MSLT data in narcoleptics by Aguirre and Broughton (2) also revealed a strong tendency toward greatest amounts of stage 2 (but not REM and stage 1) and maximum probability of SWS occurrence (despite the brief 20-min period permitted in some MSLT schedules for sleep) during these afternoon naps. Conversely, the maximum probability of SOREMPs occurred earlier, at the 1000-hr nap, during which they were significantly more frequent (84.5% incidence) than at the 1200- (58.3%), 1400- (54.5%), 1600- (54.5%), and 1800-hr (50.5%) naps (2). In adult narcoleptic subjects with the usual hours of retiring and morning awakening, each type of sleep therefore maintains a different circadian acrophase, which is similar to that of normals.

Ambulant monitoring of spontaneous sleep/wake patterns in the home environment using portable recorders (25,26) has shown that most narcoleptics take a relatively prolonged (average duration 34.3 ± 27.3 min) daytime nap in the midafternoon (Fig. 3). This increase is evident in 2-hr daytime averages of TST both in our own data (Fig. 4) (26) and in laboratory recordings with *ad libitum* sleep (12). Moreover, this nap has the highest probability of containing deep SWS (26), and the mean interval between the onset of the first SWS peak in nocturnal sleep to that of the major afternoon nap was 13 to 14 hr—essentially identical to that reported for normals during greatly extended (51) and extended shifted (52) night sleep. The marked afternoon secondary increase in sleep probability was also evident in 34-hr in-lab recordings reported by Billiard et al. (12).

This midafternoon increase in sleep propensity in narcoleptics appears to

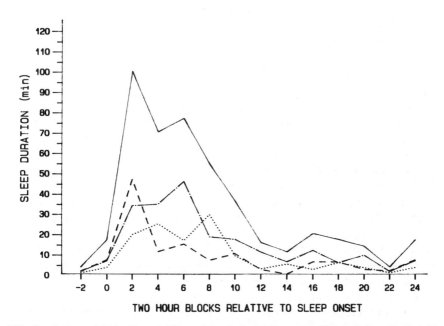

FIG. 4. Averaged 24-hr sleep patterns in ambulant home recordings of 10 patients with narcolepsy-cataplexy. Mean sleep and sleep stage durations in min were averaged in 2-hr blocks after evening sleep onset (mean time was 23:05 hr). They show poorly sustained night sleep, low amounts of morning sleep, a midafternoon increase (which peaks in the 1400–1600 hr block), and a decrease in the evening prior to sleep onset. Total sleep (—); SWS (– –); stage 2 (·—); REM (· · ·).

be followed by a period of minimal sleep pressure. This is manifested as the longest daytime sleep latencies in MSLT and MWT (64,90,91,116,117) and also is evident in the continuous in-lab recordings of Billiard et al. (12) and in the ambulant monitoring studies of Broughton et al. (25,26). Although the decreased sleep tendency around 1800 hr on MSLT has been ascribed variously to the cumulative effects of sleep in previous naps or to increased arousal related to anticipation of the end of the test, it appears more likely that it reflects a normal reduction of sleep tendency at this time (17), what Lavie (83) called the "forbidden zone" for sleep.

Although much fewer data are available in respect to idiopathic hypersomnia, sleep apnea, and other hypersomnic conditions, it certainly appears that the circasemidian sleep propensity remains evident in most or all EDS sleep disorders. Idiopathic hypersomnia, for instance, is typically characterized by a prolonged afternoon sleep episode that often contains much SWS (121). Similarly, sleep apnea patients also show more frequent afternoon, compared with morning and evening, sleep episodes (Broughton, unpublished observations).

Ultradian Aspects

It is now well documented that normal adults exhibit a variety of daytime rhythms at approximately the same period as the nocturnal NREM/REM cycle, i.e., at approximately 90 to 150 min [cf. (17,21,82; Chapter 5)]. These are particularly striking in measures reflecting alertness/sleepiness. It is, in fact, only for these variables (21) that there is any impressive evidence for an ultradian basic rest/activity cycle throughout 24 hr, as originally proposed by Kleitman (79,80). It is evident that the intermittent sampling of sleep tendency every 2 hr by MSLT is insufficiently frequent to document or analyze such fluctuations. They require either very frequent sampling or (preferably) continuous recording techniques.

That narcoleptics may show ultradian rhythmicity of recurrent daytime sleep attacks was first described by Passouant et al. (105,106) using traditional in-lab and telemetric recordings (Fig. 5, left). Their initial reports of ultradian rhythmicity dealt with attacks beginning in REM sleep. But the phenomenon also occurs for the NREM sleep episodes (Fig. 5, right), which are encountered in so-called monosymptomatic (independent) narcolepsy with sleep attacks alone (7). In these early Montpellier studies, subjects were not always recorded continuously, but rather were often re-recorded every 2 hr (Passouant, personal communication), which might have artificially entrained such periodicity. That the spontaneous ultradian rhythmicity of sleep episodes in narcolepsy is a real phenomenon is nevertheless confirmed by its frequent appearance in continuously recorded seated subjects (146).

The initial reports of repeated REM-onset diurnal sleep attacks every 2

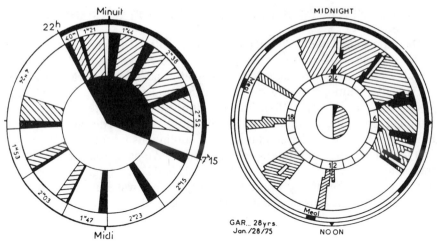

BOI... 3.Ⅲ.67

FIG. 5. Ultradian distribution of daytime sleep in laboratory recordings of narcolepsy plotted by the 24-hr clock approach (NREM sleep, *hatched*; REM sleep, *solid*; wake, *open sections*). **Left** is from a patient with narcolepsy-cataplexy. In the nocturnal sleep period a sleep-onset REM period and two major awakenings occur. They are followed by six diurnal sleep episodes repeated approximately every 2 hr in phase with night REM periods and all beginning in REM sleep. (From ref. 107a.) **Right:** sleep/wake patterns in a patient with sleep attacks alone, shows four daytime sleep episodes consisting exclusively of NREM sleep and recurring approximately every 2 hr. (From ref. 7.)

hr or so led Passouant (103) to claim that REM sleep in these patients exhibits ultradian rhythmicity throughout 24 hr. Continuous ultradian rhythmicity of REM sleep during both daytime and nighttime periods in patients with narcolepsy-cataplexy has subsequently been confirmed using formal time-series analysis by Schulz (133) and a predictive statistical technique by De Koninck et al. (44). Although this evidence exists that REM sleep in narcoleptics shows ultradian rhythmicity throughout 24 hr, there is also evidence from autocorrelation and similar analyses that (at least within nocturnal sleep) it is less rhythmic than the REM cycle in normals (72,94) and, indeed, than in monosymptomatic (NREM) narcolepsy (72).

In idiopathic hypersomnia, the daytime sleep episodes are too infrequent to show ultradian rhythmicity of repetition. Sleep apnea patients with EDS have not yet been studied in this regard.

In sum, although there is no solid evidence that any of these conditions are owing to a biorhythmic disturbance, there is definite evidence for some degree of apparently secondary disturbance of biological rhythms. The latter, however, is superimposed on relative preservation of the main features of the normal circadian, circasemidian, and ultradian distribution of sleep/wake patterns.

VIGILANCE BETWEEN DAYTIME SLEEP ATTACKS AND NAPS

Daytime sleep attacks and naps of patients with EDS typically take place within a matrix of more or less persistent excessive daytime sleepiness, which itself shows circadian, circasemidian, and ultradian variations. As previously mentioned, subjects may continuously struggle against such sleepiness [cf. also (100,101)] and develop often elaborate or even pain-inducing techniques to minimize its debilitating effects. This struggle is particularly necessary under conditions with continuous performance demands and in which either inattention or sleep could have dangerous and even life-threatening effects, such as while driving. It is this struggle, no doubt, that is reflected in the high proportion of polygraphic wakefulness containing artifacts of muscle tension and movement (so-called active wakefulness) that has been documented in seated narcoleptics compared with controls (146), although this is absent in ambulant recordings of narcoleptics (25). There is evidence in narcoleptics that this struggle comes with a later price of increased sleepiness and even facilitation of cataplexy (144).

The physiological nature of such EDS is increasingly understood and has been especially well analyzed in narcolepsy. Between overt sleep episodes, narcoleptics frequently show a more or less continuous waxing and waning of alertness (Fig. 6), mainly back and forth between EEG patterns of full wakefulness and those of light drowsiness. The latter may be subclassified (54,144) into stages 1A (slowing and diffusion of alpha, with or without slow

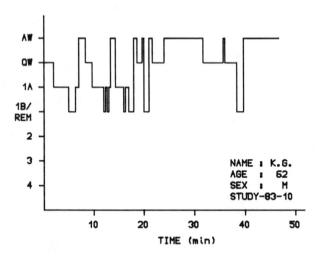

FIG. 6. Waxing and waning of alertness in a patient with narcolepsy-cataplexy recorded by ambulant home monitoring. A 50-min typical period is graphed. There are continuous fluctuations between active wakefulness (AW), with movement artifacts; quiet wakefulness (QW), without such artifacts; substage 1A, slowing and diffusion of alpha; and substage 1B, loss of alpha, medium voltage mixed frequency patterns. QW and AW follow the criteria of Volk et al. (146), and 1A and 1B those of Valley and Broughton (144).

eye movements) and stage 1B (fragmentation and disappearance of alpha, presence of medium voltage mixed frequency activity). These substages are equivalent to stages A and B of the Loomis et al. (84,85) classification, and substage 1B is equivalent to stage 1 of Rechtschaffen and Kales (112). Impressed by the great variety of EEG-polygraphic patterns of vigilance between full wakefulness and overt stage 2 or REM sleep, others have divided them into as many as seven substages (8,120).

Valley and Broughton (144) found that such continuous fluctuations of vigilance are maintained throughout prolonged boring tests like the 1-hr Wilkinson auditory vigilance task (144), although the same patients were able to remain fully awake during short challenging tasks such as the paced auditory serial addition test and digit span test (143). This also appears true for neuropsychological memory tests (3). Waxing and waning patterns are much more typical of vigilance-type tasks and similar situations than is the interruption of sustained wakefulness by brief, discrete so-called microsleeps [(60), i.e., the "momentary spells" of Ganado (53)]; the latter are in fact quite rare (144).

These relatively slow and apparently minor alertness fluctuations are far from trivial, as they have striking effects on performance abilities (144). In this study we defined sleep/wake state at the time of the signal stimuli, which were to be detected according to the 3 sec of EEG immediately preceding each stimulus. Waking patterns were further subdivided according to whether the immediately preceding 10 sec of EEG recordings contained further waking patterns (sustained wakefulness) or those of drowsiness or sleep (fragmented wakefulness). Ability to detect the signal stimuli (Fig. 7) increased from essentially nil in stage 2 to levels not significantly different from controls in sustained wakefulness. The greatest stepwise change in performance abilities with increasing EEG alertness levels was from fragmented to sustained wakefulness (17.8%–46.5% detection rates), which indicates that even recent drowsiness impairs waking performance. There is, therefore, a daytime *drowsiness inertia* effect on performance comparable with the well-described sleep inertia (sleep drunkenness) effect with awakenings from sleep itself. As well as stimulus omissions (undetected stimuli), drowsiness was associated with increased frequency of false positive responses indicating overwillingness to respond.

These results cannot be interpreted simply by the microsleep-lapse hypothesis. Substantial deficits in detection ability plus other performance changes occur independent of any recorded overt microsleeps or more prolonged sleep episodes. The markedly different capabilities on a vigilance task between sustained wakefulness, fragmented wakefulness, and substages 1A and 1B underscore the practical importance of such fine-grained analyses, rather than simply scoring wakefulness and stage 1 levels, as is generally done. Moreover, as noted, performance decreases both during light drowsiness and immediately following arousal from drowsiness. The main per-

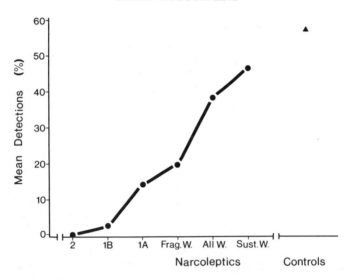

FIG. 7. Performance on the Wilkinson auditory vigilance task as a function of polygraphic sleep/wake state in patients with narcolepsy-cataplexy and matched controls. In narcoleptics, detection rates for signal stimuli are plotted separately for stages 2, 1B, 1A, and wakefulness. The latter is further separated into sustained wakefulness (Sust. W) (at least 13 sec of wakefulness prior to signal stimuli) and fragmented wakefulness (Frag. W) (with drowsiness immediately prior). Detection increases from 17.8% to 46.5% going from fragmented wakefulness to sustained wakefulness, indicating a drowsiness-inertia effect. In sustained wakefulness, narcoleptics are not significantly different from normals. [For results, see (143,144).]

formance problems therefore stem from the inability to sustain alertness over time. Stated conversely, maintenance of full wakefulness throughout a period of time is necessary for optimum levels of performance. The tendency of wakefulness to fragment in EDS patients would make the perception and retention of the chain of events necessary for processing more complex material very difficult (144). Finally, the results point to the importance of analyzing and employing state continuity for wakefulness, just as continuity is coming to be recognized as essential for assessing the relationship of nocturnal sleep to behavior (14,138).

Sleepiness, until recently, has been widely assumed to be a state that varies only in magnitude but not in kind. This may not be true. It has been postulated (20) that, just as there are the three qualitatively different mammalian and avian biological states of NREM sleep, REM sleep, and wakefulness (136), EDS might similarly show different states reflecting selective (or greater) involvement of one of these basic biological states. Specifically, one would predict three forms of sleepiness reflecting mainly pressure for REM sleep, pressure for NREM sleep, or impaired reticulocortical waking mechanisms. These might be called REM sleepiness, NREM sleepiness, and de-arousal sleepiness, respectively.

Patients with narcolepsy-cataplexy form an excellent model in which to study the possible existence of at least REM and NREM sleepiness, as they directly enter either REM and NREM sleep in daytime recordings.

We have found that REM sleepiness present immediately prior to subsequent sleep-onset REM period MSLT naps is subjectively (Stanford sleepiness scale) and objectively (MSLT latency) greater than is NREM sleepiness prior to NREM-only naps (22,23). Moreover, certain event-related potential measures have been found to differ. This is true both for simple auditory evoked potentials (110) and the more complex P300 and CNV potentials (22,23). Specifically, REM sleepiness compared with NREM sleepiness was associated in our studies of complex event related potentials with a significant amplitude increase of component P2, an amplitude decrease of component P3, and a significant near-total suppression of the CNV (Figs. 8 and 9).

Distinct differences in cerebral state therefore precede the two types of sleep, just as the post-arousal effects have long been known to differ [cf. (21)]. It seems probable (18) that narcoleptics also have impaired reticulo-cortical arousal (i.e., a so-called subvigilance syndrome or de-arousal sleepiness) coexistent with pressure for REM and NREM sleep. The fundamental

Fz

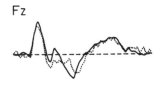

Cz

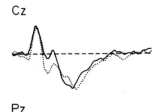

Pz

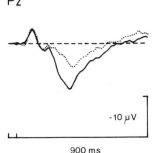

-10 μV

900 ms

FIG. 8. REM sleepiness (· · ·) and NREM sleepiness (—) prior to REM-containing and NREM-only MSLT naps distinguished by the P300 complex evoked potential paradigm. Curves represent data from narcoleptics averaged before each type of nap. REM sleepiness shows significantly higher amplitude of component P2 (peak latency of approximately 120–250 msec, measured at Cz electrode) and an almost significantly smaller P3 component at 350–400 msec latency. Positive is down. (From ref. 23.)

Fz

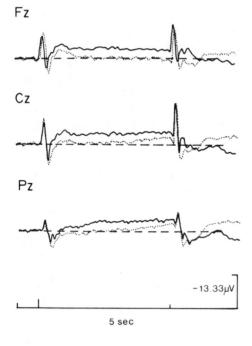

Cz

FIG. 9. REM sleepiness (· · ·) and NREM sleepiness (—) distinguished by the contingent negative variation (CNV) paradigm. In REM sleepiness there is almost total suppression of the slow negative CNV wave, which is measured at Fz. In NREM sleepiness by comparison, the CNV is of normal waking amplitude. The CNV complex confirms higher amplitude P2 components both to the S1 and S2 stimuli (From ref. 23.)

Pz

$-13.33\mu V$

5 sec

pathophysiological defect in the condition, in fact, appears to be one of general state boundary control rather than of pressure for REM sleep (35).

Although much remains to be clarified, particularly in EDS conditions other than narcolepsy, it is nevertheless clear that brain state between overt sleep attacks and voluntary naps is far from normal. Sleep attacks and naps simply represent the culmination of an abnormal propensity to enter the sleep state(s).

EFFECTS OF SLEEP ATTACKS AND NAPS

Patients with EDS often state that sleep attacks and naps are refreshing and followed by an increase in alertness. This appears particularly true for narcoleptics. This subjective improvement was documented in collaboration with Marisa Aguirre. We examined differences in Stanford sleepiness scale scores across MSLT naps and found a significant reduction in self-assessed sleepiness (2,22). Increased subjective alertness was greater for REM-containing versus NREM-only naps, an effect attributable in large part to the higher mean subjective sleepiness levels prior to REM naps (i.e., to higher levels preceding REM sleepiness compared with NREM sleepiness).

The subjective recuperative effect of daytime naps has some support in objective performance measures. Billiard (10) evaluated the effects of daytime naps on performance and cognitive functions and found improvement after naps. This effect was somewhat (but nonsignificantly) better after NREM naps; but these naps were not controlled for either duration or time-of-day effects. Recently, Godbout and Montplaisir (59) studied possible recuperative effects of naps on the sleepiness of patients with narcolepsy-cataplexy. They confirmed impaired performance levels in narcoleptics on a 4-choice RT test. On intrasubject comparisons of narcoleptics for days with MSLT naps to days without naps, there was a clear trend toward improvement on speed and accuracy on days with naps. Performance was not studied after, compared with immediately prior to, individual naps.

The effects of naps on sleepiness have also been investigated by the analysis of subsequent MSLT sleep latencies. Roehrs et al. (118) compared narcoleptics with a group of patients with other DOES conditions, mainly OSA. They found intergroup differences in the effects of altering the duration of the fourth nap at 1600 hr (either 15 or 30 min) on a subsequent fifth latency test, the interval to which was also altered (15 or 30 min later). In narcoleptics, the shorter 15-min naps led to just as great an increase in alertness as did the 30-min naps. Moreover, the recuperative effects were not sustained for a long time, as 30 min later the sleep latency had returned to baseline levels. As napping in narcoleptics did not lead to sustained alertness and increased duration of naps gave no further benefit, the taking of naps did not appear to be a particularly useful form of treatment for narcolepsy.

Despite this experimental evidence, it is nevertheless true that a significant portion of narcoleptics believe they can only get through the day if they take voluntary naps at the time(s) they have found most propitious. This is usually earlier in the afternoon, a time of day not studied in the Roehrs et al. (118) protocol. Conversely, in the other DOES patients the longer nap gave a significantly greater improvement in alertness, and this improvement was sustained 30 min later (118). At least in sleep apneics, therefore, it appears that daytime naps definitely help reverse their sleepiness, which may be an effect of sleep debt related to their fragmented, and often shortened, night sleep. This effect would be similar to the well-documented recuperative value of naps in sleep-deprived normals (86,87) (see Chapter 11).

The effects of daytime sleep on the quality of subsequent nocturnal sleep do not appear to have been studied in DOES patients.

The existence of daytime sleep attacks and marked sleepiness directly engenders a number of secondary major symptomatic and psychosocial effects. Secondary symptoms include memory problems (especially of recent memory), visual problems (mainly defocusing and diplopia), difficulties with balance, and sensitivity to the effects of alcohol. These symptoms have been shown in controlled studies to be significantly more frequent than in controls in both narcolepsy-cataplexy (27,28) and idiopathic hypersomnia (34). That sleepiness, rather than possible organic pathology, is the cause appears particularly certain for their complex memory and attentional problems, as we documented that even narcoleptics selected because of major subjective memory complaints in everyday life were able to perform a comprehensive memory test battery (involving attention, immediate recall, recent and remote memory, verbal and nonverbal material, both visual and auditory modes of presentation) at levels comparable with normals in a structured laboratory situation (3). As for short challenging performance tests like the PASAT or Knox Cube test, during which polygraphic recordings were done, narcoleptic patients are able to rally and sustain alertness and performance at normal levels during testing (143).

Marked effects on socioeconomic parameters also occur that appear attributable to the daytime sleep and sleepiness and have been documented for both narcolepsy-cataplexy (27,28) and idiopathic hypersomnia (34). These include effects on work (decreased income, threat of job loss, actual job loss, disability insurance), education (lowered achievement, problems with teachers), interpersonal relationships (including marriage break ups), driving (accidents, near accidents, increased insurance rates), and household and occupational accidents. Indeed, we found (29) the socioeconomic effects of narcolepsy-cataplexy to be, if anything, even greater than for comparable forms of epilepsy, another neurological condition well-known for its often devastating impact on the quality of life.

TREATMENT OF SLEEP ATTACKS AND EDS

Treatment of the daytime sleep attacks and EDS of sleep disorder patients has proved to be the most difficult part of their management.

In narcolepsy-cataplexy, the REM-based symptoms of cataplexy, sleep paralysis, hypnagogic hallucinations, and nightmares normally respond well to tricyclic medication such as imipramine (4,68,124), clomipramine (62,104), MAO inhibitors (151), or to other more recently introduced substances such as zimelidine (97). They also respond well to nocturnal gamma-hydroxy-butyrate (32,130), the main effect of which is to increase the continuity of sleep, especially the very fragmented REM sleep of narcolepsy (33,96,130).

On the other hand, sleep episodes and especially daytime sleepiness, although responding partially to daytime stimulants such as methylphenidate (76,153), pemoline (75,92), or amphetamines (102,107,111), are typically much more refractory to treatment. They usually remain a therapeutic problem after the REM-base symptoms have been suppressed. As stated elsewhere (20), this may reflect combined involvement in the pathogenesis of the EDS of narcolepsy of NREM sleep, REM sleep, and waking physiological and neurochemical mechanisms; current medication may simply be inadequate to produce the appropriate balance of effects. If the condition is fundamentally one of the state boundary control rather than of pressure for particular sleep/wake states (35), therapeutic progress would be expected to be particularly likely from further research into the fundamental mechanisms involved in the integration and continuity of sleep/wake states.

Scheduling of naps to minimize drowsiness has been recommended by some (114,137) but appears still to be of uncertain value. As we have seen, there is little objective evidence for any durable improvement of alertness in these patients from napping, at least in the late afternoon.

It is essential to remember that some 15% to 20% of patients with narcolepsy-cataplexy have coexistent clinically significant amounts of either sleep apnea (obstructive or central) or periodic movements in sleep. These associated conditions would alone be capable of inducing EDS and sleep attacks. Personal clinical experience with such patients indicates that treatment of these conditions (by, for example, weight loss, protriptyline, surgery; clonazepam for PMS) must be done along with prescribing the usual stimulants for narcolepsy, if optimum alertness levels are to be attained.

Idiopathic hypersomnia appears to reflect extreme sleep need, unlike narcolepsy-cataplexy in which total sleep per 24 hr remains essentially normal. The depth and amount of sleep are both increased, often on a genetic basis. Such patients typically require 12 to 16 hr or more of sleep (121). In these patients, there is some possibility that sleep satiation might lead to greater alertness during the remainder of the day. But for many sufferers of the condition the consequent low number of residual available waking hours would simply not be sufficient to permit anything like a normal life. Most patients must therefore be treated by stimulant medication with its attendant risks and side effects (101). Sleep drunkenness, when present, may be minimized by having a patient briefly awaken approximately 1 hr prior to final morning awakening, take 10 to 20 mg methylphenidate, and then return to sleep for the hour (121).

In symptomatic hypersomnias, the treatment should be oriented, whenever possible, toward the causal factor. In practice, this approach must often be supplemented by stimulant medication.

The sleep attacks and sleepiness of sleep apnea syndrome should also be treated causally insofar as is possible. For patients with OSA, this may involve weight reduction, removal of tonsils and adenoids, uvulopalatopharyngoplasty, mandibular advancement surgery, or other such means of reducing the upper airway obstruction. Continuous positive pressure to the upper airway (i.e., nasal CPAP) during sleep may also relieve the symptoms (141). Central apnea may respond best to protriptyline, medroxyprogesterone, or other respiratory stimulants. In all forms of sleep apnea, it would appear that reversing the fragmentation and lightening of night sleep and reducing the nocturnal hypoxemia produced by these sleep-related respiratory problems will help to lead to reduced daytime sleep pressure and the associated neuropsychological deficit. Central nervous system depressants such as sedative hypnotics and evening alcohol should be totally avoided in all such patients; they are at risk to develop respiratory arrest from such medication.

SUMMARY

Numerous approaches to the investigation of pathological daytime sleep and sleepiness exist and more are currently being developed. They have clarified the behavioral features and physiological nature of voluntary naps, overt sleep attacks, and the intercalated high levels of sleepiness in a number of sleep disorders. Just as for nighttime sleep states, the various expressions of daytime alertness have been separated for investigational purposes into levels, while acknowledging that these represent artificial divisions of a continuum. The postulate of qualitatively different forms of EDS has found experimental support and would appear to have pragmatic implications for treatment. The effects of napping on physiological and performance measures of sleepiness, sometimes beneficial sometimes not, have begun to be investigated, although much remains to be done in this regard. None of the classical hypersomnias to date has been found to be attributable to a primary disturbance of sleep/wake biorhythms. They all show relative preservation of the main circadian, circasemidian, and ultradian sleep/wake features characterizing normal sleep. The potentially beneficial effects of napping will no doubt be optimized by scheduling naps in phase with endogenously generated moments of enhanced sleep pressure. In all DOES conditions the treatment of sleepiness and sleep attacks remains the major therapeutic challenge.

ACKNOWLEDGMENTS

The author wishes to thank the Medical Research Council of Canada for support of research by term grant "The Investigation of Neurological Sleep Disorders" (MT-3291) and for personal support as a recipient of an MRC Career Investigator Award. He also thanks his many past and present graduate students and research associates, much of whose work is summarized herein. They include Marisa Aguirre, Dave Barker, Keith Busby, Bernardo da Costa, Pierre Duchesne, Wayne Dunham, Quais Ghanem, Tom Healey, Joel Herscovitch, Steven Liddiard, Roberto Low, Kathy Lutley, Claude Manseau, Jagdish Maru, Janice Newman, Martin Rivers, Janet Roberts, Claudio Stampi, Wlodeck Suwalski, and Victoria Valley. Barbara Reynolds kindly typed the manuscript.

REFERENCES

1. Aguirre, M., and Broughton, R. (1984): Objective and subjective measures of (REM and NREM) sleepiness in narcolepsy-cataplexy. *Sleep Res.,* 13:128.
2. Aguirre, M., and Broughton, R. (1987): Complex event-related potentials (P300 and CNV) and MSLT in the assessment of excessive daytime sleepiness in narcolepsy-cataplexy. *Electroencephalogr. Clin. Neurophysiol.,* 67:298–316.

3. Aguirre, M., Broughton, R., and Stuss, D. (1985): Does memory impairment exist in narcolepsy-cataplexy? *J. Clin. Exp. Neuropsychol.,* 7:14–24.
4. Akimoto, H., Honda, Y., and Takahashi, Y. (1960): Pharmacotherapy in narcolepsy. *Dis. Nerv. Syst.,* 21:1–3.
5. Aserinsky, E., and Kleitman, N. (1953): Regularly occurring periods of eye motility and concomitant phenomena during sleep. *Science,* 118:273–274.
6. Association of Sleep Disorders Centers (1979): Diagnostic classification of sleep and arousal disorders, 1st ed., prepared by the Sleep Disorders Classification Committee. *Sleep,* 2:1–137.
7. Baldy-Moulinier, M., Arguner, A., and Besset, A. (1976): Ultradian and circadian rhythms in sleep and wakefulness. In: *Narcolepsy,* edited by C. Guilleminault, W. C. Dement, and P. Passouant, pp. 485–497. Spectrum, New York.
8. Bente, D. (1964): *Die Insuffizienz des Vigilitätstonus Habilitations-Schrift.* Erlangen.
9. Berti-Ceroni, G., Coccagna, G., and Lugaresi, E. (1968): Twenty-four hour polygraphic recordings in narcoleptics. In: *The Abnormalities of Sleep in Man,* edited by H. Gastaut, E. Lugaresi, G. Berti-Ceroni, and G. Coccagna, pp. 235–238. Aulo Gaggi, Bologna.
10. Billiard, M. (1976): Competition between the two types of sleep and the recuperative function of REM versus NREM sleep in narcoleptics. In: *Narcolepsy,* edited by C. Guilleminault, W. C. Dement, and P. Passouant, pp. 77–96. Spectrum, New York.
11. Billiard, M. (1980): Recurring hypersomnias. In: *Sleep 1978,* edited by L. Popoviciu, B. Asgian, and G. Badiu, pp. 233–238. Basel, Karger.
12. Billiard, M., Quera Salva, M., De Koninck, J., Besset, A., Touchon, J., and Cadilhac, J. (1986): Daytime characteristics and their relationships with night sleep in the narcoleptic patient. *Sleep,* 9:167–174.
13. Blake, H., Gerard, R. W., and Kleitman, N. (1939): Factors influencing brain potentials during sleep. *J. Neurophysiol.,* 2:48–60.
14. Bonnet, M. H. (1985): The effect of sleep disruption on performance, sleep and mood. *Sleep,* 8:11–19.
15. Broughton, R. (1972): A proposed classification of sleep disorders. *Sleep Res.,* 1:146.
16. Broughton, R. (1974): Narcolepsy (letter to the editor). *Can. Med. Assoc. J.,* 110:1007.
17. Broughton, R. (1975): Biorhythmic variations in consciousness and psychological functions. *Can. Psychol. Rev.,* 16:217–230.
18. Broughton, R. (1976): Discussion. In: *Narcolepsy,* edited by C. Guilleminault, W. C. Dement, and P. Passouant, p. 667. New York, Spectrum.
19. Broughton, R. (1982): Human consciousness and sleep/waking rhythms: A review and some neuropsychological considerations. *J. Clin. Neuropsychol.,* 4:193–218.
20. Broughton, R. (1982): Performance and evoked potential measures of various states of daytime sleepiness. *Sleep,* 2:S135–S146.
21. Broughton, R. (1985): Three central issues concerning ultradian rhythms. In: *Ultradian Rhythms in Physiology and Behavior,* edited by H. Schulz and P. Lavie, pp. 217–231. Fischer-Verlag, Berlin.
22. Broughton, R., and Aguirre, M. (1985): Evidence for qualitatively different types of excessive daytime sleepiness. In: *Sleep '84,* edited by W. P. Koella, E. Ruther, and H. Schulz, pp. 86–87. Fischer-Verlag, Stuttgart.
23. Broughton, R., and Aguirre, M. (1987): Differences between REM and NREM sleepiness measures by event related potentials (P300, CNV), MSLT and subjective assessment in narcolepsy-cataplexy. *Electroencephalogr. Clin. Neurophysiol.,* 67:317–326.
24. Broughton, R., De Koninck, J., Gagnon, P., Dunham, W., and Stampi C. (1988): Chronobiological aspects of SWS and REM sleep in extended sleep of normals. *Sleep Res.,* 17:361.
25. Broughton, R., Dunham, W., Rivers, M., Lutley, K., and Duchesne, P. (1988): Ambulatory 24-hour sleep-wake monitoring in narcolepsy-cataplexy compared to controls. *Electroencephalogr. Clin. Neurophysiol.,* 70:473–481.
26. Broughton, R., Dunham, W., Suwalski, W., Lutley, K., and Roberts, J. (1986): Ambulant 24-hour sleep/wake recordings in narcolepsy-cataplexy. *Sleep Res.,* 15:109.
27. Broughton, R., and Ghanem, Q. (1976): The impact of compound narcolepsy and the life of the patients. In: *Narcolepsy,* edited by C. Guilleminault, W. C. Dement, and P. Passouant, pp. 201–220. Spectrum, New York.

28. Broughton, R., Ghanem, Q., Hishikawa, Y., Sugita, Y., Nevsilamova, S., and Roth, B. (1981): Life effects of narcolepsy in 180 patients from North America, Asia and Europe compared to matched controls: *Can. J. Neurol. Sci.,* 8:299–304.
29. Broughton, R., Guberman, A., and Roberts, J. (1984): Comparison of psychosocial effects of epilepsy and of narcolepsy-cataplexy: A controlled study. *Epilepsia,* 25:423–433.
30. Broughton, R., Low, R., Valley, V., da Costa, B., and Liddiard, S. (1981): Auditory evoked potentials compared to EEG and performance measures of impaired vigilance in narcolepsy-cataplexy. *Sleep Res.,* 10:184.
31. Broughton, R., Low, R., Valley, V., da Costa, B., and Liddiard, S. (1982): Auditory evoked potentials compared to performance measures and EEG in assessing daytime sleepiness in narcolepsy-cataplexy. *Electroencephalogr. Clin. Neurophysiol.,* 54:579–582.
32. Broughton, R., and Mamelak, M. (1979): The treatment of narcolepsy-cataplexy with nocturnal gamma-hydroxybutyrate. *Can. J. Neurol. Sci.,* 6:1–6.
33. Broughton, R., and Mamelak, M. (1980): Effects of gamma-hydroxybutyrate on sleep/waking patterns in narcolepsy-cataplexy. *Can. J. Neurol. Sci.,* 7:23–31.
34. Broughton, R., Nevsimalova, S., and Roth, B. (1980): The socioeconomic effects of idiopathic hypersomnia—comparison with controls and with compound narcoleptics. In: *Sleep 1978,* edited by L. Popoviciu, B. Asgian, and G. Badiu, pp. 103–111. Basel, Karger.
35. Broughton, R., Valley, V., Aguirre, M., Roberts, J., Suwalski, W., and Dunham, W. (1986): Excessive daytime sleepiness and the pathophysiology of narcolepsy-cataplexy: A laboratory review. *Sleep,* 9:205–215.
36. Browman, C. P., Gujavarty, K. S., Sampson, M. G., and Mitler, M. M. (1983): REM sleep episodes during the maintenance of wakefulness test in patients with sleep apnea syndrome and patients with narcolepsy. *Sleep,* 6:23–28.
37. Buchsbaum, M. S., Mendelsohn, W. B., Duncan, W. C., Coppola, R., Kelsoe, J., and Gillin, J. C. (1982): Topographic cortical mapping of EEG sleep stages during afternoon naps in normal subjects. *Sleep,* 5:248–255.
38. Cadilhac, J. (1976): Classification des troubles du sommeil. *Thérapie,* 31:7–25.
39. Cadilhac, J., Tomka, R., and Passouant, P. (1971): Les hypersomnies paroxystiques essentielles (intérêt de la télémétrie). *Rev. Electroencephalogr. Neurophysiol.,* 1:309–313.
40. Carskadon, M. A., and Dement, W. C. (1977): Sleep tendency: An objective measure of sleep loss. *Sleep Res.,* 6:200.
41. Carskadon, M. A., and Dement, W. C. (1982): The multiple sleep latency test: What does it measure? *Sleep,* 5:S67–S72.
42. Cohen, R., and Cruvant, B. A. (1944): Relation of narcolepsy to the epilepsies: A clinical-electroencephalographic study. *Arch. Neurol. Psychiatr. (Chicago),* 51:163–170.
43. Daly, D. D., and Yoss, R. E. (1957): Electroencephalogram in narcolepsy. *Electroencephalogr. Clin. Neurophysiol.,* 9:109–120.
44. De Koninck, J., Quera Salva, M., Besset, A., and Billiard, M. (1986): Are REM cycles in narcoleptics governed by an ultradian rhythm? *Sleep,* 9:162–168.
45. Dement, W. C. (1976): Daytime sleepiness and sleep 'attacks'. In *Narcolepsy,* edited by C. Guilleminault, W. C. Dement, and P. Passouant, pp. 17–42. Spectrum, New York.
46. Dement, W. C., Rechtschaffen, A., and Gulevich, G. (1966): The nature of the narcoleptic sleep attack. *Neurology (Minneap.),* 16:18–33.
47. Dickens, C. (1837): *The Posthumous Papers of the Pickwick Club.* Chapman and Hall, London.
48. Dynes, J. B., and Finley, K. H. (1941): The electroencephalograph as an aid in the study of narcolepsy. *Arch. Neurol. Psychiatr. (Chicago),* 46:598–612.
49. Endo, S., Kobayashi, T., Yamamoto, T., Fukuda, H., Sasaki, M., and Ohta, T. (1981): Persistence of the circadian rhythm of REM sleep: A variety of experimental manipulations of the sleep-wake cycle. *Sleep,* 4:319–328.
50. Etevenon, P. (1987): *Du Rêve à l'Éveil: Bases Physiologiques du Sommeil.* Albin Michel, Paris, pp. 141–168.
51. Gagnon, P., and De Koninck, J. (1984): Reappearance of EEG slow waves in extended sleep. *Electroencephalogr. Clin. Neurophysiol.,* 58:155–157.
52. Gagnon, P., De Koninck, J., and Broughton, R. (1985): Reappearance of electroencephalographic slow waves in extended sleep with delayed bedtime. *Sleep,* 8:118–128.
53. Ganado, W. (1958): The narcolepsy syndrome. *Neurology (Minneap.),* 7:197–221.

54. Gastaut, H., and Broughton, R. (1965): A clinical and polygraphic study of episodic events during sleep. *Recent Adv. Biol. Psychiatr.* 7:197–221.
55. Gastaut, H., Duron, B., Papy, J. J., Tassinari, C. A., and Waltregny, A. (1966): Étude polygraphique comparative du cycle nyctémèrique chez les narcoleptiques, les Pickwichiens, les obèses et les insuffisants respiratoires. *Rev. Neurol. (Paris)*, 115:456–462.
56. Gastaut, H., and Roth, B. (1957): À propos des manifestations électroencéphalographiques de 150 cas de narcolepsie avec ou sans cataplexie. *Rev. Neurol. (Paris)*, 97:388–397.
57. Gastaut, H., Tassinari, C.-A., and Duron, B. (1965): Étude polygraphique des manifestations épisodiques (hypniques et respiratoires), diurnes et nocturnes, du syndrome de Pickwick. *Rev. Neurol. (Paris)*, 112:568–579.
58. Gélineau, J. (1880): De la narcolepsie. *Gazette de l'Hôpital (Paris)*, 53:626–628.
59. Godbout, R., and Montplaisir, J. (1986): All-day performance variations in normal and narcoleptic subjects. *Sleep*, 9:200–204.
60. Guilleminault, C., Billiard, M., Montplaisir, J., and Dement, W. C. (1975): Altered states of consciousness in disorders of daytime sleepiness. *J. Neurol. Sci.*, 26:377–393.
61. Guilleminault, C., and Dement, W. C. (1977): Pathologies of excessive sleep. In: *Advances in Sleep Research, Vol. 3*, edited by E. Weitzman, pp. 345–390. Spectrum, New York.
62. Guilleminault, C., Raynal, D., Takahashi, S., Carskadon, M., and Dement, W. C. (1976): Evaluation of short-term and long-term treatment of the narcolepsy syndrome with chlomipramine hydrochloride. *Acta Neurol. Scand.*, 54:71–87.
63. Guilleminault, C., Raynal, D., Wilson, R., and Dement, W. C. (1973): Continuous polygraphic recording in narcoleptic patients. *Sleep Res.*, 2:151.
64. Hartse, K. M., Roth, T., and Zorick, F. J. (1982): Daytime sleepiness and daytime wakefulness: The effect of instruction. *Sleep*, 5:S107–S118.
65. Hartse, K. M., Roth, T., Zorick, F. J., and Zammit, G. (1980): The effect of instruction upon sleep latency during multiple daytime naps of normal subjects. *Sleep Res.*, 9:123.
66. Healey, T., Maru, J., and Broughton, R. (1975): A 4-channel portable recording system. *Sleep Res.*, 4:253.
67. Hess, R. (1949): Electroencephalographische Beobachtungen beim kataplektischen. *Anfall. Arch. Psychiatr. Nervenkr.*, 18:132–141.
68. Hishikawa, Y., Ida, H., Nakai, K., and Kaneko, Z. (1966): Treatment of narcolepsy with imipramine (Tofranil) and desmethylimipramine (Pertofran). *J. Neurol. Sci.*, 3:453–461.
69. Hishikawa, Y., and Kaneko, Z. (1965): Electroencephalographic study on narcolepsy. *Electroencephalogr. Clin. Neurophysiol.*, 18:249–259.
70. Hishikawa, Y., Nan'no, H., Tachibana, M., Furuya, E., Koida, H., and Kaneko, Z. (1968): The nature of the sleep attack and other symptoms of narcolepsy. *Electroencephalogr. Clin. Neurophysiol.*, 24:1–10.
71. Hishikawa, Y., Tabushi, K., Ueyama, M., Hariguchi, S., Fujiki, A., and Kaneko, Z. (1963): Electroencephalographic study in narcolepsy: Especially concerning the symptoms of cataplexy, sleep paralysis and hypnagogic hallucinations. Proc. Jpn. EEG Soc. (Tokyo), pp. 52–55.
72. Hishikawa, Y., Wakamatsu, H., Furuya, E., et al. (1976): Sleep satiation in narcoleptic patients. *Electroencephalogr. Clin. Neurophysiol.*, 41:1–18.
73. Hoddes, E., Dement, W. C., and Zarcone, V. (1972): The history and use of the Stanford Sleepiness Scale. *Psychophysiology*, 9:150.
74. Hoddes, E., Zarcone, V., Smyth, H. R., and Dement, W. C. (1973): Quantification of sleepiness: A new approach. *Psychophysiology*, 10:431–436.
75. Honda, Y., and Hishikawa, Y. (1979): Effectiveness of pemoline on narcolepsy. *Sleep Res.*, 8:192.
76. Honda, Y., Hishikawa, Y., and Takahashi, Y. (1979): Long term treatment of narcolepsy with methylphenidate (Ritalin). *Curr. Ther. Res.*, 25:288–298.
77. Jung, R., and Kuhlo, W. (1965): Neurophysiological studies of abnormal night sleep and the Pickwickian syndrome. In: *Sleep Mechanisms*, edited by K. Akert, C. Bally, and J. P. Schade, pp. 140–159. Elsevier, Amsterdam.
78. Kamei, B., Davidson, H., Gross, S., et al. (1978): Sleep parameters of narcoleptic females who were isolated from all time cues. *Sleep Res.*, 7:235.
79. Kleitman, N. (1961): The nature of dreaming. In: *The Nature of Sleep*, edited by G. E. W. Wolstenholme and M. O'Connor, pp. 349–364. Churchill, London.

80. Kleitman, N. (1963): *Sleep and Wakefulness.* University of Chicago Press, Chicago.
81. Kripke, D. F. (1976): Biological rhythm disturbances might cause narcolepsy. In: *Narcolepsy,* edited by C. Guilleminault, W. C. Dement, and P. Passouant, pp. 475–483. Spectrum, New York.
82. Lavie, P. (1982): Ultradian rhythms in human sleep and wakefulness. In: *Biological Rhythms, Sleep and Performance,* edited by W. B. Webb, pp. 239–272. Wiley, New York.
83. Lavie, P. (1986): Ultrashort sleep-waking schedule. III. "Gates" and "forbidden zones" for sleep. *Electroencephalogr. Clin. Neurophysiol.,* 63:414–425.
84. Loomis, A. L., Harvey, E. N., and Hobart, G. A. (1937): Cerebral states during sleep as studied by human brain potentials: *J. Exp. Psychol.,* 21:127–144.
85. Loomis, A. L., Harvey, E. N., and Hobart, G. A. (1938): Distribution of disturbance patterns in the human electroencephalogram with special reference to sleep. *J. Neurophysiol.,* 1:413–430.
86. Lubin, A., Hord, D. J., Tracy, M. L., and Johnson, L. C. (1976): Effects of exercise, bed rest and napping on performance decrement during 40 hours of sleep deprivation. *Psychophysiology,* 13:334–339.
87. Lumley, M., Roehrs, T., Zorick, F., Lamphere, J., Wittig, R., and Roth, T. (1985): Alerting effects of naps in normal sleep deprived subjects. *Sleep Res.,* 14:99.
88. Manseau, C., and Broughton, R. (1983): Ultradian variation in human daytime EEGs. A preliminary report. In: *Sleep 1982,* edited by W. P. Koella, pp. 196–198. Karger, Basel.
89. Manseau, C., and Broughton, R. (1984): Bilaterally synchronous ultradian EEG rhythms in awake adult humans. *Psychophysiology,* 21:265–273.
90. Mitler, M. M., Gujavarty, S., and Browman, C. P. (1982): Maintenance of wakefulness test: A polysomnographic technique for evaluating treatment efficacy in patients with excessive somnolence. *Electroencephalogr. Clin. Neurophysiol.,* 53:658–661.
91. Mitler, M. M., Gujavarty, K. S., Sampson, M. G., and Browman, C. P. (1982): Multiple daytime nap approaches to evaluating the sleeping patient. *Sleep,* 5:S119–S127.
92. Mitler, M. M., Shafer, R., Hajdukovick, R., Timms, R. M., and Browman, C. P. (1986): Treatment of narcolepsy: Objective studies on methylphenidate, pemoline and protriptylene. *Sleep,* 9:260–267.
93. Mitler, M. M., van den Hoed, J., Carskadon, M. A., et al. (1979): REM sleep episodes during the Multiple Sleep Latency Test in narcoleptic patients. *Electroencephalogr. Clin. Neurophysiol.,* 46:479–481.
94. Montplaisir, J. (1976): Disturbed nocturnal sleep. In: *Narcolepsy,* edited by C. Guilleminault, W. C. Dement, and P. Passouant, pp. 43–56. Spectrum, New York.
95. Montplaisir, J., Billiard, M., Takahashi, S., Bell, L., Guilleminault, C., and Dement, W. C. (1978): 24-hour polygraphic recordings in narcoleptics with special reference to nocturnal sleep disturbance. *Biol. Psychiatr.,* 13:73–89.
96. Montplaisir, J., and Godbout, R. (1986): Nocturnal sleep of narcoleptic patients revisited. *Sleep,* 9:159–161.
97. Montplaisir, J., and Godbout, R. (1986): Serotonergic reuptake mechanisms in the control of narcolepsy. *Sleep,* 9:280–284.
98. Mosko, S. S., Holowach, J. B., and Sassin, J. F. (1983): The 24-hour rhythm of oral temperature in narcolepsy. *Sleep,* 6:137–146.
99. Okawa, M., Matousek, M., and Petersen, L. (1984): Spontaneous vigilance fluctuations in the daytime. *Psychophysiology,* 21:207–211.
100. Parkes, D. (1981): Day-time drowsiness. *Lancet,* 2:1213–1218.
101. Parkes, D. (1985): *Sleep and Its Disorders,* W. B. Saunders, London.
102. Parkes, D., and Fenton, G. W. (1973): Levo (−) amphetamine and dextro (+) amphetamine in the treatment of narcolepsy. *J. Neurol. Neurosurg. Psychiatry,* 36:1076–1081.
103. Passouant, P. (1974): REMs ultradian rhythm during 24 hours in narcolepsy. In: *Chronobiology,* edited by L. E. Scheving, F. Halberg, and J. E. Pauly, pp. 495–498. Igaku Shoin, Tokyo.
104. Passouant, P., Baldy-Moulinier, M., and Aussilloux, C. (1970): État de mal cataplectique au cours d'une maladie de Gélineau: Influence de la chlorimipramine. *Rev. Neurol. (Paris),* 123:56–60.
105. Passouant, P., Halberg, F., Genicot, R., Popoviciu, L., and Baldy-Moulinier, M. (1969): La périodicité des accès narcoleptiques et le rhythme ultradien du sommeil rapide. *Rev. Neurol. (Paris),* 121:155–164.

106. Passouant, P., Popoviciu, L., Velok, G., and Baldy-Moulinier, M. (1968): Étude poly-graphique des narcolepsies du cours du nycthémère. *Rev. Neurol. (Paris)*, 118:431–441.
107. Passouant, P., Schwab, R. S., Cadhilac, J., and Baldy-Moulinier, M. (1964): Narcolepsie-cataplexie. Étude du sommeil de nuit et du sommeil du jour. Traitement par une am-phetamine lévogyre. *Rev. Neurol. (Paris)*, 111:415–426.
107a. Passouant, P., Cadilhac, J., and Baldy-Moulinier, M. (1967): Physio-pathologie des hy-persomnies. *Rev. Neurol. (Paris)*, 116:585–629.
108. Pond, D. A. (1952): Narcolepsy: A brief critical review and study of eight cases. *J. Ment. Sci.*, 98:595–604.
109. Pressman, M. R., Spielman, A. J., and Korczyn, A. (1980): Pupillometry in normals and narcoleptics throughout the course of a day. *Sleep Res.*, 9:218.
110. Pressman, M. R., Spielman, A. J., Pollock, C. P., and Weitzman, E. (1982): Long-latency auditory evoked responses during sleep deprivation and in narcolepsy. *Sleep* 5:S147–156.
111. Prinzmetal, M., and Bloomberg, W. (1935): The use of benzedrine in the treatment of narcolepsy. *JAMA*, 105:2051–2054.
112. Rechtschaffen, A., and Kales, A. eds. (1968): *A Manual of Standardized Terminology, Techniques and Scoring System for Sleep Stages of Human Subjects*. Institute of Health Publication No. 204. U.S. Government Printing Office, Washington, D.C.
113. Rechtschaffen, A., Wolpert, E. A., Dement, W. C., Mitchell, S. A., and Fisher, C. (1963): Nocturnal sleep of narcoleptics. *Electroencephalogr. Clin. Neurophysiol.*, 15:599–609.
114. Regestein, Q. R., Reich, P., and Mufson, M. J. (1983): Narcolepsy: A clinical approach. *J. Clin. Psychiatr.*, 44:166–172.
115. Reynolds, C. F., Coble, P. A., Kupfer, D. J., and Holzer, B. C. (1982): Application of the multiple sleep latency test in disorders of excessive sleepiness. *Electroencephalogr. Clin. Neurophysiol.*, 53:443–452.
116. Richardson, G. S., Carskadon, M. A., Flagg, W., et al. (1978): Excessive daytime sleep-iness in man: Multiple sleep latency measurements in narcoleptic and control subjects. *Electroencephalogr. Clin. Neurophysiol.*, 45:621–627.
117. Richardson, G. S., Carskadon, M. A., Orval, E. J., and Dement, W. C. (1982): Circadian variations of sleep tendency in elderly and young adult subjects. *Sleep*, 5:S82–S94.
118. Roehrs, T., Zorick, F., Wittig, R., Paxton, C., Sickelsteel, J., and Roth, T. (1986): Alerting effects of naps in patients with narcolepsy-cataplexy. *Sleep*, 9:194–199.
119. Roth, B. (1962): *Narkolepsie und Hypersomnie vom Standpunkt des Schlafes*. VEB Verlag Volk Gesundheit, Berlin.
120. Roth, B. (1964): L'EEG dans la narcolepsie-cataplexie. *Electroencephalogr. Clin. Neu-rophysiol.*, 16:170–190.
121. Roth, B. (1980): *Narcolepsy and Hypersomnia*, Karger, Basel.
122. Roth, B., and Bruhova, S. (1969): Dreams in narcolepsy, hypersomnia and dissociated sleep disorders. *Exp. Med. Surg.*, 27:187–208.
123. Roth, B., Bruhova, S., and Lehovsky, M. (1968): On the problem of pathophysiological mechanisms of narcolepsy, hypersomnia and dissociated sleep disturbances. In: *The Ab-normalities of Sleep in Man*, edited by H. Gastaut, E. Lugaresi, G. Berti Ceroni, and G. Coccagna, pp. 191–203. Aulo Gaggi, Bologna.
124. Roth, B., Faber, J., Nevsimalova, S., and Tosovsky, J. (1971): The influence of imipra-mine, dexphenmetrazine and amphetaminesulphate upon the clinical and polygraphic pic-ture of narcolepsy-cataplexy. *Arch. Suisse Neurol. Psychiatr.*, 198:251–260.
125. Roth, B., Nevsimalova, S., and Rechtschaffen, A. (1972): Hypersomnia with "sleep drun-kenness." *Arch. Gen. Psychiatry*, 26:456–462.
126. Roth, B., Nevsimalova, S., Sonka, K., and Docekal, P. (1984): A quantitative polygraphic study of daytime somnolence and sleep in patients with excessive diurnal sleepiness. *Arch. Suisse Neurol. Neurochir. Psychiatr.*, 135:265–272.
127. Roth, B., Nevsimalova, S., Sonka, K., and Docekal, P. (1986): An alternative to the multiple sleep latency test for determining sleepiness in narcolepsy and hypersomnia: A polygraphic score of sleepiness. *Sleep*, 9:243–245.
128. Roth, T., Hartse, K. M., Zorick, F., and Conway, W. (1980): Multiple naps and the evaluation of daytime sleepiness in patients with upper airway sleep apnea. *Sleep*, 3:425–439.
129. Ruther, E., Meier-Ewert, K., and Gallitz, A. (1972): Zur Symptomatologie des narkolep-tischen Syndromes. *Nerverartz*, 43:640–643.

130. Scharf, M. B., Brown, D., Woods, M., Brown, L., and Hershowitz, J. (1985): The effects and effectiveness of 8-hydroxy-butyrate in patients with narcolepsy. *J. Clin. Psychiatry*, 46:222–225.
131. Schmidt, H. (1982): Pupillometric assessment of disorders of arousal. *Sleep*, 5:S157–S164.
132. Schmidt, H. S., and Fortin, L. D. (1982): Electronic pupillography in disorders of arousal. In: *Sleep and Waking Disorders: Indications and Techniques*, edited by C. Guilleminault, pp. 127–143. Addison-Wesley, Menlo Park, CA.
133. Schulz, H. (1985): Ultradian rhythms in the nycthemeron of narcoleptic patients and normal subjects. In: *Ultradian Rhythms in Physiology and Behavior*, edited by H. Schulz and P. Lavie, pp. 164–185. Springer-Verlag, Berlin.
134. Schulz, H., Wilde-Frenz, J., Simon, O., Weber, R., and Ruther, E. (1983): Sleep-wake rhythms in a narcoleptic patient under normal entrained conditions and during isolation from time cues. In: *Sleep 1982*, edited by W. P. Koella, pp. 336–338. Karger, Basel.
135. Schwartz, B. A., Suguy, M., and Escande, J. P. (1967): Correlations EEG, respiratoires, oculaires et myocloniques dans le 'syndrome pickwickien' et autres affections paraissant apparentées. *Rev. Neurol. (Paris)*, 117:145–152.
136. Snyder, F. (1963): The new biology of dreaming. *Arch. Gen. Psychiatry*, 8:381–391.
137. Soldatos, C. R., Kales, A., and Cadieux, R. J. (1983): Treatment of sleep disorders. II: Narcolepsy. *Ration. Drug Ther.*, 17:1–7.
138. Stepanski, E., Lamphere, J., Badia, P., Zorick, F., and Roth, T. (1984): Sleep fragmentation and daytime sleepiness. *Sleep*, 7:18–26.
139. Stepanski, E., Zorick, F., Roehrs, T., Young, D., and Roth, T. (1988): Daytime alertness in patients with chronic insomnia. *Sleep*, 11:54–60.
140. Stoupel, M. N. (1950): Étude électroencéphalographique de sept cas de narcolepsie-cataplexie. *Rev. Neurol. (Paris)*, 83:563–570.
141. Sullivan, C. E., Berthon-Jones, M., Issa, F. G., and Eves, L. (1981): Reversal of obstructive sleep apnea by continuous positive airway pressure applied through the nares. *Lancet*, 1:862–865.
142. Takahashi, Y., and Gimbo, M. (1963): Polygraphic study of the narcolepsy syndrome with special reference to hypnagogic hallucinations and cataplexy. Proc. Joint Meet. Jpn. Soc. Psychiatr. Neurol. and Am. Psychiatr. Assoc. (Tokyo), pp. 343–347.
143. Valley, V., and Broughton, R. (1981): Daytime performance deficits and physiological vigilance in untreated patients with narcolepsy-cataplexy compared to controls. *Rev. Electroencephalogr. Neurophysiol. Clin.*, 11:133–139.
144. Valley, V., and Broughton, R. (1983): The physiological (EEG) nature of drowsiness and its relation to performance deficits in narcoleptics. *Electroencephalogr. Clin. Neurophysiol.*, 55:243–251.
145. Vogel, G. (1960): Studies in the psychophysiology of dreams: III. The dream of narcolepsy. *Arch. Gen. Psychiatry*, 3:421–428.
146. Volk, S., Simon, O., Schulz, H., Hansert, E., and Wilde-Franz, J. (1984): The structure of wakefulness and its relationship to daytime sleepiness in narcoleptic patients. *Electroencephalogr. Clin. Neurophysiol.*, 57:119–128.
147. von Economo, G. (1919): Grippe-encephalitis und Encephalitis lethargica. *Wien Klin. Wochenschr.*, 32:393–396.
148. von Economo, G. (1930): Sleep as a problem of localization. *J. Nerv. Ment. Dis.*, 71:249–259.
149. Walsh, J. K., Smitson, S. A., and Kramer, M. (1982): Sleep-onset REM sleep: Comparison of narcoleptic and obstructive sleep apnea patients. *Clin. Electroencephalogr.* 13:57–60.
150. Wilson, S. A. K. (1928): The narcolepsies. *Brain*, 51:63–109.
151. Wyatt, R. J., Fram, D. H., Buchbinder, R., and Snyder, F. (1971): Treatment of intractable narcolepsy with a monoamine oxidase inhibitor. *New Engl. J. Med.*, 25:987–991.
152. Yoss, R. E., and Daly, D. D. (1957): Criteria for the diagnosis of narcolepsy syndrome. *Proc. Mayo Clin.*, 32:320–328.
153. Yoss, R. E., and Daly, D. D. (1959): Treatment of narcolepsy with Ritalin. *Neurology (Minneap.)*, 9:171–173.
154. Yoss, R. E., Mayer, N. J., and Ogle, K. N. (1969): The pupillogram and narcolepsy. *Neurology*, 19:921–928.

Sleep and Alertness: Chronobiological, Behavioral, and Medical Aspects of Napping, edited by
D. F. Dinges and R. J. Broughton.
Raven Press, Ltd., New York © 1989.

14

The Significance of Napping:
A Synthesis

*David F. Dinges and †Roger J. Broughton

*Unit for Experimental Psychiatry, The Institute of Pennsylvania Hospital and
University of Pennsylvania, Philadelphia, Pennsylvania 19139-2798, USA, and
†Division of Neurology, University of Ottawa and Ottawa General Hospital,
Ottawa, Ontario, Canada K1H 8L6*

The preceding chapters leave little doubt that the study of naps, napping, and nappers challenge current concepts of the manner in which sleep is most appropriately obtained. A relatively inflexible, monophasic sleep/wake pattern in adult humans may not necessarily be the optimal or most efficient pattern. When the variety of nap patterns is considered in humans and across species, sleep/wake behavior takes on a conceptually richer and more subtle functional architecture that is especially evident in the seascape of wakefulness—rolling waves of alertness marked by a daily prominent midafternoon tidal wave of sleepiness. Having reviewed the accumulated data, it seems abundantly clear that napping is much more than an amusing, unusual, or aberrant event in the sleep/wake pattern. There are more fundamental factors at work in its expression than have generally been considered.

At the beginning of this volume we have referred to napping as "sleep's orphan." In examining sleep's orphan, we have found a lost progeny. But if it is certain that napping is a fundamental human behavior, it is equally uncertain how or why it came to be. What account can be made for the presence of naps in our species?

NAP TIME

Data bearing on the temporal patterning of sleepiness and napping are presented in no fewer than seven chapters of this volume. No matter what kind of empirical or experimental paradigm is used, there is ample evidence that napping is an integral feature of sleep/wake fluctuations. The timing and infrastructure of naps appear governed by the same biological and behavioral systems that determine the nature and timing of nocturnal sleep. As Tobler's review showed, its pervasiveness among mammals is evidenced by the fact that most mammalian species have a sleep/wake period length rather less than 24 hr.

There is now unequivocal evidence to support the proposal of Broughton (3) that a circasemidian or biphasic sleep/wake behavioral pattern is characteristic of humans and reflects an endogenous two per day rhythm of sleep facilitation. In fact, the bulk of the data from a number of approaches, including nap patterns at all ages, sleep tendency as assessed by sleep latency, ultradian sleepability, napping during temporal isolation, and napping among shift workers, strongly supports the existence of a biphasic sleep pattern, with each sleep phase separated by a zone of enhanced wakefulness (or diminished sleep pressure). Although the two wake zones (late morning and early evening in sidereal time) have comparable durations, the two sleep zones (nocturnal sleep and midafternoon nap times) have very unequal durations. The latter is probably owing to the different circadian phases at which they occur, although this point remains controversial. Whether the fundamental biphasic rhythm is one of sleep, or one of wakefulness (as suggested by Broughton), remains to be determined, as must the issue of whether the biphasic rhythm is a separate circa 12-hr rhythm (5) or an inherent quality of the circadian system (1).

If a robust two-per-day rhythm of sleep tendency exists, why is the midafternoon nap so idiosyncratic and prone to be a "now-you-see-it, now-you-don't" phenomenon? As Campbell and Zulley put it, "what kind of rhythmic system includes one component which, in normal adults, is most conspicuous by its frequent lack of occurrence?" Clearly, the expression of this two-per-day pattern is not so infrequent as to be discounted. Apart from its common occurrence in populations that have the opportunity or inclination to nap such as children, college students, shift workers, persons engaged in prolonged activity, siesta cultures, and hypersomnic patients, napping is also remarkably common among subjects living in temporal isolation, even when instructed not to take naps. In none of these groups or situations, however, is napping an everyday occurrence in every person.

The probability of nap behavior, in short, is overwhelmingly less than that of major (nocturnal) sleep behavior. On the other hand, the potential for nap sleep may be as great as the potential for nocturnal sleep. That is, sleep tendency, as assessed by both sleep latency (Carskadon) and sleep infrastructure (Broughton, Lavie), shows evidence of increased sleep pressure in the midafternoon, regardless of the overall level of sleepiness (or lack of it) resulting from the amount of prior wakefulness or other factors. This occurs at the same time that naps are prevalent (Dinges). It is reasonable to suggest, therefore, that this transient period of midafternoon sleepiness is the main factor underlying napping.

Whether an individual actually takes a nap (i.e., the probability of nap behavior), however, depends on other factors. Because the nap zone is of shorter duration than the zone for night sleep and because it takes place during a circadian phase when humans are typically active and environmental and social forces are urging wakefulness, nap behavior is often suppressed. Although this midafternoon period of increased sleep tendency (enforced rest) may have evolved in response to dominant environmental stimuli [e.g., heat], as Webb (20) has speculated, powerful social forces can suppress nap behavior at this time (Webb and Dinges). These same factors can also affect the duration and quality of nocturnal sleep. The paradoxical net result is an increase in daytime sleepiness and pressure for sleep during both the nocturnal and nap sleep zones in societies pressing to have more people awake at different times of the 24-hr day. Whereas many chapters provided consistent evidence for the chronobiological regulation of napping, just as many have concluded that overall daily sleep pressure or sleep need is associated with nap probability.

By not sleeping or not sleeping enough during the biorhythmically regulated nocturnal and diurnal times of increased sleep tendency, sleepiness is increased across all periods of the day. Ultradian, circasemidian, and circadian endogenous sleep/wake rhythms continue to shape the intensity of this sleepiness. In fact, their influence can become even more prominent when overall sleep pressure is increased (Lavie, Broughton). Under these circumstances, the potential for sleep (reflected in the chronobiological dynamics of sleep) and the probability of sleep (reflected in the tendency to go to sleep despite environmental contingencies) become very close. The greater the sleep pressure, the more likely the person is to nap, especially at chronobiologically appropriate times. It is not that sleep pressure selectively enhances sleepiness during the nap zone (although there is some evidence that this may occur), but rather that sleepiness throughout the day is increased and chronobiological regulation results in its being greatest during the nap zone. Thus, the potential to nap is a permanent feature of human sleep/wake cyclicity, but its behavioral expression is variable.

Superimposed on the midafternoon increase in sleep propensity are documented 90 to 120 min ultradian oscillations reflected in "sleepability" in ultrashort sleep/wake schedules (Lavie) and in various indices of alertness reviewed in the sleep model and sleep disorders chapters (Broughton). Like the midafternoon period of sleep facilitation, these more rapid oscillations are enhanced by ongoing sleepiness from prior sleep deprivation or by pathology (5). They appear to represent the daytime component of Kleitman's (10) basic rest activity cycle (BRAC). Much remains to be clarified concerning these oscillations, not the least of which is to demonstrate that they are in phase with nocturnal NREM/REM cycles, as the BRAC hypothesis predicts.

The concept that napping reflects a chronobiologically regulated sleep tendency that is amplified by sleep pressure and gated by environmental opportunity accounts for the fact that naps are not a daily feature of everyone's sleep/wake patterns. It does not, however, explain why there should be a major tendency to sleep approximately midway in the roughly 16-hr period of wakefulness in our species and in many others. Moreover, there is even evidence for some degree of qualitative differences in sleepiness and nap behavior as a function of whether they express pressure mainly for REM sleep (normally maximal in the morning) or for SWS (maximum later in the day) (4,6). These are among the most interesting aspects of nap behavior and ones that challenge us theoretically.

THEORIES AND MODELS

Theories of sleep have historically focused on why organisms sleep, (9,10,14,16,21), not why they sleep or nap at particular times of the day. More recently, mathematical models of sleep/wake regulation have been developed to account for the patterning of major sleep episodes observed during temporal isolation (2,7,13). Again, however, napping has rarely been considered. This is probably because it was deemed to account for too small a proportion of total sleep time or its occurrence within the circadian cycle was judged to be too infrequent to warrant attention. Although it is understandable that theorists and modelers would seek first to explain the regular circadian reappearance of major sleep episodes, the data in this volume make it unlikely that any theory of sleep function or model of sleep/wake regulation can henceforth be considered adequate unless it also accounts for the presence and timing of naps.

Another reason napping has not been discussed in most theories of sleep and models of sleep/wake regulation is no doubt because it is not as readily accounted for and its very existence challenges some of the constructs that current models seek to explain. For example, naps are frequently taken by subjects living in temporal isolation, even if prohibitions against them exist (Campbell and Zulley). When such naps are evaluated, spontaneous internal desynchronization, which is considered the "Rosetta stone" for chronobiological control of sleep, may not occur (26). Although the effect of naps on internal desynchronization during temporal isolation remains controversial [cf. (22)], it now appears that at the very least some mathematical models of sleep/wake regulation do not account for the existence of the nap phase in temporal isolation studies (18). In fact, as has been reviewed, models of sleep/wake regulation that involve either two oscillatory processes (13) or both oscillatory and recovery processes (7) do not predict either napping itself (Broughton) or the existence of a privileged nap zone (Broughton, Lavie).

Although Kronauer (12) recently expanded his chronobiological model to account for napping during temporal isolation, only Broughton (3,5,6) has consistently attempted to construct a theoretical framework for understanding the temporal regulation of naps in the broader context of multiple interrelated sleep-facilitating rhythms. His conceptual model, articulated in Chapter 5, incorporates the evidence for ultradian, circasemidian, and circadian sleep-facilitating rhythms as follows:

An early postulate [3] was that humans generally tend to exhibit biological rhythms that may be considered as harmonics of the basic circadian (cosmic solar) day. These include both the 48-hr bicircadian . . . subharmonic and the approximately 12-hr (circasemidian), 6-, 3-, and 1.5-hr superharmonic rhythms. Assuming that these apparently privileged periodicities occur in the raw data, a more plausible current hypothesis is that there exists an evolutionary natural selection for those periodicities that show fixed integer ratios to each other and to the basic 24-hr solar day [5]. Briefly stated, although oscillators with all conceivable periodicities might have appeared at some point in evolution, there would be survival value (more economic use of available energy resources) to maintain rhythms that can be consistently phase related to each other and to the major external zeitgeber (viz., the solar day).

Whereas this model emphasizes the chronobiological regulation of daytime sleepiness leading to naps, Webb's (20) theory of napping, as an evolved adaptation to the torrid heat of midday, stresses the influence of environment on the biorhythmicity of nap behavior. Webb argues that napping evolved as an endogenous rhythm to ensure forced rest during the midday heat, when activity could be detrimental to health. There are other reasons environmental temperature may be associated with napping (Dinges, Webb and Dinges), but there is little doubt that napping depends, in part at least, on a variety of behavioral pressures and opportunities for sleep. In this regard, Webb's theory reminds us of the importance of the context in which the sleeper lives.

Although these theories emphasize the endogenous chronobiological and environmental or contextual factors that influence napping, they make little mention of the role of daily sleep need in accounting for naps. Yet naps often seem to be associated with increased sleep need, and theories that stress the homeostatic aspects of sleep must account for the relationship of napping to basal daily sleep need (Dinges). For example, to the extent that naps account for less than 25% of daily total sleep time, it might not appear on the surface that they could be considered as "core" sleep in theories that posit a need for a basal amount of sleep (9,24). In particular, according to Horne's (9) theory, core sleep serves primarily to restore cerebral function, whereas "optional" sleep occupies time (i.e., enforced time-out). In this framework, naps can be categorized as optional sleep since they occur after nocturnal (core) sleep is obtained and, according to Webb (20), they function as enforced time-out. On the other hand, midafternoon naps preferentially contain high amounts of SWS, which appears to be under homeostatic control each 24-hr period (Broughton, Dinges) and which Horne (9) postulates is the primary sleep of restitution. In this sense, naps containing SWS may provide core sleep.

Whether regarded as core or optional sleep, napping often appears to play a significant role in satisfying basal daily sleep need, regardless of the function(s) served by that need. Theorists who choose to ignore the role of sleep need in sleep/wake regulation or sleep function must offer, at the very least, some explanation of how sleep loss, even modest sleep loss, sensitizes or otherwise enhances the sleep-promoting influence of endogenous chronobiological and exogenous contextual factors on sleep behavior, including napping.

HEALTH AND FUNCTIONING

If napping evolved to ensure an enforced time-out at a circadian phase of increased behavioral risk, as Webb (20) and others have conjectured, there is no reason to believe that such environmental risk is relevant to the vast majority of humans alive today. Should napping therefore be discarded as an archaic adaptation that no longer serves a useful purpose? Certainly many modern cultures have taken this view of daytime sleep and actively prohibit napping (Webb and Dinges). But what if napping continues to offer adaptational advantages in the context of modern environmental pressures? There is considerable evidence throughout this volume that this may indeed be the case. Modern cultures are plagued by increasing pressure to have more people awake at more times of the 24-hr day, making occupationally induced sleepiness a serious public health and safety issue (8,15). A critical function of napping in such societies could be to reduce sleepiness and maintain alertness.

The data strongly support the belief that napping, like nocturnal sleep, is inextricably linked to both endogenous oscillations of sleep ability and sleepiness resulting from inadequate sleep. It is the relief of sleepiness engendered by a daytime nap, either through the passage of time (to a chronobiologically controlled phase of greater alertness) or reduced sleep pressure (as a function of having slept) that produces the positive relationship between a nap and subsequent mood and performance. The more pronounced the sleepiness, the more likely that the nap will enhance behavioral alertness, especially performance capability and the ability to sustain subsequent wakefulness (Stampi, Dinges, Åkerstedt et al., Naitoh and Angus). To the extent that naps facilitate functioning in situations involving limited sleep opportunities during circadian phases of increased sleep pressure, they have a potentially important role to play in sleep scheduling to optimize alertness.

There appear to be few risks associated with taking a daytime nap in such situations. The major one is transient sleep inertia (Stampi, Dinges, Åkerstedt et al., Naitoh and Angus), which Piéron (16) regarded as impaired "critical reactivity" and consists of a period of reduced ability to function optimally immediately upon awakening. A minor risk of long naps (e.g., 2 hr or more) is the possible reduction of sleep depth on the subsequent night (Dinges). Aside from these two potential and seldom serious disadvantages, there are no other known adverse consequences of a daytime nap. In short, the advantages of napping in advance of occupationally induced sleep loss, [i.e., prophylactic napping (Dinges)], generally outweigh any disadvantages.

Whether brief naps can be the only sleep obtained during sustained work scenarios remains an important theoretical and applied question (Stampi, Naitoh and Angus). In order for nap patterns or polyphasic sleep/wake cycles to be effective in humans, it may be that at least one sleep episode must be long enough (e.g., 50% of usual nocturnal sleep duration) to serve as an "anchor" sleep to maintain internal synchrony among biological rhythms that influence mood and behavior.

In addition to the benefits that naps may offer for situations in which the daily sleep need is not being fully met by the major (usually nocturnal) sleep, there are reasons a daily nap may be advantageous, even when nocturnal sleep is routinely obtained. Whether the reasons are compelling enough to suggest that maintaining a biphasic sleep/wake cycle is important for health and functioning remains, however, uncertain.

There is little evidence that not following the body's endogenous tendency for a daily nap will in any significant way either impair an individual's ability to perform or lead to dysphoric mood. But this may have more to do with the limitations of our current measurement devices than with lack of effects per se. Few studies of napping in fact have ever measured performance and mood repeatedly after a nap, or done so over repeated days with comparisons of monophasic and biphasic sleep/wake patterns (Dinges). As reliable ambulatory monitoring devices become more practical and behavioral assessments become more portable, such studies can be undertaken. If a biphasic sleep/wake pattern, or for that matter a polyphasic pattern, has behavioral advantages, these will be evident in field studies.

A socially very significant issue that requires more study concerns the epidemiology of napping as it pertains to health. What are the long-term consequences for health of acute or chronic sleep loss, displaced sleep, highly variable sleep/wake schedules, and sustained biphasic or polyphasic sleep/wake patterns? Despite the importance of such issues to public health in a world that increasingly values wake time and the widespread belief that disturbed sleep, sleep loss, shift work, and jet lag can have adverse medical effects, there have been few studies of the relationships between sleep patterns and either morbidity or mortality. What little has been done suggests that in the United States at least, chronic short (6 hr or less) and chronic long (9 hr or more) nocturnal sleep durations are associated with increased mortality and with a heightened incidence of coronary heart disease (11,25). These studies, however, have not included daytime napping in the analyses.

An intriguing recent study by Trichopoulos et al. (19) of 81 patients with coronary heart disease (CHD) and 71 control patients in Greece, where the siesta is being legislated out of existence (17), reported the following results:

> The association between duration of night sleep and rate of CHD events is slight and non-significant. By contrast there is a strong, duration dependent and significant (two-tail $p < 0.04$) association between afternoon sleep with the occurrence of (at least) non-fatal CHD episodes, indicating that a half-hour siesta may be related to an almost 30% reduction in CHD incidence. Among patients reporting a siesta the association of its duration with CHD rate was even stronger: the point estimate of the rate ratio for a 30 min increase in the duration of siesta is 0.53, indicating that psychosocially conditioned self-selection in the afternoon rest group is not a likely confounder in the observed association.

Such tantalizing observations require independent confirmation using much larger sample sizes. They do not prove that a siesta would prevent or diminish the risk of coronary heart disease, but they do suggest, for the first time, that a monophasic versus biphasic sleep/wake pattern may be an important health risk factor for some patients. If these findings can be replicated in larger studies and/or in other populations, we may begin to formulate more specific experiments to determine which aspects of a biphasic sleep/wake pattern (e.g., total sleep time per 24 hr, sleep architecture, sleep timing) may be most important to optimize health and survival.

Clearly, there is much yet to be learned about the significance and value of both single daily naps and polyphasic sleep patterns for health and efficient functioning. This volume has only begun that effort. By accumulating knowledge from diverse areas of research, we hope that it will provide a more comprehensive understanding of sleep and alertness, and in particular of the role of naps. Whether as a sign of chronobiological regulation of sleep tendency, increased sleep need, a change in behavioral restraints, a deliberate effort to ward off anticipated later sleepiness, or any combination of these factors, napping is most evidently an adaptive option for the sleep/wake system. When we fully understand why polyphasic sleep is so prevalent among those species (including our own) that sleep, it is likely that we will unlock the very mystery of sleep itself.

ACKNOWLEDGMENTS

The substantive evaluation on which this chapter is based was supported by grants MH19156 and MH44193 from the National Institute of Mental Health, U.S. Public Health Service, a grant from the Institute for Experimental Psychiatry Research Foundation (D.F.D.), and grant MT3219 and a

Career Investigator Award from the Medical Research Council of Canada (R.J.B.).

REFERENCES

1. Aschoff, J., and Gerkema, M. (1985): On diversity and uniformity of ultradian rhythms. In: *Ultradian Rhythms in Physiology and Behavior*, edited by H. Schulz and P. Lavie, pp. 321–334. Springer-Verlag, Berlin/Heidelberg/New York/Tokyo.
2. Borbély, A. A. (1982): A two process model of sleep regulation. *Hum. Neurobiol.*, 1:195–204.
3. Broughton, R. J. (1975): Biorhythmic variations in consciousness and psychological functions. *Can. Psychol. Rev.*, 16:217–230.
4. Broughton, R. J. (1982): Performance and evoked potential measures of various states of daytime sleepiness. *Sleep*, 5(suppl. 2):S135–S146.
5. Broughton, R. J. (1985): Three central issues concerning ultradian rhythms. In: *Ultradian Rhythms in Physiology and Behavior*, edited by H. Schulz and P. Lavie, pp. 217–233. Springer-Verlag, Berlin/Heidelberg/New York/Tokyo.
6. Broughton, R. J., DeKoninck, J., Gagnon, P., and Dunham, W. Circadian, circasemidian and ultradian sleep-wake rhythms: A brief review and further data from extended sleep studies. In: *Sleep and Biological Rhythms: Basic Mechanisms and Applications to Psychiatry*, edited by J. Montplaisir and R. Godbout. Oxford Press, Oxford/New York (*in press*).
7. Daan, S., Beersma, D. G. M., and Borbély, A. A. (1984): The timing of human sleep: A recovery process gated by a circadian pacemaker. *Am. J. Physiol.*, 246(Regulatory, Integrative Comp. Physiol. 12):R161–R178.
8. Dinges, D. F. (1989): The nature of sleepiness: Causes, contexts and consequences. In: *Perspectives in Behavioral Medicine: Eating, Sleeping, and Sex*, edited by A. Stunkard and A. Baum. Lawrence Erlbaum, Hillsdale, NJ.
9. Horne, J. (1988): *Why We Sleep: The Functions of Sleep in Humans and Other Mammals*. Oxford University Press, Oxford.
10. Kleitman, N. (1963): *Sleep and Wakefulness*. University of Chicago Press, Chicago.
11. Kripke, D. F., Simons, R. N., Garfinkel, L., and Hammond, E. C. (1979): Short and long sleep and sleeping pills. *Arch. Gen. Psychiatry*, 36:103–106.
12. Kronauer, R. E. (1987): Temporal subdivision of the circadian cycle. *Lect. Math. Life Sci.*, 19:63–120.
13. Kronauer, R. E., Czeisler, C. A., Pilato, S. F., Moore-Ede, M. C., and Weitzman, E. D. (1982): Mathematical model of the human circadian system with two interacting oscillators. *Am. J. Physiol.*, 242(Regulatory Integrative Comp. Physiol. 11):R3–R17.
14. Mayes, A. ed. (1983): *Sleep Mechanisms and Functions in Humans and Animals: An Evolutionary Perspective*. Van Nostrand Reinhold, England.
15. Mitler, M. A., Carskadon, M. A., Czeisler, C. A., Dement, W. C., Dinges, D. F., and Graeber, R. C. (1988): Catastrophes, sleep and public policy: Consensus report of a committee for the Association of Professional Sleep Societies. *Sleep*, 11:100–109.
16. Piéron, H. (1913): *Le Problème Physiologique du Sommeil*. Masson, Paris.
17. Soldatos, C. R., Madianos, M. G., and Vlachonikolis, I. G. (1983): Early afternoon napping: A fading Greek habit. In: *Sleep '82*, edited by W. P. Koella, pp. 202–205. Karger, Basel.
18. Strogatz, S. H. (1986): *The Mathematical Structure of the Human Sleep-Wake Cycle*. Lecture Notes in Mathematics No. 69. Springer-Verlag, Berlin/Heidelberg/New York/London/Paris/Tokyo.
19. Trichopoulos, D., Tzonou, A., Christopoulos, C., Havatzoglou, S., and Trichopoulou, A. (1987): Does a siesta protect from coronary heart disease? *The Lancet*, Aug. 1, 270–271.
20. Webb, W. B. (1978): Sleep and naps. *Specul. Sci. Tech.*, 1:313–318.
21. Webb, W. B. (1979): Theories of sleep functions and some clinical implications. In: *The Functions of Sleep*, edited by R. Drucker-Colin, M. Shkurovich, and M. B. Sterman, pp. 19–36. Academic Press, New York.

22. Wever, R. A. (1986): Characteristics of circadian rhythms in human functions. *J. Neural Transm.*, 21(suppl.):323–373.
23. Wilkinson, R. T. (1970): Methods for research on sleep deprivation and sleep function. In: *Sleep and Dreaming*, edited by E. Hartmann, pp. 369–381. Little, Brown, Boston.
24. Wilkinson, R. T., Edwards, R. S., and Haines, E. (1966): Performance following a night of reduced sleep. *Psychonom. Sci.*, 5:471–472.
25. Wingard, D. L., and Berkman, L. F. (1983): Mortality risks associated with sleeping patterns among adults. *Sleep*, 6:102–107.
26. Zulley, J., and Campbell, S. S. (1985): Napping behavior during "spontaneous internal desynchronization": Sleep remains in synchrony with body temperature. *Hum. Neurobiol.*, 4:123–126.

Subject Index